ACS SYMPOSIUM SERIES 487

Biosensors and Chemical Sensors

Optimizing Performance Through Polymeric Materials

Peter G. Edelman, EDITOR
Ciba Corning Diagnostics Corporation

Joseph Wang, EDITOR
New Mexico State University

Developed from symposia sponsored
by the Divisions of Polymeric Materials: Science and Engineering
and of Analytical Chemistry
at the 201st National Meeting
of the American Chemical Society,
Atlanta, Georgia,
April 14–19, 1991

American Chemical Society, Washington, DC 1992

Library of Congress Cataloging-in-Publication Data

Biosensors and chemical sensors: optimizing performance through polymeric materials / Peter G. Edelman, Joseph Wang, eds.

p.   cm.—(ACS symposium series, 0097–6156; 487)

"Developed from symposia sponsored by the Divisions of Polymeric Materials: Science and Engineering and of Analytical Chemistry at the 201st National Meeting of the American Chemical Society, Atlanta, Georgia, April 14–19, 1991."

Includes bibliographical references and indexes.

ISBN 0–8412–2218–5

1. Biosensors—Congresses.   2. Membranes   (Technology)— Congresses. 3. Polymers—Congresses.

I. Edelman, Peter G., 1956–   . II. Wang, Joseph, 1948–   . III. American Chemical Society. Division of Polymeric Materials: Science and Engineering. IV. American Chemical Society. Division of Analytical Chemistry. V. American Chemical Society. Meeting (201st: 1991: Atlanta, Ga.) VI. Series.

R857.B54B552   1992
681′.2—dc20   92–6330
                                                CIP

The paper used in this publication meets the minimum requirements of American National Standard for Information Sciences—Permanence of Paper for Printed Library Materials, ANSI Z39.48–1984.   ∞

ACS Symposium Series

M. Joan Comstock, *Series Editor*

1992 ACS Books Advisory Board

Foreword

THE ACS SYMPOSIUM SERIES was founded in 1974 to provide a medium for publishing symposia quickly in book form. The format of the Series parallels that of the continuing ADVANCES IN CHEMISTRY SERIES except that, in order to save time, the papers are not typeset, but are reproduced as they are submitted by the authors in camera-ready form. Papers are reviewed under the supervision of the editors with the assistance of the Advisory Board and are selected to maintain the integrity of the symposia. Both reviews and reports of research are acceptable, because symposia may embrace both types of presentation. However, verbatim reproductions of previously published papers are not accepted.

Contents

ELECTROPOLYMERIZED THIN FILMS

INDEXES

Preface

BIOSENSORS CONTINUE TO BE a field of exciting research. This fact was exemplified at the symposium on which this book is based, where 80 biosensor presentations were spread through seven symposia. Many challenges have been conquered in this field of research, and yet these successes seem overshadowed by the challenges that remain. The widespread optimism of 10 or 15 years ago has been replaced by savvy realism. Many significant transducer developments have been realized, but transducer-enhancing membranes remain the Achilles heel limiting practical applications. Developments in miniaturization of transducers may have progressed, but technology to apply polymers to modify these tiny interfaces has lagged behind.

Research in this field has escalated recently because of a growing awareness of the inherent complexities and limitations of polymers at interfaces. The polymer must fulfill stringent design rules:

- The polymer must possess a desired permselectivity or other attributes.

- The permselectivity or other attributes must not change with time.

- The polymer must be applied in reproducible fashion to the transducer.

- This membrane must remain adhered to the surface to extend the useful life of the sensor.

- The polymer should be compatible with the incorporated biological entity.

- For sensors in contact with blood, not only must the membrane not be fouled in the complex matrix of proteins and lipids, but also the membrane must protect the transducer from fouling.

There are further membrane requirements for a nonthrombogenic in vivo device.

Whereas the focus of this book is polymers for sensor applications, this book is also about interfaces:

- the interface between two multidisciplinary fields (analytical chemistry and polymer chemistry),
- the interface between sample and membrane, and
- the interface between membrane and transducer.

To impart added functionality and reactivity to a sensitive transducer, the interface between transducer and sample can be modified with a polymer. The success of all these interfaces is critical to sensor development and performance.

Research in this discipline is important because of the potential to accelerate the materials development for sensors by applying what has already been learned in other fields, such as electronics, aerospace, and biomaterials. By combining an understanding of sensor issues with a broad understanding of how polymer problems have been solved in other fields, polymer development for specific sensor applications will advance more rapidly. The logical progression from materials development by trial and error is to tailor new materials systematically for each specific application, based on understanding of material science.

In putting the symposium together for the Division of Polymeric Materials: Science and Engineering (PMSE), it was anticipated that the authors would discuss in detail correlations of polymer properties with sensor performance. However, there is still incomplete understanding of material properties, making correlations to sensor performance difficult. This fact makes it even more difficult to go one step further and tailor new materials to meet the design requirements better. This lack of understanding is undesirable, but it creates opportunities for exciting possibilities for biosensor-related materials research. This book captures a moment in time and delineates the current state of the art. It will be interesting to look back at this book in 10–15 years to measure our progress.

This book is intended for two audiences. The first includes those who are involved in sensor research who have polymer-related problems and need to find some potentially elegant remedies. The second audience includes polymer scientists who are looking for a challenging discipline in which to practice their arts. Many of the solutions already worked out for polymers used in electronics or biomedical applications may be adapted to polymers for sensors. As these scientists become more directly involved with sensor-related materials development, better solutions to problems may be realized.

Many other volumes on sensors have been compiled that deal with novel aspects of sensor-related research. This collection is unique in that it is the first sensor book that focuses strictly on polymer and membrane-related research for sensor fabrication.

There are four main goals in putting this book together:

1. to bring together in one volume the current state of materials development for biosensors,

2. to heighten the awareness in the polymer science community of the challenges and opportunities inherent in biosensor development,

3. to emphasize the importance of materials development for biosensor advancement, and

4. to emphasize the necessity for interdisciplinary efforts.

Most of the chapters in this book were presented in symposia organized by two divisions. The majority are from the PMSE division, and most of the remainder are from the Division of Analytical Chemistry. In addition, there is one chapter each from the Divisions of Agriculture and Food Chemistry and of Biological Chemistry. We are fortunate that many excellent authors have agreed to include their works. Three chapters not presented at this symposium were added to round out the volume.

We gratefully acknowledge financial support from Ciba Corning Diagnostics Corporation and the PMSE division. We also thank all those without whose efforts this book would not be possible, especially the authors, reviewers, and the very supportive staff of ACS Books.

PETER G. EDELMAN
Ciba Corning Diagnostics Corporation
Medfield, MA 02052–1688

JOSEPH WANG
New Mexico State University
Las Cruces, NM 88003

January 21, 1992

Chapter 1

Overview of Biosensors

Elizabeth A. H. Hall

Institute of Biotechnology, University of Cambridge, Tennis Court Road, Cambridge CB2 1QT, United Kingdom

Biosensor models have been investigated utilising enzymes, antibodies/antigens, and other cellular components, or even whole cells, to provide a recognition surface for an analyte of diagnostic interest. The realisation of the model brings together many technologies, the components of which must often be studied individually before their union can be achieved. This overview considers the transduction of the analytical parameter and some factors which influence the materials and reagents of the transduction.

Biosensors . . . what does this imply? Traditional chemical and biological analytical methods involve a reaction which takes place in solution following the addition of reagents and sample. A biosensor is frequently described as a 'reagentless' system, but strictly speaking, it is more correct to say that the reagents are already part of the reaction chamber and do not therefore have to be added by the user. It follows from the implications of this statement that Biosensors will most likely be concerned with immobilised reagents - that is to say with reactions at surfaces and their interrogation. To adopt a well utilised definition however, biosensors comprise an analyte selective interface in close proximity or integrated with a transducer (figure 1), whose function it is to relay the interaction between the surface and analyte either directly or through a chemical mediator.

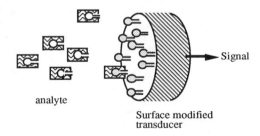

analyte

Surface modified
transducer

Figure 1. The Biosensor concept

0097–6156/92/0487–0001$06.00/0

The nature of the transducer and the transduced parameter will depend on the type of bioanalytical event concerned with detection of the analyte - that is to say, that a system designed for immunoassay where antibody-antigen binding must be detected is unlikely to be appropriate to an enzyme linked redox reaction! However, since several parameters can often alter during an analytical reaction pathway, the choice of device is not necessarily restricted to a single transducer. This book is concerned primarily with the developments of materials appropriate for the realisation of biosensing transduction devices in major areas of this ever broadening field. The nature of the parameters concerned are summarised in figure 2.

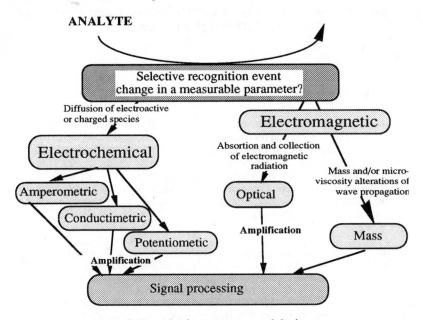

Figure 2. Transduction parameter and device type

Choice of Transduction Parameter

Amperometric. Discounting the centuries old use of live birds or fish to detect the sudden presence of toxins in air or water respectively (the species dies, thus 'transducing' an alarm signal!) the first report of a biosensor can probably be attributed to Clark & Lyons (1) who described an 'enzyme electrode' using glucose oxidase as a selective biorecognition molecule for glucose. The enzyme was held next to a platinum electrode in a membrane sandwich. The Pt electrode, polarised at +0.6V vs SCE, responded to hydrogen peroxide, produced by the enzyme reaction (**Scheme 1**):

$$\text{Glucose} + O_2 \xrightarrow{\textit{glucose oxidase}} \text{Gluconic acid} + H_2O_2$$

Scheme 1

This lead to the development of the first glucose analyser for the measurement of glucose in whole blood (Yellow Springs Instrument 23 YSI, 1974).

In the early days of this first device, the major effort in biosensor research was probably concerned with amperometric systems utilising redox enzymes. This is perhaps hardly surprising, since nowhere better in available biosensor formats, is the reaction parameter, namely the electron transfer concerned with the redox enzyme, better matched with the transduction parameter - i.e. electron transfer at the electrode. Nevertheless, as can be deduced from the scheme above this can hardly be a straightforward electrode transfer between enzyme and electrode, since the signal that was monitored is that due to the oxidation of a product of the transformation - hydrogen peroxide. Looking at the scheme more closely, the electrode fails to compete with the natural oxidant (molecular oxygen) and the transduced electrode signal is therefore an indirect one, rather than one due to the direct oxidation of enzyme.

In practical terms, as far as a biosensor is concerned, this may be of no consequence. However, utilising H_2O_2 as a measurand requires a constant supply of oxygen (hardly ideal for low or fluctuating O_2 environments) and a measuring potential (+0.6V *vs* SCE) where many interferents (eg Vitamin C) can add to the signal.

Mediators. Since direct electrode transfer between enzyme and electrode can rarely be performed with repeated efficiency, it was obvious that a synthetic replacement must be found for oxygen, and there thus began the hunt for modified electrodes and enzyme mediators. Many electron acceptor molecules and complexes have been considered for the role of mediator. Amongst the favourites ferrocene (η^5-bis-cyclopentadienyl-iron) and its derivatives still attracts attention - almost as a redox mediator standard (*2,3*).

The cyclic reaction processes which result from the use of such systems (scheme 2b) make them particularly attractive for deployment in a biosensor, since the only molecule which is now consumed by the sensor is the analyte itself and the reagents are regenerated within the assay.

Optimisation of one enzyme linked assay in this manner can also open doors to adjacent schemes and have implications beyond the analyte of original interest. For example schemes 2a/b show how the original system can be applied in a 'competition' for glucose with other enzymes using glucose as a substrate, and thus leading to an assay for say creatine, hexokinase or creatine kinase (*4*). Where the mediator is also an antigen label it can be employed in an enzyme linked immunoassay (scheme 2c). This latter scheme (*5*) depends on electrochemical and mediator inactivity of the ferrocene labelled antigen-antibody complex and the activity of the uncomplexed labelled antigen.

Such mediator-linked assays, even at the simplest original one enzyme level can seem an elegant solution for a biosensor. Nevertheless, optimisation of the reagents alone is not the complete solution. As discussed earlier, the reagents must be immobilised to interact with both analyte and transducer as a self contained system, before the 'biosensor' label is attached. This prerequisite is far from trivial and is a major preoccupation in all branches of biosensor development.

Cells and Cellular Components. Even in this narrow amperometric biosensor field the reagents may be of widely different natures. In particular, the very nucleus of the analyte selective reaction is not always an isolated enzyme, but may be a whole cell (*6*) or cellular component. Microbial biosensors have been described with a specificity for analytes ranging from CO_2 to Vitamin B. Tissue preparations have also been described utilising sources which appear limited only by the imagination: banana pulp, for example contains a high concentration of polyphenol oxidase which can be used in the estimation of dopamine and catechols (*7*). In the majority of instances these cell based devices have been based on a change in the O_2 tension monitored at an electrode mounted behind the cell layer. Like the H_2O_2 measurement already described, this is an indirect signal. Also like the isolated enzyme counterparts,

(a)

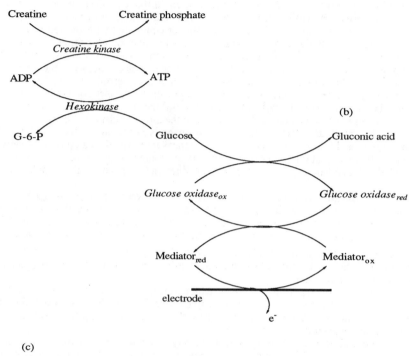

(b)

(c)

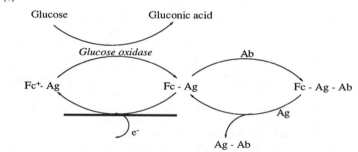

Scheme 2. Reaction pathways utilising a redox enzyme linked amperometric measurement via a redox mediator. (b) assay for glucose; (a)+(b) assay for creatine, creatine kinase or hexokinase; (c) competitive immunoassay using Fc-labelled antigen.

artificial redox mediators have been investigated, but in this instance transport is restricted by the cell walls and membranes. A considerable variation in efficiency of mediation has often been noted, even between similar mediators and the same organism, or related organisms and the same mediator; factors such as lipophilicity and surface interactions can also play a more important role than in enzyme preparations.

Cell preparations are also of use in some instances where the specificity of an enzyme linked assay is not required or the nature of the analyte makes it an unsuitable solution.

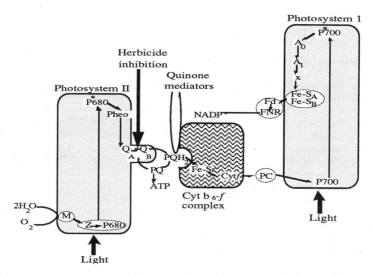

Figure 3. The Photosynthetic electron transport chain (PET)

An elegant example of this is the monitoring of herbicide residues via the photosynthetic electron transport (PET) pathway by utilising cyanobacteria or thylakoid membranes (*8*). For many herbicides the mode of action is as inhibitors of PET, often acting between the 2 photosystems as indicated in figure 3, and the result is a decrease in the photocurrent.

Within the boundaries of a biosensor, immobilisation of both the bacteria and thylakoid membranes together with a redox mediator, capable of charge transfer between PET and electrode, has allowed detection of some herbicides <10ppb (Martens, N. and Hall, E. A. H., Institute of Biotechnology, University of Cambridge, unpublished result). The advantage of this system is that it is not specific to a single herbicide but for any PET active toxin. In an area such as pesticide monitoring where the compounds have not been developed with predetermined properties, but are more the result of random screening, such a solution is more attractive than one which requires the development of many individual, unrelated devices in order to detect several compounds . . . and then probably finally fails to detect the actual herbicide in use!

Potentiometric. Not all analytes can be readily assayed via a redox enzyme and in these instances the assay schemes suggest parameters other than electron transfer which may be probed. Indeed, even where a redox enzyme is available for the analyte in question, it is not always desirable to deploy an amperometric technique.

Potentiometric measurements have been most frequently developed around pH sensitive electrodes and the same analytical reagents have been used in pH-FETs. Any of the enzyme pathways which result in a change in H^+ can be applicable here, but the most studied routes are those involving penicillinase or urease:

$$Penicillin \xrightarrow{\quad penicillinase \quad} Penicilloate + H^+$$

$$Urea \xrightarrow{\quad urease \quad} NH_4^+ + HCO_3^-$$

The final response of the pH sensor depends on the balance of all the equilibria involving H^+: protonation and deprotonation reactions of the enzyme reaction products; buffering capacity etc. A recent theoretical kinetic model, considering all these associated transport phenomena predicts the steady state response of an enzyme based pH sensor (9).

A possibly more novel application of potentiometric sensors and FET devices is in immunoassay involving antibody-antigen complex formation (10). The proper assignment of the theoretical basis for these measurements is not always clear. For example, Aizawa et al (11) showed that a membrane specific for the Wassermann antibody, prepared by casting a lipid antigen of cardiolipin, phosphatidylcholine and cholesterol in triacetylcellulose, mounted in a potentiometric cell, would respond to antibody. It was however suggested that this response resulted from an incidental secondary effect due to changed ion-exchange properties in the membrane when the antibody was bound. This being the case, then it is hardly surprising that static measurements of d.c. potential shifts are prone to drift and non reproducibility.

Schasfoort et al (12) have recently reported a means of overcoming these problems, by applying a series of step changes in electrolyte concentration and monitoring the transient membrane potential. An attempt has also been made to rationalise a theoretical treatment of the mode of response.

Optical. Optical sensors have been considered in two modes, according to the optical configuration. In the intrinsic mode the incident wave is not directed through the bulk sample, but propagates along a wave guide and interacts with the sample at the surface within the evanescent field. In the extrinsic mode, the incident light passes through the sample phase and interacts directly with the sample.

   Extrinsic.   Many of the traditional bioassays are based on optical measurements, but these are solution assays without immobilised reagents. Optical biosensors are concerned with interrogation of reactions at surfaces, but often the reagents are the same; the optical fibres acting, in the extrinsic mode as light guides to transmit information to and from a remote target reaction at its terminus.

The most sensitive optical sensors of this type are based on a fluorescent measurement and although many are the result of a detection of evolved products (eg H^+), fluorescent labels also offer an attractive strategy for probing binding events, such as antibody antigen complex formation. A typical sensor developed to monitor pH might use two fibres to guide the light to and from the distal end of the fibre (13), where the sensing chemistry is localised. Indicators have generally to be chosen to have long wavelength absorption and emission maxima, since most commonly available fibres have poor transmittance below λ_{420nm} and in any case the simple cheaper light sources are at wavelengths above λ_{400nm}. The choice of pH indicator is influenced by the pH range of interest and thus tuned to the pK_a of the immobilised fluorophore (which may be different to that in solution). Like all optical pH indicators, the sensitivity of the equilibrium to ionic strength can have a considerable influence on the assay and must be taken into account in interpretation of the results.

The most widely employed dyes for immunoassay have been fluorescein derivatives, but they are less than ideal since the short excitation λ limits the optical materials that can be employed. Rhodamine derivatives with longer λ_{ex} and an adequate Stokes shift are often favoured as are an increasing catalogue of alternatives ranging from pyrene to porphyrin derivatives. Fluorescent lanthanide metal ions and their chelates have also raised a lot of interest, particularly due to their long delay times, making them ideal for time resolved fluorometric immunoassay (14).

Several configurations for the sensor are possible. An especially viable alternative would seem to be the competitive displacement of fluorescent label. Since this is an equilibrium, fouling or contamination of the surface should not alter the absolute result. Krull *et al* (*15*) have reported the reproducible immobilisation of a stable phospholipid membrane containing fluorophore in this context. Concurrent fluorescence polarisation measurements can offer the possibility of multidimensional analysis (*16*) and are in any case experiencing a rejuvenation of interest as a highly selective technique, when the effective molecular weight of the antibody is increased relative to the antigen, by immobilisation on a latex or metal particle (*17*)

Intrinsic. In this mode, the equivalent of the extrinsic optical sensors already discussed is the waveguide where the reagents are immobilised on the surface within the evanescent field. Obviously, although this configuration has the advantage that the excitation light does not pass through the bulk sample, it has the disadvantage that in normal multimode fibres, the interaction between the light and surface immobilised reagents is less. Considering the ray model for the propagation of light through a guide, it can be seen that interaction with the surface only occurs at the 'points' of total internal reflection within the wave guide (figure 4).

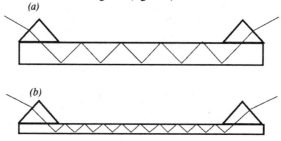

Figure 4. Ray diagram model of light propagating through (a) 'thick' film and (b) 'thin' film waveguides.

Thin film wave guides increase the interaction (*18*), but their fabrication is far from trivial. Ion exchange at the surface of a glass slide, giving a layer of higher refractive index is one solution (*19*), but it does not give a 'step' wave guide, but a gradient refractive index. Low temperature cure phosphate glasses have also been investigated and have shown promise (*20*). The fabrication of suitable thin film waveguides would allow many of the extrinsic mode assays to be converted and lead to a considerable leap forward in optical devices.

Other Surface Methods of Optical Interrogation.   Several other optical methods have been studied to give signal enhancement, many based on modulation of the field excited at an interface between different materials due to incident light. Of these surface plasmon resonance has probably received more coverage. This technique probes changes within the field propagating along the surface of a thin metal (plasmon) film. The earlier reports of SPR in the detection of protein binding investigated unlabeled interactions with, for example, and antibody immobilised on the metal film (*21, 22*). It soon became clear that although bound antigen could be detected, there was often no means of distinguishing this signal from any other non specific interaction with the surface (*23*). Current developments in this area therefore investigate refractive index labels and other optical labels and promise new low detection limits.

Similar conclusions have been arrived at in an SPR model designed as a DNA probe (24). The DNA probe sequence was immobilised to a thiol modified silver film deposited on a waveguide in a SPR format. The probe showed a characteristic change in the SPR excitation on binding complementary DNA. In this instance the use of an antibody to double stranded DNA could be used in a subsequent step to check the specifity of the result and allow an independent estimation of the level of detection. Without the addition of high molecular weight or refractive index labels however, this technique could not achieve a desirable level of detection, nor eliminate non specific interactions.

Acoustic waves. The microbalance is generally based on quartz crystals which are mechanically distorted when subjected to an electrical potential. The source is usually an alternating frequency in the MHz region and will be perturbed by minor changes at the surface. The basic equation relates a frequency change ΔF to <u>mass</u> change ΔM according to the Sauerbry equation:

$$\Delta F = -K\Delta M$$

In its simplest form, the technique was demonstrated first by Shons (25), who recorded the response of an antigen, attached at the surface of a bulk 9MHz crystal, to solution antibody. The major limitations of these devices are seen as non specific adsorption and surface occlusion..........the same problems as those encountered in the unlabeled optical assays mentioned above. The equivalent solution here must therefore be a 'mass' label!

The simple piezoelectric mass detection systems and the more sensitive surface acoustic wave (SAW) devices operate well in a dry or constant humidity gaseous environment (26,27), but they suffer from loss of signal in aqueous media. In the former case, this is largely a non specific response to water at the surface, and in the latter case it is associated with serious loss of the surface Rayleigh wave to the bulk solution.

Bulk acoustic wave (BAW) devices can be deployed with some success in the aqueous phase (28) - as can an ever increasing number of ingenious manipulations of format and control modes. Recent approaches have included the Lamb wave device discussed by White *et al*, which used a thin piezoelectric film, supported on a silicon substrate to carry low velocity shear vertical waves at a frequency around 10MHz (29,30). The analyte recognition surface is prepared, on the reverse side of the device, in the field present at the substrate interface below the thin piezoelectric film. In this manner the interaction between the Rayleigh wave and the solution is avoided. Nevertheless, the fragility of this thin film device makes its fabrication difficult. A plate mode device, proposed by Martin *et al* (31) uses higher frequencies and a thicker waveguide to propagate shear horizontal waves, creating a field at the upper and lower surfaces of the device, which may be easier to fabricate.

Better sensitivity has been reported (32,33) with a surface guided wave, known as a Love wave. A shear horizontal wave is directed along the surface; on top of the device a layer is deposited in which the acoustic waves are carried at a lower velocity.

Immobilisation of the Analyte Recognition Species.

In general the immobilisation matrix may function purely as a support, or else it may also be concerned with mediation of the signal transduction mechanism associated with the analyte. Techniques can be loosely divided into 4 groups:

(i)   Physical entrapment by an inert membrane

(ii)   Adsorption   -   physical
                              -   chemical

(iii)   Binding to a functionalised support

(iv)   Entrapment in an 'active' membrane

Adsorption.

Physical adsorption. Physical adsorption of the biorecognition molecule on the surface of the transducer is the simplest of the immobilisation processes. It may also perhaps be the least satisfactory since there often seems to be little that one can do to direct the orientation or site of attachment without altering the surface itself, and that the surface can have a major influence on the activity of the adsorbed species! This has been well illustrated by Duschl and Hall (*34*), who showed that although human immunoglobulin G (hIgG) could be adsorbed at a SiO_2 or an amino silane treated SiO_2 surface in a manner to give an apparently similar mono-layer, the reaction of the antibody, anti-hIgG, with these two layers was quite different, the former binding about 3 antibody molecules per molecule of adsorbed hIgG, while the latter could support 5 or 6. This was interpreted to indicate different interactions between the adsorbed protein and the surface in the two models.

The attachment of protein to the transducer surface for immunoassay is a difficult problem, since it must be achieved without interfering with the active site. If an unlabeled assay is to be performed it must also be such that subsequent non specific interaction with the surface can be inhibited. This requirement is often contrary to those for effective antibody-antigen complex formation. A surface close packed with antibody will be sterically hindered and its reaction with antigen inhibited. On the other hand, a suitably spaced packing allows non-specific interactions to occur and large false positive signals to be recorded. As mentioned earlier, this was demonstrated by Cullen and Lowe who used the surface plasmon resonance technique to probe specific and non-specific protein interactions at metal surfaces. (*23*).

Chemisorption. A key step in the development of a successful membrane modified FET is the adherence and longevity of the membrane and the success of the encapsulation procedure. Sudhölter *et al* (*35*) have proposed a method for the attachment of the ion selective membrane to a silylated SiO_2 gate oxide. An organofunctional silane, for example methacryloxypropyl trimethoxysilane, forms a bond between the surface and a photocrosslinkable polymer (eg polybutadiene). The ionophore would be directly bound to the polymer backbone, thus eliminating the need for plasticiser. Without any further modification with ionophore the membrane is sensitive to pH and shows a long life time (some months). Similar approaches have also been proposed by other workers, eg (*36*)

Methods for chemical attachment have also been exhaustively explored for protein immobilisation. No single preferred method has evolved, although coupling most frequently involves a surface $-NH_2$ group on the protein, but results in very varied surface coverages. These methods appear to be most useful for structures with large surface areas, whereas direct attachment to the 'transducer' would frequently involve a more planar lower surface area, so that it is particularly relevant in flowing systems (eg Flow Injection Analysis, FIA) where the reagent phase is immobilised onto beads in a packed column.

Physical retention in polymer matrices. One of the favourite methods for entrapment of enzymes and whole cells is by cross linking to give a gel matrix where the receptor species is physically retained. The methods of achieving this 'gel' have been numerous. For those with endless patience a low temperature cured sol-gel silicon glasses may be the answer (*37,38*), but when glasses takes days, weeks or even months to 'gel' this method may not be the one ideal for the less stable biological molecules. Nevertheless, 'bioactive glasses' are certainly an interesting concept and low temperature glasses other than those from the silanes may require less patience.

A typical matrix material for ion selective sensors and sensors utilising the ion selective response, is PVC. Macrocyclic receptor molecules are dissolved in a plasticiser as solvent and retained within the polymeric matrix. The ratio of membrane components is critical (39) and deviation leads to excess leaching of the plasticiser or insufficient receptor molecule to obtain a response. The PVC matrix is only able to retain a limited proportion of plasticiser and the ionophore has limited solubility in the plasticiser. In fact the most stable membranes have the lowest proportion of plasticiser, but these are also the least sensitive and selective, since they contain a low ionophore ratio; some stability must be lost at the expense of sensitivity (Duschl C and Hall E A H Hall, unpublished data).

These PVC based ionophore carrying matrices are also membranes which can be adapted for immunoassay systems. Keating and Rechnitz (40) have described an antigen digoxin coupled to one of the potassium ionophores (eg cis-dibenzo-18 crown-6) and the resulting conjugate immobilised in the PVC membrane.

Polyvinylalcohol, Polyvinylpyridene, gelatine, agar, polyacrylamide and many other gels have all been thoroughly investigated. They are probably none of them ideal. Nevertheless, a dialdehyde crosslinked polymer of polyacrylamide hydrazine has shown some promise as an enzyme immobilisation matrix on a pH electrode (41). Since most of these polymers can be clear and permeable they also have obvious applications in other types of electrochemical sensors and even optical ones. Real development of these kinds of polymer however needs to show an improvement in the storage characteristics, particularly in its stabilisation of the immobilised biorecognition molecule and, for the aqueous based or water containing matrices, the hydration properties.

Immobilisation matrices with 'transduction properties'. When it comes to membranes for amperometric biosensors an insulating membrane deposited over the electrode will hardly assist the electron transport and signal sensitivity, although may still be appropriate if analyte and electroactive measurand can permeate the membrane. Such membrane covered biosensors may trap the enzyme behing the membrane in an electrolyte film next to the electrode, or may entrap the enzyme within the membrane. However, the potential benefits of an enhanced charge transport across the membrane was probably the original driving force for the investigation of electrochemically deposited conducting polymers as immobilisation matrices. The first reports of glucose oxidase entrapped in polypyrrole (42) opened the door to a large family of polymers, but at the same time the constraints imposed by the desire for coimmobilisation of a biomolecule have mostly restricted the emphasis to the polymers which can be electrodeposited from aqueous media, since immobilisation of the biomolecule is achieved simply by its physical entrapment due to its presence in the polymerisation solution.

Perhaps the original hope for these polymers was that they would act simultaneously as immobilisation matrix and mediator, facilitating electron transfer between the enzyme and electrode and eliminating the need for either O_2 or an additional redox mediator. This did not appear to be the case for polypyrrole, and in fact while a copolymer of pyrrole and a ferrocene modified pyrrole did achieve the mediation (43), the response suggested that far from enhancing the charge transport, the polypyrrole acted as an 'inert' diffusion barrier. Since these early reports, other mediator doped polypyrroles have been reported (44,45) and curiosity about the actual role of polypyrrole or any other electrochemically deposited polymer, has lead to many studies more concerned with the kinetics of the enzyme linked reactions and the film transport properties, than with the achievement of a 'real' biosensor.

One such study has been directed towards polyaniline (Pani), since it was observed that this polymer did not appear to adopt an 'inert' role when glucose oxidase (GOD) was entrapped in its structure. In fact, in the presence of glucose, the modified electrode polarised at +0.6V *vs* SCE showed a current response which increased with time, not reaching a steady state even after 12 hours. This di/dt measurement was proportional to glucose concentration (figure 5) (*46,* Cooper J C and Hall E A H Hall submitted paper) and gave the fascinating indication that it must be related to some direct or indirect interaction with the polymer and not a H_2O_2 linked signal, since it appeared to be independent of O_2 concentration and was not reduced even in the complete absence of O_2. Attempts to simulate the response, have excluded a direct H^+ effect from the gluconic acid formed by the enzyme reaction. However, a continuing investigation of polyaniline has revealed a complex and potentially useful polymer for charge transduction.

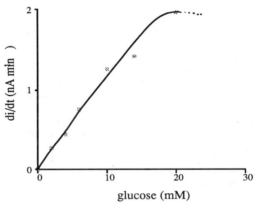

Figure 5. di/dt response to glucose at +0.65V *vs* SCE at a Pani/GOD modified electrode

This finding, while not yet fully explained gives some hope for designing a polymeric matrix which can interact efficiently with the redox enzymes. However, let us not expect that this is a straightforward process. Initial studies of rate determining processes and reaction layers suggest that our understanding must be considerably advanced before we are able to predict and design the action of these films.

Similar words of caution belong with redox polymers. Polyvinylferrocene typifies the type of polymeric compound where the redox centre is an integral part of the polymer. In view of ferrocene's well tried mediation talents this would seem an ideal choice of model, but the heterogeneous charge transfer rate constant for this redox polymer is typically 10^2 - 10^3 lower than the solution species (*47*) and although catalytic activity can usually be partially retained the kinetics of electron transfer are undesirably slow..

Hale *et al* (*48*) have proposed a 'redox polymer' based on a general structure of a polysiloxane polymer containing ferrocene modified side chains, then mixed with a carbon paste to form an electrode. The efficiency of the mediation is sensitive to changes in the length of the side chain and the relative loading of ferrocene, but it does not yet totally 'out-compete' O_2. However, the decrease in the mediated signal is less at high glucose concentrations.

This result is contrary to that found in the solution proposed by Heller and coworkers, of attaching the mediator to the enzyme on a flexible chain (49), and adsorbing the modified species on an electrode in a polyelectrolyte. In this instance the O_2 induced decrease in signal was constant with glucose concentration.

Other so called 'redox polymers' have been shown to have a high degree of catalytic efficiency, namely the ion exchange polymers which have been equilibrated with ionic redox species, so that the mediator is immobilised electrostatically through the surface bound ion exchanger. Unfortunately, although the charge transfer mediation kinetics are proficient, so is the leaching of mediator out of the polymer into the solution!

Comments, Challenges and Conclusions

The early driving force for the development of Biosensors came mostly from the medical diagnostic field, where rapid diagnosis and treatment is seriously inhibited by the long time delay between sample taking and the results from a central laboratory. Without doubt, in terms of the size of the potential market for rapid 'desk top' 'reagentless' assay devices, this must still remain an attractive market. Nevertheless, the ever increasing awareness of the state of our environment, and the demand for control and monitoring of effluent, air and ground species is providing new incentives for the Biosensor Technologists.

Whatever the market, wherever the application, the development of the Sensor Device requires separate and linked investigation at various levels. Even without a particular final goal, our basic understanding of immunoassay, enzyme-linked assay, recognition proteins, catalytic active sites and their 'electronic transduction' will continue to occupy the field, in addition to more 'downstream' considerations such as life-time levels of detection etc.; the list could be unending, these for example, are just some of the considerations:

(a)   nature of the analyte and identification of a specific recognition pathway and transduction parameter.

(b)   identification of the physico-chemical method for transduction of that parameter and its optimisation.

(c)   optimisation of the transducer technology.

(d)   linking the recognition reaction with the transduction.

(e)   immobilisation of the recognition species and optimisation of its recognition pathway.

(f)   immobilisation of any other 'transduction' species and their optimisation.

(g)   assessment of levels and range of detection

(h)   assessment of interferents

(i)   consideration of needs of particular application:
  biocompatibility?
  quantitative or qualitative (alarm)?
  operation in 'real' samples?
  specific interferents?
  required working life?
  required shelf life?
  appropriate technology?

(j)   ease of fabrication

(k)   market forces

(l)   many other considerations!!

There appears to be no simple summary of areas which should be targeted for further investigation, or statement of what might be involved. Perhaps a suitable description might be that we are concerned with the matching of natural and synthetic materials and technologies to allow interference free communication between analyte and a data handling circuit. This book focuses on the development of materials which may play a role in this communication pathway.

Since the first stages of biosensor technology the field has become more multidisciplinary and more multinational! Any technology may suddenly appear, being exploited in a biosensor format and achieving varying degrees of success. It all serves to keep every participant in biosensor research fully alert and ready to acknowledge both major and minor advances in many fields.

As we complete the third decade since the first report of a Biosensor (*1*), we could pause and reflect on the number of Biosensors which have been realised in the commercial market. However, assessments of the success of Biosensors, judged in this manner, should be made in conjunction with a general consideration of the goals, achievements and failures. The Biosensor concept is a simple one, its realisation is far from simple! The integration of naturally occuring molecules with synthetic elements has not been straightforward and while the nature of the possible measurement parameters for a given analyte can often be easily identified, the mechanism of their transduction has yet to be optimised. In a field such as Biosensors, it is obviously necessary to keep the target of their application well in sight. This should not however be allowed to be so dominant that short term goals are achieved at the expense of the development of our fundamental understanding of the mechanisms and technologies.

If after 3 decades of 'preliminary' achievements, we now fail to make the investment in a fundamental investigation and understanding of the systems that we are trying to develop as practical devices, then we must not be surprised to find our future progress seriously inhibited!

Literature cited.

1. Clark Jr, L.C.; Lyons, C. *Ann. N. Y. Acad. Sci.* **1962**, *102*, 29.
2. Cass, A.E.G.; Davis, G.; Francis, G.D.; Hill, H.A.O.; Aston, W.J.; Higgins, I.J.; Plotkin, E.V.; Scott, L.D.L.; Turner, A.P.F., *Anal. Chem.*, **1984**, *56*, 667 - 671.
3. Löffler, U.; Wiemhöfer, H.D.; Göpel, W., *Biosensors and Bioelectronics,* **1991**, *6*, 343 - 352.
4. Davies, P; Green, M.J.; Hill, H.A.O., *Enzyme Microb. Technol.*, **1986**, *8*, 349 - 352.
5. diGleria, K.; Hill, H.A.O.; McNiel, C.J.; Green, M.J., *Anal. Chem.*, **1986**, *58*, 1203.
6. Wang J.; Naser N., *Anal Chim Acta*, **1991**, *242*, 259.
7. Deshpande, M.K.; Hall, E.A.H., *Biosensors and Bioelectronics*, **1990**, *5*, 431 - 448.
8   Carpentier R.; Lemieux, S.; Mimeault, M.; Purcell, M.; Goetze, D.C., *Bioelectrochemistry and Bioenergetics*, **1989**, *22*, 391 - 401.
9   Glab, S.; Koncki, R.; Hulanicki, A., *Electroanalysis*, **1991**, *3*, 361 - 364.
10. Schasfoort, R. B. M.; Bergveld, P.; Bomer, J.; Kooyman, R. P. H.; Greve, J., *Sensors and Actuators*, **1989**, *17*, 531.
11 Aizawa, M.; Kato, S.; Suzuki, S., *J. Membrane Sci.*, **1977**, *2*, 125 - 130.
12 Schasfoort, R. B. M.; Bergveld, P.; Kooyman, R. P. H.; Greve, J., *Biosensors and Bioelectronics*, **1991**, *6*, 477 - 489
13. Offenbacher, H.; Wolfbeis, O. S.; Fürlinger, E., *Sensors and Actuators*, **1986**, *9*, 73 - 79
14. Bannwarth, W.; Schmidt, D.; Stallard, R. L.; Hornung, C.; Knorr, R.; Mueller, F., *Helv. Chim. Acta.*, **1988**, *71*, 2085
15. Tedesco, J. L.; Krull, U. J.; Thompson, M., *Biosensors*, **1989**, *4*, 123.
16. Krull, U. J.; Brow, R. S.; Hougham, B. D.; McGibbon, G.; Vandenberg, E. T., *Talanta*, **1990**, *37*, 561.

17.   Tsurnoka, M.; Tamiya, E.; Karube, I., *Biosensors and Bioelectronics*, **1991**, *6*, 501 - 505.
18.   Nellen, Ph. M.; Lukosz, W., *Biosensors and Bioelectronics*, **1991**, *6*, 517 - 525.
19.   Carlyon, E.E.; Lowe, C.R., Reid, D.; Bennion, I, *Biosensors and Bioelectronics*, **1991**, in press.
20.   Sloper, A. N., *PhD Thesis, University of London*. **1991**.
21   Liedberg, B.; Nylander, C.; Lundström I., *Sensors and Actuators*, **1983**, 299 - 306.
22.   Cullen, D.C.; Brown, R.G.W.; Lowe, C.R., *Biosensors*, **1987/88**, *3*, 211 - 225.
23.   Cullen, D.C.; Lowe, C.R., *Sensors and Actuators*, **1990**, *B1*, 576 - 579.
24.   Hall, E. A. H.; Duschl, C.; Liley, M. J., *J. Cell Biochem.*, **1990**, *14B*, 359.
25.   Shons, A.,*J Biomed. Mater. Res.* **1972**, *6*, 565 - 570.
26.   Nieuwenhuizen, M. S.; Arnold, A. J., *Anal Chem*, **1988**, *60*, 236 - 240.
27.   Nieuwenhuizen, M. S.; Arnold, A. J.; Barendsz, A. W., *Anal Chem*, **1988**, *60*, 230 - 236.
28.   Thompson, M.; Dhaliwal, G. K.; Arthur, C. L.; Calabrese, G. S., *IEEE Trans Ultrasonics, Ferroelect and Freq Control*, **1987**, *34(2)*, 127 - 135
29.   Wenzel, S. W.; White, R. M., *IEEE Trans Electron Dev*, **1988**, *35(6)*, 735 - 743.
30.   Zellers, E. T.; .Wenzel, S. W.; White R M, *Sensors and Actuators*, **1988**, *14*, 35 - 45.
31   Ricco, A. J.; Martin, S. J., *Appl. Phys. Lett.* **1987**, 21, 1474 - 1476.
32.   Gizeli, E.; Stevenson, A. C.; Goddard, N.; Lowe, C. R., *Proc Transducers '91*, **1991**, San Francisco USA.
33.   Gizeli, E.; Stevenson, A. C.; Goddard, N.; Lowe, C. R., *IEEE Control Syst.*, **1991**, April.
34.   Duschl, C.; Hall, E. A. H., *J Coll Int Sci*, **1991**, *144*, 368 - 380.
35.   Sudholter, E. J. R.; Van Der Wal, P. D.; Skowronska-Ptasinska, M.; Van Den Berg, A.; Reinhoudt, D. N. *Sensors and Actuators* **1989**, *17*, 189-194.
36.   Battilotti, M.; Cololli, R.; Giannini, M.; Giongo, M., *Sensors and Actuators* **1989**, *17*, 209 - 215
37.   Hench, L.L.; West, J.K., *Chem. Rev.* **1990**, *90*, 33 - 72.
38.   Braun, S.; Rappoport, S.; Zusman, R.; Avnir, D.; Ottolenghi, M., *Materials Lett.*, **1990**, *10(1,2)*, 1 - 5.
39.   Telting-Diaz, M.; Diamond, D.; Smyth, M. R.; Seward, E. M.; Mc Kervey, A. M., *Electroanalysis*, **1991**, *3*, 371 - 375.
40.   Keating, M. Y.; Rechnitz, G. A., *Anal. Chem.*, **1984**, *57*, 1998.
41.   Tor, R.; Freeman, A., *Anal Chem*, **1986**, *58*, 1042.
42.   Foulds, N.C.; Lowe, C.R., *J. Chem. Soc., Faraday Trans. 1*, **1986**, *82*, 1259 - 1264.
43.   Foulds, N.C.; Lowe, C.R., *Anal. Chem.*, **1988**, *60*, 2473 - 2478.
44.   Kajiya, Y.; Tsuda, R.; Yoneyama, H., *J. Electroanal. Chem.*, **1991**, *301*, 155 - 164.
45   Kajiya, Y.; Sugai, H.; Iwakura, C.; Yoneyama, H., *Anal. Chem.*, **1991**, *63*, 49 - 54
46.   Hall, E. A. H.; Cooper, J.C.; Deshpande, M.K., *J. Cell Biochem.*, **1990**, *14B*, 364.
47.   Leddy, J.; Bard, A. J., *J. Electroanal. Chem.*, **1985**, *189*, 203 - 219.
48.   Hale, P. D.; Boguslavsky, L. I.; Inagaki, T.; Karan, H. I.; Lee, H. S.; Skotheim, T. A.; Okamoto, Y., *Anal Chem*, **1991**, *63*, 677 - 682.
49.   Gregg, B. A.; Heller, A., *Anal Chem*, **1990**, *62*, 258 - 262.

RECEIVED January 21, 1992

PERMSELECTIVE MEMBRANES AND IMMOBILIZATION FOR ENZYME SYSTEMS

Chapter 2

Multienzyme Sensors

F. Scheller, A. Warsinke, J. Lutter, R. Renneberg, and F. Schubert

Central Institute of Molecular Biology, O-1115 Berlin-Buch, Germany

The interplay of mass transfer, partition and enzymatic substrate conversion determines the dynamic measuring range, response time, and accessibility towards interferences of enzyme sensors. New principles for designing the analytical performance by coupled enzyme reactions are presented in this paper:

(1) signal augmentation by accumulation of an intermediate in a reaction sequence

(ii) elimination of interfering signals by "stripping" an intermediate of the analyte conversion

(iii) signal separation by preconcentration of a cosubstrate.

Coupled Enzyme Reactions in Biosensors

This line has been initiated in the late 70s by Rechnitz' group (1), who introduced enzyme sequences into enzyme electrodes. The field has been considerably widened by introducing the competitive (parallel), recycling and accumulation sensors. The following main merits of such sensors shall be pointed out:

(i) Expansion of the scope of analytes is possible by sequential and parallel coupling of different enzymes with each other, within and with subcellular organelles, and with tissue slices. Examples include families of sensors based on glucose oxidase, lactate monooxygenase and glucose-6-phosphate dehydrogenase. The combination of lactate converting enzymes has led to enzyme elec-

0097–6156/92/0487–0016$06.00/0

trodes for the determination of 14 substrates and 4 enzyme activities (2).
(ii) Shifting of the measuring range to high analyte
concentrations by parallel coupling of enzymes (3).
(iii) The sensitivity of biosensors can be enhanced
up to four orders of magnitude (leading to picomolar
detection limits) by amplification using enzymatic analyte recycling (3).
(iv) Factors disturbing the response of biosensors can
be eliminated, i.e., by using antiinterference layers.

Sequences with accumulation

For a high functional stability an excess of the analyte
converting enzyme is generally fixed in front of the
sensor. In this way, in spite of the time-dependent deactivation of the immobilized biocatalyst, complete analyte conversion is maintained over a long operation
time. Under these conditions the sensitivity and the
response time are determined by the rate of mass transfer both outside and inside the enzyme layer.
Highest sensitivity is reached when high enzyme activity within a thin layer is used and effective external
mass transfer is provided. Under these conditions, substrate measurement can be managed down to the range of
1 micromolar with imprecision below 2 %. Therefore,
owing to their limited sensitivity, "normal" enzyme
electrodes are applicable only to metabolites present
in the micro and millimolar concentration range.
An increase of sensitivity is gained if a reaction product of the analyte conversion is accumulated. Kulys et
al.(4) demonstrated the formation of a peak current up
10 times higher when reduced mediator was accumulated
at an oxidase-loaded organic metal electrode and the
charge was stripped by applying an oxidation potential.
Alternatively, the signal transfer was interrupted on
level of a sequential enzymatic reaction, i.e., an electrode inactive intermediate formed in the analyte conversion by the "accumulator" enzyme E_1 is accumulated in
front of the polarized indicator electrode (5). The
"generator" (E_2) converts the accumulated intermediate upon addition of a cosubstrate, which results in an
electrochemically detectable concentration change.
The principle may be explained for the glucose determination with a sensor using the sequence glucose dehydrogenase (GDH) /lactic dehydrogenase (LDH) /lactate monooxygenase (LMO). The oxidation of glucose in the presence of an excess of NAD^+ generates the intermediate
NADH. The low diffusivity of NADH within the enzyme
layer results in a low outflux into the bulk solution.
Since during the accumulation time of NADH no pyruvate
is present, the NADH formation does not trigger an oxygen consumption via the sequence LDH-LMO. The accumulated NADH is stripped by addition of an excess of

pyruvate. Due to the high concentration and diffusivity
of pyruvate the reaction rate considerably exceeds that
for the steady state current which is determined by the
glucose influx from the bulk. Ten seconds after the
start with pyruvate the current reaches a peak. Since
the NADH is consumed in the LDH-LMO reaction the current
decreases and approaches a steady state identical to the
one without accumulation of NADH. The peak current rises
with accumulation time. A limit is set by the equilibrium of the reaction. Therefore the peak current levels
off at long accumulation time. The peak current at an
accumulation time of 10 min is almost 15 times higher
than the steady state current. In the kinetic mode
(di/dt) the sensitivity is even increased by a factor of
40.
In analogy to the enzymatic stripping in the amplified
glucose determination glycerol has been determined using
the sequence glycerol dehydrogenase (GlyDH/lactate dehydrogenase (LDH)/lactate oxidase (LOD).

In this system molecular weight-increased NAD$^+$ (PEG-
NAD$^+$, MW 20.000 Dalton) has been coentrapped within the
enzyme layer behind a dialysis membrane (molecular
cutoff 8.000). In the normal procedure the sensitivity
for glycerol was not decreased as compared with the measurement in the presence of 2 mmol/l (unmodified) NAD$^+$
in the bulk solution. This behaviour reveals a sufficiently high reaction rate between the macromolecular
NAD(H) and the enzymes GlyDH and LDH.
In the stripping mode the signal is generated after
accumulation of NADH by injecting an excess of pyruvate.
After 6 minuts of PEG-NADH accumulation an increase of
the sensitivity in the differential mode of 64 is found.
The linear concentration dependence extends up to 10
/umol/l and the peak current is reached 10 s after injecting the pyruvate.
The concept of enzymatic stripping has been extended to
glucose-6-phosphate and formate (Tab. 1). The amplification factors are determined by the equilibrium constant
and by the difference in the permeability of analyte and
initiator. Thus the measurement of lactose is not amplified after accumulation of glucose and subsequent addition of NAD$^+$, because the molecular weight of the india-
tor (NAD$^+$) is much higher than that of the accumulated
species.

Elimination of Interfering Signals

Another aspect of the signal generation by adding the
initiator substance is the elimination of interfering
contributions of the sample to the measured signal. Since
the basic current includes the contribution of all electrochemically active sample components the measuring
signal reflects only the analyte concentration. This advantage has been used in the determination of triglyce-
rides (Fig. 1). In the measuring procedure the triglyce-

Table 1: Accumulation Sensors

Analyte	E_1/E_2	Intermediate	Initiator	Amplification factor
Glucose	GDH/LDH-LMO	NADH	pyruvate	18
Glucose-6-phosphate	G6PDH/ "	"	"	17
Formate	FDH/ "	"	"	6
Glycerol	GlyDH/LDH-LOD	NADH-PEG	"	64
Lactose	ß-Gal/GDH	glucose	NAD+	1

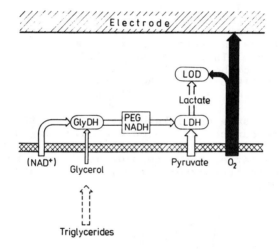

Fig. 1: Scheme of the sequential conversion of glycerol to give an oxygen depletion

rides are hydrolyzed in alkaline methanol. After completion of this reaction the sample is injected into the
buffered measuring solution. Of course the addition of
methanol causes a large signal of the basic oxygen sensor
resulting from the different oxygen solubility in the
sample and in the background solution. Therefore the
addition of pyruvate is carried out after the establishment of a stable baseline thus avoiding this disturbance.
In this way triglyceride concentrations have been determined in serum samples with good correlation to the
established method (r = 0.993).

Signal Separation for Different Analytes

Multi-layer enzyme sensors offer the opportunity to separate the signals of different analytes. For this purpose
a cofactor dependent enzyme converting the analyte A is
fixed in front of a layer which contains enzymes converting both analytes B and A under the formation of an
electrode-active product. Two procedures for separating
the signals for both analytes have been established.

(i)   In the absence of the cofactor both analytes reach
      the second layer and are transformed therein to
      create a concentration change of an electrode-active
      species. In this case the signal represents the sum
      of analytes A and B. On addition of the cofactor,
      analyte A is eliminated in the first layer. There-
      fore only analyte B generates the signal. By combi-
      ning both measured values the concentration of both
      analytes is accessible.
      This principle has been demonstrated for the deter-
      mination of glucose and maltose using a layer of
      hexokinase in front of a glucoamylase-glucose oxi-
      dase layer. In the absence of ATP the current change
      represents the sum of both sugars whilst the addi-
      tion of ATP prevents the glucose from reaching the
      second layer, i.e., only the maltose is indicated.

(ii) In the same multi-layer arrangement the signal of
      glucose can be influenced depending on the concen-
      tration of the cofactor with the couple glucose and
      ATP the latter has a considerably lower permeabili-
      ty due to its higher molecular weight and negative
      charges.

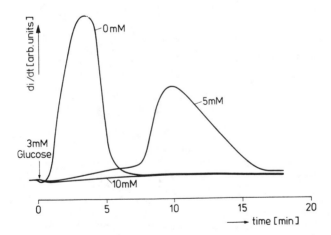

Fig. 2: Kinetic signal-time curve of a hexokinase/GOD
         sensor on addition of glucose in the presence
         of different concentrations of ATP

This effect is demonstrated by the concentration dependence of glucose in the presence of different concentrations of ATP (Fig. 2). With rising ATP concentration the dependence is parallelly shifted indicating the increasing conversion of glucose in the hexokinase layer (6). Extrapolation of the respective linear dependences shows that an almost threefold excess of ATP is necessary to achieve complete conversion of glucose, i.e. zero signal. At a medium ATP concentration the glucose signal is not only decreased (as compared with that in the absence of the cofactor) but also delayed. This behaviour allows to separate the glucose signal from that generated by maltose.

The arrangement of multi-layer sensors containing coupled or independently acting enzymes offers new prospects in designing potent biosensors.

Literature Cited

(1) Jensen,M.A., Rechnitz,G.A.; J.Membrane Sci. 1979, 5, 117

(2) Scheller,F., Schubert,F., Weigelt,D., Mohr,P., Wollenberger,U.; Makromol.Chem. 1988, 17, 429

(3) Scheller,F., Schubert,F.; Biosensoren, Birkhäuser Verlag, Basel, Boston, Berlin, 1989

(4) Kulys,J.J., Cenas,N.K., Svirmickes,G.J.S., Svirmickiene,V.P.; Anal.Chim.Acta, 1982, 138, 39

(5) Schubert,F., Lutter,J., Scheller,F.; Anal.Chim.Acta, 1991, 243, 17

(6) Warsinke,A., Renneberg,R., Scheller,F.; Anal.Letters, 1991, 24, 1363

RECEIVED March 10, 1992

Chapter 3

Biosensors Based on Entrapment of Enzymes in a Water-Dispersed Anionic Polymer

Guy Fortier, Jian Wei Chen, and Daniel Bélanger

Groupe de Recherche en Enzymologie Fondamentale et Appliquée, Département de Chimie et Biochimie, Université du Québec à Montréal, C.P. 8888, Succursale A, Montreal, Quebec H3C 3P8, Canada

Biosensors using choline and glucose oxidase have been prepared by deposition at the surface of platinum electrodes of an aliquot containing a blend of amorphous polyester cationic exchangers (AQ 29D:AQ 55D; ratio 1:1) dispersed in water and the enzyme. The polymer-enzyme films were prepared according to the following three protocols: 1) the enzymatic film is dried at room temperature, 2) same as 1 and covered by a Nafion overlayer or 3) the enzymatic film is dried by heating at 50°C for 30 min. The evaluation of the biosensors was based on the amperometric detection of enzymatically generated hydrogen peroxide in presence of choline chloride or glucose. Enzyme electrodes prepared using protocols 2 & 3 are stable but those dried by heating at 50°C for 30 min gave the higher amperometric responses. Also, the AQ-enzymatic film keeps the same permselectivity against small anionic redox species than the AQ film alone of the same thickness.

During the last decade, immobilization of oxidase type enzymes by physical entrapment in conducting or ionic polymers has gained in interest, particularly in the biosensor field. This was related to the possibility for direct electron tranfer between the redox enzyme and the electroconducting polymers such as polypyrrole *(1,2)*, poly-N-methyl pyrrole *(3)*, polyindole *(4)* and polyaniline *(5)* or by the possibility to incorporate by ion-exchange in polymer such as Nafion *(6)* soluble redox mediators that can act as electron shuttle between the enzyme and the electrode.

Previously, Nafion was used in electroanalysis as electrode covering membrane to immobilize cationic species *(7,8)* and as a barrier to anionic species *(9,10)*. In biosensor field, the main limitation in the use of Nafion for enzyme entrapment is related to the solubility of the Nafion in lower aliphatic alcohols which are not compatible with the enzyme activity *(11)*. Until now, only glucose oxidase was successfully immobilized in Nafion.

0097–6156/92/0487–0022$06.00/0
© 1992 American Chemical Society

More recently, a new series of water dispersed anionic polymers, the AQ 29D, 38D and 55D polymers were released by Eastman Kodak. Since that time, these polymers were used as electrode modifier *(12, 13)*, as covering membrane *(14)* and as support for enzyme immobilization *(15, 16)*. AQ polymers are high molecular weights (14,000 to 16,000 Da) sulfonated polyester type polymers *(17, 18)*. Their possible structures have been recently presented *(18)*. The AQ polymer serie shows many interesting characteristics useful for the fabrication of biosensors. They are water dispersed polymers and thus compatible with enzymatic activity. They have sulfonated pendant groups similar to Nafion and they can act as a membrane barrier for anionic interferring substances and they offer the possibility to immobilize redox mediators by ion exchange.

In this paper, we have evaluated three different protocols for the preparation of AQ-enzyme film using choline and glucose oxidases and mixture of AQ 29D and AQ 55D (1:1). This AQ mixture is recommended by Eastman Kodak to increase the adherency of the film to a surface such as a platinum electrode *(17)*. The values 29 and 55 represent the glass transition temperature of each polymer. Also, the main structural difference between the two polymers is that, in the case of AQ 55D, an aliphatic glycol moiety replaces the cycloaliphatic glycol moiety found in the AQ29 *(17, 18)*.

These biosensors transform their respective substrates following the reactions 1 and 2.

$$\text{Choline} + O_2 \xrightarrow{\text{Choline oxidase}} \text{Betaine aldehyde} + H_2O_2 \quad (1a)$$

$$\text{Betaine aldehyde} + O_2 \xrightarrow{\text{Choline oxidase}} \text{Betaine} + H_2O_2 \quad (1b)$$

$$\text{ß-d-Glucose} + O_2 \xrightarrow{\text{Glucose oxidase}} \text{d-Gluconolactone} + H_2O_2 \quad (2)$$

Material and Methods

Reagents.   Choline oxidase and glucose oxidase and their respective substrates, choline chloride and glucose were purchased from Sigma Chemical Co. (St.Louis, USA). The cationic exchangers AQ 29D and 55D were kindly supplied by Eastman Chemical Inc. (Kingsport, USA) and were obtained as dispersed polymer solutions at concentration of 30 and 28% (w/v) in water, respectively. A blend of the AQ polymer solutions was prepared by mixing the AQ 29D with the AQ 55D in a ratio (1:1), and was further diluted with water to a final concentration % (w/v) indicated in the text. Nafion (equivalent mass 1100 g) 5% (w/v) in a mixture of lower aliphatic alcohols and 10% of water was obtained from Aldrich (St.Louis, USA) and diluted with methanol to yield a stock solution of 0.5% (w/v).

Film Fabrication.   The platinum electrode (0.28 cm^2 area) was fabricated and cleaned as previously described *(19)*. Thin films of AQ-enzyme were prepared by dissolving an amount of the enzyme, as indicated below, in 10 µl of 1.5% AQ polymers solution at room temperature. Two aliquots of 5 µl were deposited atop the platinum electrode and the first aliquot was allowed to dry before the second addition. This procedure corresponds to the first protocol. In addition, for the second protocol, 10 µl of the 0.5% Nafion solution was casted atop the dried AQ-enzyme film and the methanol was allowed to evaporate at room temperature. The third protocol consisted in the deposition of 10 µl of a 1% of AQ solution containing the enzyme, atop the platinum electrode followed by heating in an oven at 50°C during 30 min. In each case, 2 U of glucose oxidase were used.

All the enzyme electrodes were held at 0.7 V *vs* SCE in 5 ml of 0.1 M phosphate buffer, pH 7, for one minute before addition of their respective substrate. The steady-state values recorded after 1 min were used to obtain the calibration curves. The electrode was removed from the solution and washed with water prior to be used for another assay.

Instrumentation.   All electrochemical experiments were carried out in a conventional one-compartment cell. Potentials were applied to the cell with a bipotentiostat (Pine Instruments Inc., USA) model RDE4. Current-time responses were recorded on a XYY' recorder model BD 91 (Kipp & Zonen, USA) equiped with a time base module. All potentials were measured and quoted against a saturated calomel electrode (SCE).

Kinetic Analysis.   The kinetic parameters were obtained by iterative non-linear curve fitting of raw data (current generated versus the substrate concentration).The data fitted a modified Michaelis-Menten equation:

$$Is = (Imax \ [S])/ (Km'' + [S]) \qquad (3)$$

where   Is is the observed current and Imax is the maximal value for the current response. Km'' is an effective apparent constant that represents the substrate concentration giving one half of the Imax in air saturated solution. The Km'' is compounded by the mass transport rates of both substrates because all the currents used were obtained at steady-state. Also, we have made the assumption that the structure of the enzyme is not modified following the immobilization.

Results and Discussion

Three different protocols have been evaluated for the enzyme-AQ film preparation to obtain the enzymatic film that will be the most stable in the assay solution. These protocols consisted i) in depositing the mixture of glucose oxidase and AQ atop the platinum electrode and allowing the solvent to evaporate at room temperature, ii) in covering this AQ-glucose oxidase film with a Nafion layer or iii) in heating the AQ-glucose oxidase film at 50°C in an oven during 30 min. The reproducibility of the time-current traces for 4 consecutive assays of 20 mM of glucose was evaluated as probe of the film stability for each protocol. Figure 1 depicts the stability of the glucose oxidase-AQ film obtained following the different protocols. When the glucose biosensor is prepared by mixing 10 μl of AQ 1.5% mixture with 2 U of glucose oxidase (Fig.1A), the value of the steady-state current decreases by 25% after each glucose assay. For the 5th assay, no response was observed indicating that the film has been slowly dissolved in the assay solution during each assay.

To aleviate the dissolution of the enzymatic film during the glucose assay procedure, a covering membrane made by addition of 10 μl of Nafion 0.5% was used. The amperometric current increases until a constant value of 20 μA is reached on the fourth glucose assay (Fig.1B). The addition of Nafion film atop of the AQ-enzyme layer doesn't seem to denaturate the enzyme entrapped in the AQ film because the current reached a similar value as the current observed in absence of Nafion. On the other hand, the use of Nafion as covering membrane gives a thicker film that leads to mass transport limitation of the substrates in the AQ film *(20)*. It results in a longer time to obtain the steady-state current.

The last protocol, similar to the one use by Wang and co-workers *(16)*, consisted in heating the enzymatic film in an oven at 50°C for 30 min. After this period, a glassy and resistant enzymatic film is obtained. An enzyme loading of 2 U

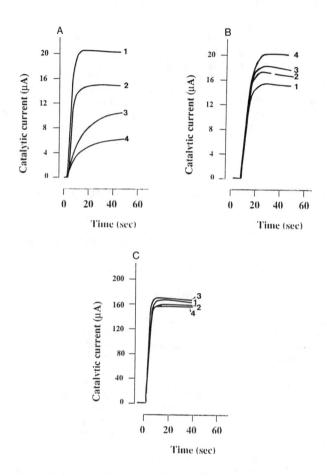

Figure 1. Current-time curves of glucose biosensors prepared A) in depositing the mixture of glucose oxidase and AQ atop the platinum electrode and allowing the solvent to evaporate at room temperature, B) in covering this AQ-glucose oxidase film with a Nafion layer or C) in heating the AQ-glucose oxidase film at 50°C in an oven for 30 min. The assays were done with 20 mM of glucose at pH 7.

was used with 10 μl of 1% AQ solution. As seen in Fig.1C, the current reaches 170 μA for the first assay in presence of 20 mM of glucose. This current value is 8 times larger than the value obtained above using the two others preparation techniques. This higher value seems to be related to a better organization of the film structure leading to enhanced mass transport of the substrates and products.

The effect of GOD concentration on the amperometric response to 20 mM of glucose of the biosensor prepared with the third protocol was evaluated for 3 glucose oxidase concentrations. As shown in Figure 2, the amperometric current response increases non-linearly from 120 μA to 180 μA in presence of oxygen as electron acceptor when the glucose oxidase concentration increases from 2 U to 12 U in 25 μl of AQ polymer. This is to be contrasted with AQ-GOD films coated with Nafion for which no increase of current was observed when the GOD concentration in the film is increased from 2 to 70 U (14). It is important to note that a minute amount of GOD, i.e. 2 U gives high catalytic responses. We cannot conclude for the moment if AQ has any stabilizing effect on the structure of glucose oxidase during the heating treatment even if a high catalytic current is generated. Effectively, recent studies on thermoinactivation of glucose oxidase in aqueous solution (21) have shown that less than 15% of glucose oxidase is inactivated when the glucose oxidase is heated at 50°C during 30 min.

The calibration curve for the AQ-choline oxidase electrode, prepared by casting 25 μl of 1% AQ solution and 3 U of enzyme, is given in Figure 3. The amperometric response of the biosensor reaches a maximum value of 18 μA when 2.5 mM of choline is added to the air saturated solution assay. The calibration curve is linear up to 1 mM of choline. An effective apparent Michaelis-Menten constant of 0.46 mM was calculated for choline and is to be compared to the value of 1.2 mM choline reported for soluble choline oxidase (22) obtained in air saturated solution. The Imax of the enzymatic film is 21.2 μA which is higher than the value of 8 μA obtained when the biosensor is covered with a Nafion layer (15). The time required to obtain the steady-state diminishes with the increase of choline concentration in the assay solution. This is explained by the neutralization of AQ negative charges when the concentration of the positively charged choline at pH 7 increases. As no information about the mass transfer coefficients are available, we cannot say if the small Km'' observed is due to the preconcentration of choline inside the film or to oxygen depletion inside the film.

The calibration curve for the AQ-glucose oxidase electrode, prepared by casting 25 μl of 1% AQ solution and 12 U of enzyme, is given in Figure 4. The amperometric response of the biosensor reaches a maximum value of 180 μA when 25 mM of glucose is added to the air saturated solution assay. The calibration curve is linear up to 10 mM of glucose as shown in the inset of Fig. 4. An effective apparent Michaelis-Menten constant for glucose was calculated and was 11.8 mM glucose which is to be compared to the value of 33 mM glucose reported for soluble glucose oxidase (23) obtained in air saturated solution. The Imax of the enzymatic film is 269 μA which is higher than the value of 45 μA obtained when the biosensor is covered with Nafion layer (15). These differences can be related to a decrease in the film permeability to glucose and oxygen when the GOD-AQ film is covered with a layer of Nafion.

One important limitation of a biosensor that is to be used for the evaluation of the concentration of an analyte in biological fluids is related to the amperometric current generated by anionic interferents in the biological sample. So, we have evaluated the effect of enzyme in the AQ film on its permeability to anionic species. This was performed by following the cyclic voltammetry of ferrocyanide 5 mM in 100 mM KCl solution on bare platinum, on platinum covered with 10 μl of 1% AQ solution containing 4 U of glucose oxidase and covered by 30 μl of 1%AQ solution containing 12 U of glucose oxidase. The cyclic voltammograms are presented in

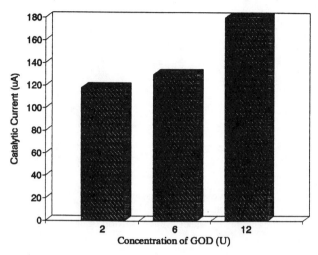

Figure 2. Effect of glucose oxidase concentration on the amperometric response to 20 mM glucose for biosensors prepared with different amounts of enzyme mixed with 25 μl of AQ 1%. The AQ-glucose oxidase films were dried at 50°C during 30 min in an oven.

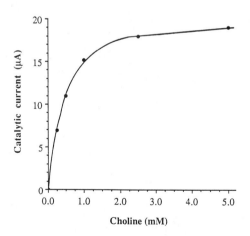

Figure 3. Calibration curve for AQ-choline oxidase electrode, prepared by casting 25 μl of 1% AQ solution and 3 U of enzyme. The film was dried at 50°C during 30 min in an oven.

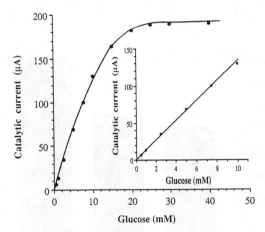

Figure 4.   Calibration curve for AQ-glucose oxidase electrode, prepared by casting 25 µl of 1% AQ solution and 12 U of enzyme. The film was dried at 50°C during 30 min in an oven. Inset: Linear portion of the calibration curve.

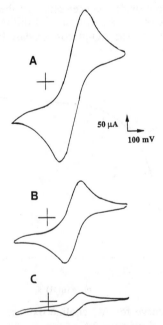

Figure 5. Cyclic voltammetry of 5 mM Fe(CN)$_6^{3-}$ in 0.1M KCl solution at (A) bare platinum electrode, (B) at a platinum electrode covered by 10 µl or (C) 30 µl of a 1% solution of AQ containing 2U/10 µl of glucose oxidase.The films were dried at 50°C during 30 min in an oven. The scan rate was 100 mV/s.

Figure 5. At a bare platinum electrode, Figure 5A, the height of the anodic peak current is 240 µA. With 10 µl (Fig.5B) and with 30 µl of AQ 1% solution containing 4 and 12 U of glucose oxidase respectively, the value of the anodic peak decreases to 105 and 45 µA. Indeed with the presence of glucose oxidase in the AQ film, the enzymatic film keeps a similar permselectivity against these anionic redox species than the one observed for the AQ film alone (results not shown).

The long-term stability of the Pt/AQ-GOD electrode stored at room temperature in air was studied by testing the reponse of the biosensor to 10 mM glucose during a period of three weeks. The biosensor was submitted to about 15 wet-dry cycles. During this period of time, no noticeable variation of the biosensor response was observed. The operational stability of the glucose biosensor was evaluated by determining the number of consecutive assay to 10 mM glucose that was possible to perform with the same biosensor. More than 100 samples analysis have been performed before the enzymatic film pelled off from the platinum electrode. During the first 40 assays, a decrease of less than 10% of the response of the biosensor to glucose was noticed. During the subsequent assays, the response of the biosensor was diminishing slowly to reach at the last assay a value of about 70% of the initial response. This response decrease was observed and correlated to the detachment of the enzymatic film from the platinum electrode.

Conclusion

In this study, it was demonstrated that AQ polymer is a very good polymer to entrap oxidase type enzyme since a small amount of enzyme is needed to obtain high catalytic currents. Glucose and choline biosensors having high catalytic current and good linearity have been obtained using a protocol which consisted in drying the film at 50°C during 30 min in an oven. This approach has given better results than the one using a Nafion overlayer since the redissolution of AQ-enzyme film in the assay solution is prevented *(15)*. Using AQ, it should be possible to prepare a wide variety of biosensors composed with enzymes of various structural stability. Experiments are in progress in our laboratories to determine if the protocol (50°C/30min) can be used with other enzymes more sensitive to heat than glucose or choline oxidases. Also, as Nafion *(6)*, AQ offers the possibility to incorporate by ion exchange redox mediator that will enable the use of a lower potential for the biosensor.

Acknowledgments

We thank the Natural Sciences and Engineering Research Council of Canada and the FCAR, programme d'établissement de nouveaux chercheurs of the Quebec Government for their financial support. J.W.C. would also like to thank UQAM for a post-doctoral fellowship. We also thank Prof. J. Wang for providing a reprint of the Ref. 16 prior to its publication.

Literature Cited

1) Fortier, G.; Brassard, E.; Bélanger, D. *Biosensors & Bioelectronics* **1990,** *5 ,* 472.
2) Foulds, N.C.; Lowe, C.R. *J. Chem. Soc. Faraday Trans. 1* **1986,** *82,* 1259.
3) Bartlett, P.N.; Whitaker, R. G. J. *Electroanal. Chem.* **1987,** *224,* 37.

4) Pandey, P.C. *J. Chem. Soc., Faraday Trans. 1,* **1988,** *84,* 2259.
5) Barlett, P.N.; Whitaker, R.G. *Biosensors,* **1987/88,** *3,* 79.
6) Bélanger, D.; Vaillancourt, M.; Chen, J.W.; Fortier, G. *in preparation.*
7) Faulkner, L.R. *C & EN,* **1984,** *62,* 28.
8) Murray, R.W., in *Electroanalytical Chemistry,* Marcel Dekker, A.J. Bard (ed), N.Y., NY, 1984, pp. 191-368.
9) Nagy, G.; Gerhardt, G.A.; Oke, A.F.; Rice, M.E.; Adams, R.N. J. *Electroanal. Chem.,* **1985,** *188,* 85.
10) Kristensen, E.W.; Werner, G.K.; Wightman, R.M. *Anal. Chem.,* **1987,** *59,* 1752.
11) Fortier, G.; Vaillancourt, M.; Bélanger, D. *Electroanalysis,* **1992,** in press.
12) Wang, J.; Lu, Z. *J. Electroanal. Chem.,* **1989,** *266,* 287.
13) Wang, J.; Golden, T. *Anal. Chem.,* **1989,** *61,* 1397.
14) Gorton, L.; Karan, H.I.; Hale, P.D.; Inagaki,T.; Okamoto, Y.; Skotheim, T.A. *Anal. Chim. Acta.,* **1990,** *228,* 23.
15) Fortier, G.; Béliveau, R.; Leblond, E.; Bélanger, D. *Anal. Lett.,* **1990,** *23,* 1607.
16) Wang, J.; Leech, D.; Ozsoz, M.;Martinez, S.; Smyth, M.R. *Anal. Chim. Acta,* **1991,** 245, 139.
17) Eastman Chemicals Products Inc. Publication No. GN-389A, **1989.**
18) Gennett, T.; Purdy, W.C. *Anal. Chem.,* **1990,** *62,* 2155.
19) Fortier, G.; Brassard, E.; Bélanger, D. *Biotechnol. Tech.,* **1988,** *2,* 177.
20) Harrison, D.J.; Turner, R.F.B.; Baltes, H.P. *Anal. Chem.,* **1988,** 60, 2002.
21) Fortier, G.; Bélanger, D. *Biotechnol. Bioengin.,* **1991,** *37,* 854.
22) Ikuta, S.I.; Imamura, S.I.; Misaki, H.; Horiuti, Y. *J. Biochem.,* **1977,** *82,* 1741.
23) Swoboda, B.E.P.; Massey, V. *J. Biol. Chem.,* **1965,** *240,* 2209.

RECEIVED October 22, 1991

Chapter 4

Electrochemical Characterization of Ferrocene Derivatives in a Perfluoropolymer Glucose Oxidase Electrode

Jian Wei Chen, Daniel Bélanger, and Guy Fortier

Groupe de Recherche en Enzymologie Fondamentale et Appliquée, Département de Chimie et Biochimie, Université du Québec à Montréal, C.P. 8888, Succursale A, Montreal, Quebec H3C 3P8, Canada

Glucose oxidase, GOD, was immobilized into Nafion film, into which artificial electron transfer mediators were incorporated. The mediators were ferrocene derivatives with various hydrophobicity obtained by varying the alkyl chain length. It was demonstrated that they were incorporated in the film with different maximum loadings and at different rates depending on the hydrophobicity. The diffusion coefficients for ferrocene derivatives in Nafion-GOD films decreased when the hydrophobicity of the redox couples increased. The rapid decrease of the catalytic current of glucose oxidation is related to the leaching of the ferrocene derivatives from the Nafion-GOD layer into the solution. However, hexadecanyldimethylaminomethyl ferrocene bromide, the most hydrophobic mediator used, gave a low but stable catalytic current in presence of glucose.

It is well demonstrated that glucose oxidase can not directly exchange electrons with a metallic electrode because its redox center is too far from the glucose oxidase's surface (*1*). Therefore, during the last decade, many considerations were directed to the redox mediators that can be used for the transport of electrons from the active center of the enzyme to the electrode surface. The first generation of amperometric biosensors uses the oxygen as the electron acceptor. In this particular case, the enzymatic reaction rate corresponds to the consumption of oxygen or to the electrochemical oxidation of hydrogen peroxide. The fluctuation of oxygen pressure and the high anodic potential required to oxidize hydrogen peroxide (0.7 V vs SCE) severely limited the application, especially in presence of blood interferents. Consequently, it has been proposed to substitute oxygen by artificial electron acceptors such as ferrocenes (*2-12*), quinones (*13*), ruthenium hexaamine (*14*), osmium complexes (*15-17*) and components of organic salts (*18*) which are characterized by lower redox potentials.

The approaches that have been proposed to immobilize artificial mediators include the adsorption of the redox mediator (*9*), the immobilization in carbon paste (*13*), the covalent linkage on electroinactive (*15*) or conducting polymer backbone (*10*), the covalent attachement to the enzyme structures (*3*) and

0097–6156/92/0487–0031$06.00/0

entrapment in ion-exchange polymer (19-22). DuPont's Nafion perfluoronated sulfonated ion-exchange polymer has been extensively used as the immobilization matrix of cationic species (23) and as the covering membrane for enzyme electrodes (24-26).

We have recently reported our work on the entrapment of glucose oxidase in Nafion film on a platinum electrode in an aerobic environment (19) and with dimethylaminomethyl ferrocene as an electron acceptor for reduced glucose oxidase (27). In the latter, incorporation of mediators into Nafion film without leakage to the aqueous solution is essential for long term operation of the Nafion-glucose oxidase-mediator systems. Considering the structure of Nafion, which possesses hydrophobic fluorocarbon chains and negatively charged sulfonated groups, we synthesized a serie of positively charged ferrocene derivatives with variable hydrocarbon length chain with the aim of improving their retention in the Nafion film (28) containing the enzyme. Their electrochemical behavior inside Nafion film has been studied and characterized in terms of incorporation and desorption rates, diffusion coefficients and, amperometric current generated in presence of glucose.

EXPERIMENTAL

Enzyme and substrate: Glucose oxidase, type VII-S(E.C. #1.1.3.4) was supplied from Sigma Chemicals Co. A stock solution of glucose oxidase (24,000 U/ml) was prepared by dissolving the appropriate amount of the enzyme in water. A stock solution of glucose (2M) (BDH Chemicals) solutions was prepared 24 hours prior to use in order to obtain the α anomer.

Mediators: Dimethylaminomethyl ferrocene (DMAFc) from Strem Chemicals was used as received. Trimethylaminomethyl ferrocene (C_1Fc^+) chloride was obtained by metathesis of the corresponding iodide salt (Strem Chemicals) with AgCl. AgI precipitate was filtered from the solution and cyclic voltammetry was employed to verify if I^- was completely removed. Hexanyldimethylaminomethyl ferrocene bromide (C_6FcBr) and hexadecanyl dimethylaminomethyl ferrocene bromide ($C_{16}FcBr$) were synthesized from DMAFc with bromohexane and bromohexadecane, respectively (29). The crude products were recrystallized from acetone-ether. Their structure and purity were verified by ^{1}H-NMR spectroscopy. DMAFc, C_6FcBr and $C_{16}FcBr$ are lightly soluble in aqueous solution whereas C_1FcCl has relatively higher solubility.

Buffer: All experiments were performed in 0.1M K_2HPO_4-KH_2PO_4 (pH 7.0) buffer solution. Distilled water from a Barnstead water system was used throughout.

Electrode and Nafion film: Platinum disk electrode (area = 0.28 cm^2) was fabricated as previously described (27). A solution of Nafion (EW = 1100) 5% w/v in a mixture of lower aliphatic alcohols and 10% water was obtained from Aldrich and diluted with methanol to give a stock solution of 0.35% w/v. Films of Nafion-glucose oxidase were formed by syringing aliquot (20 μl) of the mixed solution (86.7 μl of Nafion stock solution and 13.3 μl of glucose oxidase solution) at the surface of the platinum electrode. The solvents were left to evaporate at room temperature for at least 30 min. The extra amount of GOD was removed from the film by immersing the electrode in 0.1 M phosphate buffer for about 10 min. The film thickness is calculated based on the density 1.58 g/cm^3 (*19*).

Glucose measurement : The amperometric response to glucose was evaluated in a deaerated 0.1 M phophate buffer solution under mild stirring conditions. Prior to glucose measurement, the enzyme electrode was potentiostated at 0.7 V for 1 min and an appropriate aliquot of the stock solution of glucose was added to the buffered solution. The reading was taken at steady state.

Apparatus: Cyclic voltammetry and amperometric current-time curves were obtained with a Pine Instrument Inc., Model RDE4 bipotentiostat and Kipp & Zonen BD 91 XYY' recorder equipped with a time base module. All measurements were performed in a conventional single-compartment cell with a saturated calomel electrode as the reference electrode and a Pt mesh as the auxiliary electrode at room temperature. Chronoamperometry was made with EG&G Princeton Applied Research potentiostat/galvanostat Model 273 equipped with Model 270 Electrochemical Analysis Software.

RESULTS AND DISCUSSION

1. Electrochemistry of mediators in the Nafion-GOD film.

1.1 Incorporation of mediators in the Nafion-GOD film.

The incorporation of ferrocene derivatives into Nafion-GOD films was carried out by immersion of a Nafion-GOD coated electrode in a solution containing the ferrocene derivative. The ferrocenes such as DMAFc, C_1Fc^+ and C_6Fc^+ in the aqueous solution, respectively, partitionned or preconcentrated into mostly hydrophobic part of the Nafion-GOD films. A continuous cyclic voltammogram was performed during the incorporation and showed an increase in the anodic and cathodic currents. The increased peak current between each scan monitored the incorporation rate. The loading of ferrocene derivatives in the film was evaluated from the peak area on the slow scanned voltammogram. The steady-state cyclic voltammograms, corresponding to an equilibrium partition concentration, were obtained after approximately 30 min depending on the mediator concentration in the soaking solution. When the same concentration of DMAFc, C_1Fc^+ and C_6Fc^+ is used in the soaking solution, the initial incorporation rate is in the order DMAFc > C_1Fc^+ > C_6Fc^+, respectively. The final mediator concentration in the film of Nafion-GOD vs. the concentration for the different mediators in the solution are compared in Table I. It is interesting to note that when the mediator concentration in the soaking solution is small (<1 x 10^{-3} M), DMAFc is incorporated in the film at the higher concentration than C_1Fc^+, but at a lower concentration than C_6Fc^+. However, when the mediator concentration in the soaking solution is increased at 5 x 10^{-4} M for these three mediators, the DMAFc and C_1Fc^+ film concentrations become similar and C_6Fc^+ film concentration is slightly larger than both DMAFc and C_1Fc^+. In all cases, the C_6Fc^+ film concentration is the highest.

Nafion is characterized by its multi-phase structure (*30*). The interfacial model (*31*) includes three domains: a Teflon-like hydrophobic fluorocarbon phase, an ionic cluster phase and an interfacial region that separates the ionic and fluorocarbon chain phases. It was pointed out (32) that uncharged Fc resided in the interfacial domain whereas Fc^+ cations resided in the ionic clusters. Also it was demonstrated that (33) $Ru(bpy)_3^{2-}$ partitioning into the ionic clusters was negligible, compared with the amount retained in the interfacial region.

As a consequence of the hydrophobic nature of the Nafion film, DMAFc which is slightly more hydrophobic than C_1Fc^+, entered into Nafion film at a higher rate than C_1Fc^+ and partitioned into the hydrophobic interfacial region. The total concentration of DMAFc in the film is higher, as shown in Table I when its concentration in the solution is less than 1×10^{-4} M. As the concentration of the mediator in the solution was increased, the hydrophobic sites became occupied, then the subsequently loaded ferrocenes were forced to occupy less hydrophobic domain (ionic clusters) in the film. This is apparently not favorable for the incorporation of the uncharged DMAFc (in the reduced state) and therefore the concentration of C_1Fc^+ is higher than that of DMAFc. The larger concentration of C_6Fc^+ in this film is in agreement with the well known strong affinity shown by Nafion for hydrophobic cations (34). The presence of the long alkyl chain on the ferrocene allows for a greater extent of hydrophobic interactions with Nafion.

TABLE I. Partition of DMAFc, C_1Fc^+ and C_6Fc^+ Between the Nafion-GOD Film and the Soaking Solutions

Mediator Concentration in the Soaking Solution (M)	Mediator Concentration in the Film (M)		
	DMAFc	C_1Fc^+	C_6Fc^+
5×10^{-5}	4.1×10^{-2}	1.5×10^{-2}	6.1×10^{-2}
1×10^{-4}	6.2×10^{-2}	3.3×10^{-2}	9.1×10^{-2}
5×10^{-4}	1.7×10^{-1}	1.5×10^{-1}	2.9×10^{-1}
2.5×10^{-3}	4.0×10^{-1}	4.4×10^{-1}	5.5×10^{-1}

1.2 Diffusion coefficients of incorporated species in the Nafion-GOD films.

When a ferrocene derivative loaded Pt/Nafion-GOD electrode is transferred to a phosphate buffer solution containing no ferrocene derivatives, the observed redox waves reflect their electrochemical behavior in the film. The scan rate dependence of the cyclic voltammetry for the Pt/Nafion-GOD incorporated with different ferrocene derivatives was studied. A linear plot of the anodic peak current against the square root of the scan rate was obtained in all cases, which is indicative of diffusion controlled redox process. Typical chronoamperometric

Cottrell plots for Pt/Nafion-GOD with loaded ferrocene derivative are shown in Figure 1 for DMAFc, C_1Fc^+ and C_6Fc^+ respectively. As predicted for a diffusion controlled process, these plots are linear with zero intercepts. Apparent diffusion coefficients were calculated from the Cottrell plot slopes. The values of the apparent diffusion coefficient for C_1Fc^+, DMAFc and C_6Fc^+ in Nafion-GOD films are 2.5×10^{-9}, 6.5×10^{-10} and 9.4×10^{-10} cm^2/s, respectively (Table II). Thus, the apparent diffusion coefficient is very sensitive to the length of the alkyl chain and the hydrophobicity of the ferrocene derivative. The apparent diffusion coefficient decreases when the length of the alkyl chain increases and when the hydrophobic nature of the ferrocene derivative increases. The latter point pertains to the fact that C_1Fc^+ is less hydrophobic than DMAFc although it bears one additional methyl substituant on the nitrogen. Such behavior has been reported for plain Nafion films loaded with the same ferrocene derivative (*28*) and for Nafion gels with several ferrocenes with variable hydrophobic nature (*35*). Table II gives their diffusion coefficients in Nafion film and in homogeneous solution. The diffusion coefficient of C_6Fc^+ in homogeneous solution is the smallest because of its size according to Stokes-Einstein equation. However, this difference becomes larger in Nafion films. The decrease of the apparent diffusion coefficient with an increase in the hydrophobic nature of the ferrocene can be attributed to the hydrophobic interactions between the alkyl chain and the perfluorinated chains (*28*) and the increase of the size of the electroactive probe as the length of the alkyl chain of the ferrocene is increased. Another factor might also be invoked to explain the observed variation of the apparent diffusion coefficient. It is the decrease in the overall content of water in the Nafion films with an increase in the hydrophobic nature of the ferrocene. The decrease of the water content limits the ions movement across the Nafion matrix (*28*).

TABLE II Comparison of Diffusion Coefficients in Homogeneous Aqueous Solution and in Nafion Film (cm^2/s)

	in 0.1 M Phos.	in Nafion-GOD
DMAFc	2.9×10^{-6}	6.5×10^{-10}
C_1Fc^+	2.4×10^{-6}	2.5×10^{-9}
C_6Fc^+	1.7×10^{-6}	9.4×10^{-10}

2. Mediation of glucose oxidation by incorporated ferrocene derivative.

In Figure 2 is shown the decrease of the anodic peak current of the cyclic voltammograms at a sweep rate of 100 mV/s, which is proportional to the mediator loading in the Nafion-GOD film, as a function of time after the fully loaded Pt/Nafion-GOD-DMAFc, Pt/Nafion-GOD-ferrocene were respectively

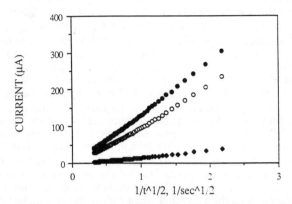

Figure 1. Cottrell plots obtained from oxidation of (◆) Pt/Nafion-GOD-DMAFc, (O) Pt/Nafion-GOD-C_1Fc^+ and (●) Pt/Nafion-GOD-C_6Fc^+ in 0.1M phosphate buffer. DMAFc, C_1Fc^+ and C_6Fc^+ were incorporated into the Nafion-GOD films from a 0.5 mM solution.

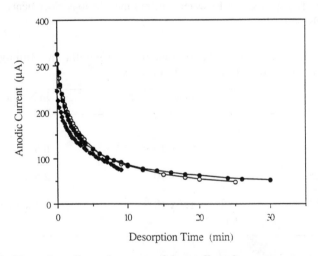

Figure 2. Plots of anodic peak current of the cyclic voltammograms recorded at 100 mV/s vs time for the Nafion-GOD films fully loaded with (O) C_1Fc^+, (●) C_6Fc^+ and (◆) DMAFc, respectively in 0.1M phosphate buffer solution containing no mediators.

brought in contact with buffer solution containing no ferrocene derivatives. In contrast with the incorporation process, the desorption process first occurs in the ionic cluster phase resulting in a sharp decay for the first minutes. The decay slowed down but continued until a low level loading corresponding to the loading in the interfacial region is reached. Following the loading decay, the catalytic current due to the glucose oxidation consequently decreased as a function of the number of glucose sample measurement. Table III presents the decrease of the glucose response obtained from Pt/Nafion-GOD-DMAFc, Pt/Nafion-GOD-C_1Fc^+ and Pt/Nafion-GOD-C_6Fc^+ electrodes, respectively. The catalytic current to 20 mM glucose was stabilized at ~9 μA for Pt/Nafion-GOD-DMAFc and ~12 μA for Pt/Nafion-GOD- C_1Fc^+. With the C_6Fc^+, the glucose response consequently decreased to 25-30% of its initial value after 13 measurements. These Nafion-GOD films contain 4-6 x 10^{-3} M of the mediators after the tenth sample measurement as evaluated by cyclic voltammetry at slow scan rate.

TABLE III Catalytic Current Response of Nafion-GOD loaded with mediator to 20 mM Glucose (μA)

Measurement Number		Mediator	
	DMAFc	C_1Fc^+	C_6Fc^+
1	30	41	23
5	15	19	12
9	9	10	8
13	9	12	6

* The applied potential was 0.7 V vs SCE.

Garcia and Kaifer (*28*) pointed out that ferrocene derivatives with longer alkyl chain were retained inside of a Nafion film more strongly. However, our results suggest that the presence of glucose oxidase in Nafion films induced changes in the Nafion film's hydrophobicity. In our hands, a longer alkyl chain, C_6Fc^+ ferrocene derivative is not retained for longer period than C_1Fc^+ in the Nafion-GOD film. It is worth noting that the initial concentration of the ferrocene derivatives in the Nafion-GOD films of Fig.2 is higher than that used by Garcia and Kaifer (*28*) for the Nafion coated electrode. However, the lack of stability for the C_6Fc^+ based electrode cannot be related to the fact that the initial concentration is higher and thus leads to a rapid leaching of ferrocene because the final concentration of ferrocene derivatives in the Nafion-GOD coating (4 - 6 x 10^{-3} or 0.56 - 0.84 x 10^{-9} mol/cm^2) is much lower than that found by Garcia and Kaifer (*28*).

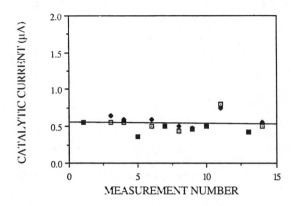

Figure 3. Current responses to anaerobic 20 mM glucose vs number of glucose measurements for two Pt/Nation-GOD-$C_{16}Fc^+$ electrodes poised at 0.7 V vs SCE.

Due to the micellization and adsorption effect, $C_{16}Fc^+$ cannot be incorporated into Nafion film by immersion of a coated Pt/Nafion-GOD electrode. Instead, a mixture of Nafion methanol solution, GOD water solution and a calculated amount of $C_{16}FcBr$ was directly deposited on the Pt electrode. The concentration of $C_{16}FcBr$ (0.28M) is calculated based on the 20% compensation of available Nafion sulfonate sites. However, the concentration of $C_{16}Fc^+$ evaluated from a slow scan cyclic voltammogram is only 0.02 M or 1.3% compensation. The difference between the amount of $C_{16}Fc^+$ deposited and measured may be due to the fact that a large fraction of $C_{16}Fc^+$ are not electrochemically accessible. However, a more likely explanation is that the ferrocene derivatives would be simultaneously removed with the enzyme from the Nafion coating when the enzyme electrode was washed in phosphate buffer during the fabrication of the electrode. The Nafion-GOD-$C_{16}Fc^+$ electrode is more stable during successive glucose measurements (Figure 3). It is worth noting that from Figure 3 and Table III ferrocenes with the longer alkyl chain yield the lowest catalytic current responses.

CONCLUSION

We used Nafion as a polymer support to immobilize glucose oxidase and to incorporate electron transfer mediators. Fully loaded Pt/Nafion-GOD film with DMAFc, C_1Fc^+ and C_6Fc^+ showed rapid leaching of the mediator from the film to the contacting solution. The catalytic currents corresponding to 20 mM glucose decrease after each glucose measurement. However, the catalytic current can be stabilized when the mediator concentration in the Nafion-GOD film is low. $C_{16}Fc^+$ can strongly be retained in the Nafion-GOD film, giving stable but small catalytic currents to 20 mM glucose. Indeed, a counterbalance of mediator efficiency vs retention inside the Nafion matrix should be considered for our present system.

ACKNOWLEDGMENTS

The financial support of the Conseil de Recherche en Sciences Naturelles et en Génie, CRSNG, du Canada and the Fonds pour la Formation de Chercheurs et l'Aide à la Recherche, FCAR, du Gouvernement du Québec is gratefully acknowledged.

Literature Cited.

1. Heller, A. *Acc. Chem. Res.*, **1990**, 23, 128.
2. Cass, A.E.G.;   Davis, G.;   Green, M.J.; Hill, H.A.O. *J. Chem. Soc., Chem. Commun.*, **1987**, 1603.
3. Degani, Y.; Heller, A. *J. Phys. Chem.*, **1987**, 91, 1285.
4. Jonsson, G.; Gorton, L.; Peterson, L. *Electroanalysis*, **1989**, 1, 49.
5. Lange, M.A.; Chambers, J.Q. *Anal. Chim. Acta*, **1985**, 175, 89.
6. Wang, J.; Wu, L.H.; Lu, Z. Li , R.; Sanchez, J. *Anal. Chim. Acta*, **1990**, 228, 251.
7. Amine, A.; Kauffmann, J.M.;   Patriarche, G.J. *Talanta*, **1991**, 38, 107.
8. Hale, P.D.; Inagaki, T.; Karan, H.I.; Okamoto, Y.;   Skotheim, T.A. *J. Am. Chem. Soc.*, **1989**, 111, 3482.
9. Cass, A.E.G.; Davis, G.; Francis, G.D.; Hill, H.A.O.; Aston, W.J.; Higgins, I.J.; Plotvin, E.V.;   Scott, L.D.L.;   Turner, A.P.F. *Anal. Chem.*, **1984**, 56, 667.
10. Foulds, N.C.; Lowe, C.R. *Anal. Chem.*, **1988**, 60, 2473.
11. Schuhmann, W.; Wohlschäger, H.; Lammert, R.; Schmidt, H.L.; Löffler, U., Wiemhöfer, M.D.; Göpel, W. *Sensors and Actuators*, **1990**, B1, 571.
12. Schuhmann, W.; Ohara, T.J.; Schmidt, H.L.; Heller, A. *J. Am. Chem. Soc.*, **1991**, 113, 1394.
13. Ikeda, T.; Hamada, H.; Miki, K.;   Senda, M. *Agric. Biol. Chem.*, **1985**, 49, 541.
14. Degani, Y.; Heller, A. *J. Am. Chem. Soc.*, **1988**, 110, 2651, .
15. Degani, Y.; Heller, A. *J. Am. Chem. Soc.*, **1989**, 111, 2357, .
16. Pishko, M.V.; Katakis, I.; Lindquist, S.E.; Ye, L.; Gregg, B.A.; Heller, A. *Angew. Chem. Int. Ed. Engl.*, **1990**, 29, 82, .
17. Gregg, B.A.;   Heller, A. *Anal. Chem.*, **1990**, 62, 258.
18. Bartlett, P.N. *J. Electroanal. Chem.*, **1991**, 300, 175.
19. Fortier, G.; Vaillancourt, M.; Bélanger, D. *Electroanalysis*, **1992**, in press.
20. Vaillancourt, M.; Bélanger, D.; Fortier, G. *177th Meeting of the Electrochemical Society*, Abstract #712 **(1990)**.
21. Rishpon, J.; Zawodzinski, T.A.; Gottesfeld, S. *177th Meeting of the Electrochemical Society*, Abstract #713 **(1990)**.
22. Weber, J.; Kavan, L.;   Sticha, M. *J. Electroanal. Chem.*, **1991**, 303, 237.
23. Murray, R.W. in *Electroanalytical Chemistry;* Bard, A.J. Ed., Marcel Dekker: New York, **1984**, Vol. 13; pp 191-368.
24. Gunasingham, H.; Beng Tan, C. *Analyst*, **1989**, 114, 695.
25. Harrison, D.J.; Turner; R.F.B.;   Baltes, H.P. *Anal. Chem.*, **1988**, 60, 2002.

RECEIVED November 26, 1991

Chapter 5

Protein Stabilization in Biosensor Systems

Timothy D. Gibson[1] and John R. Woodward[2]

[1]Department of Biochemistry and Molecular Biology, University of Leeds,
Leeds, Yorkshire LS2 9JT, United Kingdom
[2]Yellow Springs Instrument Company, 1726 Brannum Lane, Box 279,
Yellow Springs, OH 45387

One of the most important problems in the development of commercial biosensors is the stabilization of the enzymes used as the biological detection agents in the sensor. Although many different biosensors have been constructed, few have been developed in commercially successful instruments. The molecular parameters which influence the activity and stability of enzymes in the dry state, and when rehydrated in an electrode surface, have been investigated. A new and novel system has been developed, which appears to have broad application to the stabilization of enzymes. The use of this system to stabilize enzymes on electrodes is reported, together with a discussion on the mode of action of stabilization in general.

It was, perhaps, no accident that the first working enzyme electrode to be produced utilized glucose oxidase to measure glucose via hydrogen peroxide (1). The reasons for the use of glucose oxidase were twofold; firstly, glucose was an interesting and important analyte, especially in medical diagnostic tests, and secondly glucose oxidase was found to be a very stable enzyme. Constructing a multitude of enzyme based sensors since that time has not been a major problem; the problem has been to obtain stable, long lived enzyme based sensors which can stand prolonged storage and perform well in use for extended periods. Unfortunately not many enzymes are as obligingly stable as glucose oxidase. Indeed even glucose oxidase presents problems; there are perhaps dozens of patents on glucose electrodes containing this enzyme and hundreds of papers, yet only two or three successful commercial probes are produced which use glucose oxidase. One of the major problems in expanding the use of enzyme sensors beyond that for glucose has been protein stabilization, especially when it has been necessary to dry and then rehydrate on the transducer's surface before use.

This problem lies at the heart of all enzyme electrode manufacture. Proteins are biological molecules, and there is an apocryphal tale concerning Bragg when he was head of the Cavendish Laboratory, and Frances Crick told him he was going to study DNA crystal structure. Bragg's reply was gruff, and as it turned out quite apt, "thats no

0097–6156/92/0487–0040$06.00/0

good" he is reported to have said "that's a biological molecule, biological things wriggle"! Here in lies the ability of proteins to act as enzymes. The conformational structure of a protein is not fixed completely in 3-dimensional space. The protein breathes, sometimes quite literally as in the case of heamoglobin, where Max Perutz was able to show that the protein structure expanded during oxygen binding to allow the entry of a water molecule into the center of the structure. Movement is often essential to activity but excess movement can lead to unfolding and thus loss of activity. Therefore, all protein chemists and especially those working on stabilization, have the task of maintaining a delicate balance of movement in the protein, enough to ensure activity, not enough to allow unfolding and loss of activity.

The aim of this paper is to describe various methods employed to stabilize enzymes which are particularly labile. The problem of stabilization on thin films and on surfaces will also be addressed.

Methods of enzyme stabilization

Molecular Modification. A number of methods of enzyme stabilization have been tried with respect to enzyme electrode construction. Obviously where naturally stable enzymes are available, they are used in preference to unstable alternatives. Recently the study of X-ray crystal structure and DNA sequence information has led to predictive models which suggest that by changing certain amino acids in the protein sequence, via site directed mutagenisis, improvements in stability can be achieved. Workers such as Fersht are making point mutations in the DNA sequence to bring about such amino acid changes and thus improve the temperature stability of various enzymes (3,4). These experiments are beyond the scope of this paper but clearly point to exciting new technologies for enzyme stabilization in the future.

Other methods of stabilization include chemical or carbohydrate modification of enzymes. Modifications of reactive groups on proteins without insolubilization has been used to enhance stability in solution. Grafting of polysaccharides or synthetic polymers, alkalation, acetylation and amino acid modification have all been reported. (5)

It is well known that glycosylated proteins are very stable, therefore grafting polysaccharides or short chains of sugar molecules onto proteins has been used to improve enzyme stability. (6, 7, 8)

Cross linking. Enzymes can be cross linked between sub units or the domains of the protein using reagents such as gluteraldehyde, dicarboxilic acids, diisocyanates, bisimidates or diamines. The effect of cross linking is to introduce a clamping effect on the enzyme which produces a "strap" across the surface of the protein, reducing domain mobility and degradation of the 3-D structure. Cross linking together with entrapment or immobilization on a surface is often used in the preparation of enzyme electrodes.

Physical entrapment. This is often used in sandwich membranes, the enzyme is trapped in a cross linked polymer or adsorpted onto a gel matrix when constructing a sensor. It relies on weak hydrogen bonding and electrostatic interactions between the protein and the surface. Whereas this is often effective, loss of activity by protein leaching is often a severe problem.

Immobilization of enzymes. Enzymes consist of amino acids which contain reactive groups such as amino(lysine, c-terminus), thiol (cysteine), carbonyl (aspartate, glutamate and c-terminus), aromatic hydroxyl (tyrosine) and aliphatic hydroxyl (serine and threonine). Chemical, ionic or chelation reactions with such groups can enable us to attach the amino acids and hence proteins to insoluble, inert supports. Immobilization is one of the best ways of stabilizing enzymes. There is a vast literature on this subject and the reader is directed to Barker (9) and Coughlan et al (10) for further reading on specific systems, techniques and applications of immobilization.

Immobilization imparts stability to proteins by restricting the movement of the protein molecule by attachment to an inert body via chemical bonds. The various domains are therefore held in the correct orientation to retain activity at least over an extended period of time when compared with enzymes in free solution. Interaction with the surface upon which the protein is immobilized is important and has a direct effect on stability and activity, possibly by electrostatic interaction between the surface and adjacent amino acids. This phenomenon is well illustrated when immobilizing alcohol oxidase on to nylon tubing to construct tube reactors for Flow Injection Analysis (FIA) systems (11). We found that in the process of immobilizing the enzyme we could see direct effects of the distance between the enzyme and the surface of the nylon tubing. An alkane chain was used as a spacer between the derivatised nylon and the amino acid involved in the immobilization. We found that enzyme activity varied according to the length of the alkane chain (Gibson and Woodward unpublished results). Proximity to the nylon appeared to inhibit the enzyme probably due to the hydrophobic nature of the nylon surface. Polysaccharide amines were effective in reducing hydrophobic interactions when they were used as spacers, and they prolonged enzyme life in use when compared to alkane chain spacer molecules (Fig 1).

Another excellent method of enzyme immobilization is the derivitization of controlled pore glass (CPG) (12) followed by the binding of the selected enzyme. The enzyme is contained within pores in the glass using diamino alkane side chains bound on one end of the glass through silanization and to the protein via an aldehyde group. The stabilization effect of this matrix is excellent. We have immobilized alcohol oxidase onto CPG and used CPG columns in an FIA system for at least three months at room temperature (20 - 24°C). Such columns can be stored for one year at 4°C and still show high-activities. Prolonged storage is followed by a loss of activity over the first few days of use, after which the remaining activity is stable for at least thirty days (Fig. 2). Combining CPG columns with peroxide probes or separate membrane biosensors produces a very stable long lived system.

Enzyme stabilization by carbohydrates and polymers.

Stabilization with sugar additives. In general it is necessary to immobilize enzymes for two reasons. Firstly, because it helps to improve their stability and hence the lifetime of the device in which they are operating; secondly because we do not want them to escape! The enzyme needs to be held in the matrix and in the case of an elec-

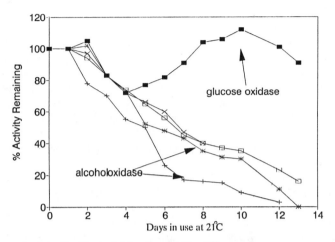

Figure 1.   Activity Profile for Nylon Immobilized Enzymes.
Spacer molecules used in this experiment were: (■) 1,2 Diaminoethane; (+)
1,2 Diaminoethane; (∗) Adipic Acid Dihydrazide; (□) Polyamine Dextran;(✕)
Pectinamine.

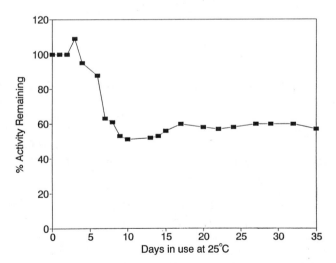

Figure 2.   Activity Profile for Control Pore Glass Immobilized Alcohol Oxidase.

trode it needs to be held in intimate contact with the electrode surface. Thus immobilization is an integral part of sensor construction in many cases. However, this is not sufficient to guarantee a good commercial product or even a good research product. The electrode must be stored. For convenience this is usually dry storage, and thereby we introduce a whole new set of problems in the preparation of electrodes and other enzyme assay systems.

One of the best (or at least, favorite) ways of drying and storing enzymes in free solution is to freeze dry them. This is often done in the presence of stabilizers. Many different compounds have been used but by far the most popular have been poly-alcohols and sugars. Many workers have reported that sugars such as trehalose, maltose, lactose, sucrose, glucose and galactose can be used to stabilize proteins, especially enzymes (13, 14, 15, and 16). However, there has been no overall pattern to the use of these compounds as stabilizers, their use has to a large extent been empirical with the general explanation that lowering water activity yet maintaining water interactions with the protein improves stability (13).

We have extended some of this work over the last five years by studying the effect of sugars and polymers on the stability of enzymes in the free state, in diagnostic dry strip tests and on electrodes. Our findings are reported in the rest of this paper using alcohol oxidase as a model unstable enzyme system.

Alcohol oxidase. We have chosen to use alcohol oxidase from Hansenula polymorpha (2) in our research. The enzyme has eight sub-units, all of which must be associated for the enzyme to exhibit activity. It is a large enzyme Mr≈600,000 and has poor stability upon freeze drying. This factor combined with an almost flat activity response to pH between 6 and 10 and a maximum temperature activity at 40-50°C made the enzyme an ideal candidate for the study of enzyme stabilization in the dry state, especially as we wished to prepare diagnostic kits for ethanol and solid phase enzyme based sensors.

Results and Discussion

Effective Stabilizers: In general, many of the stabilizers reported for other enzymes were not effective in preserving activity on drying of alcohol oxidase. The enzyme auto-oxidizes upon drying and destroys itself. Reducing sugars, added at 10% final volume, appeared to make this effect worse rather than improving stability (Fig. 3a). However, disaccharides, added as described by Gibson (11), were found to give very good results, especially trehalose, a sugar which is found in many organisms which dry out and dehydrate in natural conditions (Fig. 3b & 3c). Trehalose, inositol and cellobiose all appeared to induce a stimulation of activity. No adequate explanation exists for this phenomenon unless it represents a true stimulation of the enzyme kinetics by slight structural movements within the molecule to accommodate these sugars even after rehydration. From our results, it was immediately obvious that reducing sugars and non cyclic sugars destroyed enzyme activity, whereas cyclic polyalcohol type sugars or disaccharides with blocked functional aldehydes tended to stabilize. The stabilizing properties of these sugars appeared to be related to their structure, very slight changes in structure could lead to devastating effects. Sorbitol completely destroyed activity after drying (Fig 3c) but mannitol treated enzyme retained about

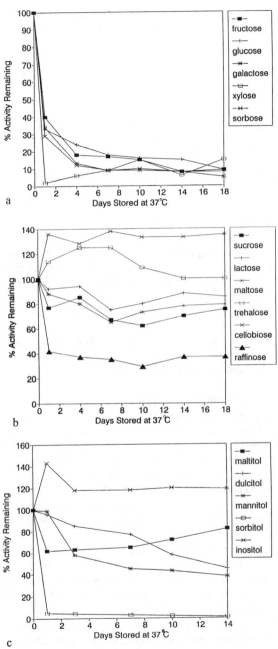

Figure 3. Alcohol oxidase stabilization by mono-, di-, and trisaccharides. Sugars were added to the enzyme at concentrations of 1–10% immediately prior to drying. The solutions were dried in shallow dishes at 30 °C under a vacuum and then harvested, ground to a powder, and stored in vials at 37 °C. Enzyme activity was assayed with methanol as a substrate using an oxygen electrode and colorimetric assay.

60% activity after three days and 40% after fourteen days. The only difference between these sugars appears to be the position of one hydroxyl group on the second carbon atom.

$$
\begin{array}{ccc}
CH_2OH & CH_2OH & CH_2OH \\
| & | & | \\
HO\text{-}C\text{-}H & H\text{-}C\text{-}OH & H\text{-}C\text{-}OH \\
| & | & | \\
HO\text{-}C\text{-}H & HO\text{-}C\text{-}H & HO\text{-}C\text{-}H \\
| & | & | \\
H\text{-}C\text{-}OH & H\text{-}C\text{-}OH & HO\text{-}C\text{-}H \\
| & | & | \\
H\text{-}C\text{-}OH & H\text{-}C\text{-}OH & H\text{-}C\text{-}OH \\
| & | & | \\
CH_2OH & CH_2OH & CH_2OH \\
\text{Mannitol} & \text{Sorbitol} & \text{Dulcitol}
\end{array}
$$

Dulcitol, another linear polyalcohol sugar gave improved stabilization with a further slight variation on the structure of sorbitol. Raffinose appears to have been a poor stabilizer due to the presence of a reducing sugar in the structure. The requirement for multiple alcohol groups on the stabilizer was further demonstrated using dextrans of various molecule weights (Fig. 4), similar results were reported by Monsan and Combes (13).

Whereas the use of disaccharides and some polysaccharides gave excellent results when drying the free enzyme, we experienced very poor results when we attempted to use the same stabilizers with immobilized enzymes or enzymes adsorbed onto solid surfaces.

Alcohol oxidase could be immobilized onto DEAE-sepharose (Pharmacia) and had very high stability even when dried and rehydrated. We therefore tried to stabilize alcohol oxidize on a solid surface and as free enzyme using DEAE dextran. However this was not successful (Fig. 5). We also knew that lactitol had been used in stabilizing various antibodies during the manufacture of antibody based tests, however lactitol also had poor stabilizing properties when dried with alcohol oxidase (Fig 5). Unexpectedly, however, we found that a combination of DEAE dextran and lactitol gave excellent stabilization of the enzyme (Fig 5.). Further experiments showed that the combination of stabilizers was generic and would be used to stabilize broad selection of enzymes (Table 1 and see ref. 17).

The DEAE dextran/lactitol stabilization system was particularly useful when examining the effect of pH and method of drying on enzyme stability. Figure 6 shows the result of drying and storing the alcohol oxidase previously disolved in buffer at various pH values. As can be seen from the graph maximum stability occurred at pH 7.87 almost exactly in the middle of the pH activity curve for alcohol oxidase (see ref. 2). However when the enzyme was co-immobilized with peroxidase on a graphite electrode the optimum pH for storage was 6.0 (Fig. 7). We observed that storage conditions such as the texture and material make up of a surface could affect the stability of the enzyme and the pH at which it must be dried. Hence, when dry strip diagnostic assays were prepared a cellulose base was used and this required a pH of 7.91 (17).

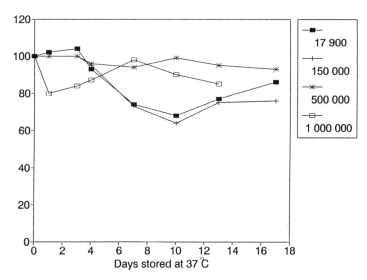

Figure 4. Effect of Dextrans on the Dry Stabilization of Alcohol Oxidase. Dextrans were added to the liquid enzyme at a concentration of 1-5%.

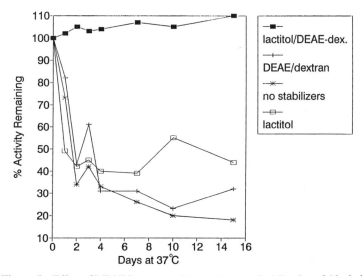

Figure 5. Effect of DEAE Dextran and Lactitol on Dry Stabilization of Alcohol Oxidase.

Table 1. Stabilization of Enzymes by Polyelectrolyte/polyhydroxyl Molecules

Enzyme	Stabilizer†	% Activity Remaining		Days at 37°C
		(a) After drying	(b) After incubation at -37°C	
Cholesterol oxidase	+	96.0	93.0	17
	-	78.0	38.0	17
Choline oxidase	+	97.7	80.6	15
	-	63.3	3.9	15
Glycerol 3 phosphate oxidase	+	98.2	111.3	15
	-	110.7	59.7	15
Lactate oxidase	+	90.0	91.6	15
	-	77.1	91.1	15
Galactose oxidase	+	125.0	117.0	16
	-	117.0	38.0	16
Uricase	+	91.0	100.0	10
	-	82.0	26.0	15
Peroxidase	+	117.0	100.0	15
	-	106.0	31.3	15
Alkaline phoshpatase	+	115.0	113.0	15
	-	112.0	64.0	15
Diacetyl reductase	+	216.0	144.0	8
	-	79.0	9.0	8
Lactate dehydrogenase	+	96.0	88.0	54
	-	64.0	46.0	54
B-galactosidase	+	115.0	113.0	36
	-	119.0	52.0	36
Maleate dehydrogenase	+	114.0	96.0	20
	-	14.0	2.0	20

† + = stabilizers added before drying
 - = stabilizers absent

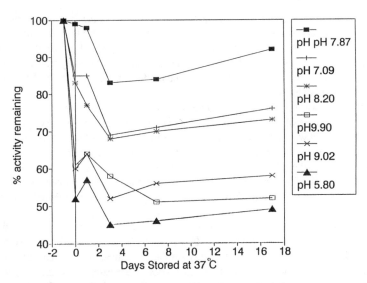

Figure 6.   Alcohol Oxidase Stabilization; pH Effect on Activity.
The enzyme was dried with a mixture of lactitol and DEAE dextran (see Gibson & Woodward, 1989) at the pH's indicated in the figure.

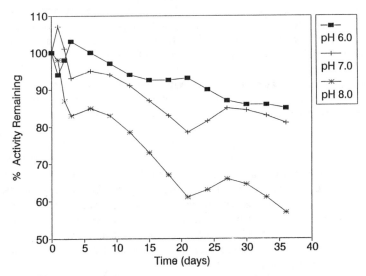

Figure 7.   Effect of pH on Storage Stability of Alcohol Electrodes.
Alcohol Electrodes were prepared by layering 20μL of a saturated solution of N-methyl phenazinium-tetracyanoquinodimethane onto a base graphite electrode. After drying, these electrodes were dipped in N-cyclohexyl-n (2-morpholinoethyl) - carbodimide methylpotoluene sulphonate for 90 minutes. They were then washed in carbonate buffer pH 9.8 and immersed in a solution of alcohol oxidase and peroxidase, and stabilized at various pH values. Electrodes were vacuum dried and stored under vacuum.

Although freeze drying has often been used to prepare stable dry enzyme preparations we found that when preparing enzyme powders and electrodes it was preferable to vacuum dry the enzyme in the presence of stabilizers (Fig. 8). However this was not true in all cases. Cholesterol oxidase was found to be stabilized best by freeze drying in stabilizers and buffer at pH 5.5 well below the pH for optimal activity (Gibson and Woodward unpublished results).

This work was further extended to the study of enzymes immobilized on graphite electrodes in the case of alcohol oxidase/peroxidase electrodes. The results illustrate the fact that sugars which stabilize one enzyme in free state do not necessarily do so on solid surfaces. Figure 9 shows that trehalose was a very poor stabilizer on graphite electrodes whereas the DEAE dextran/lactitol treated electrodes retained at least 95% of their activity after 35 days at 37°C. Mediated glutamate oxidase probes contructed on graphite gave similar results (Parker, S., University of Leeds, unpublished results).

Mechanism of action of stabilizers. Figure 10 shows the structure of some of the sugars which stabilize enzymes upon drying. It can be seen that although maltose, ß-cellobiose and trehalose have superficially similar structures, these differences cause major effects on stability. Both trehalose and cellobiose are excellent stabilizers, however, maltose gives only partial stability. The only difference between cellobiose and maltose is that cellobiose is 4-0 ßD glucopyranosyl D-glucose and maltose is 4-0-α-D-glucopyranosyle D-glucose, thus there is a stereo specific difference between the two sugars. The question therefore arises as to how these sugars effect stabilization.

The addition of polyhydroxyl compounds to enzyme solutions have been shown to increase the stabilities of enzymes, (13, 16, 19, 20). This is thought to be due to the interaction of the polyhydroxyl compound, (e.g. sucrose, polyethylene glycols, sugar alcohols, etc), with water in the system. This effectively reduces the protein - water interactions as the polyhydroxy compounds become preferentially hydrated and thus the hydrophobic interactions of the protein structure are effectively strengthened. This leads to an increased resistance to thermal denaturation of the protein structure, and in the case of enzymes, an increase in the stability of the enzyme, shown by retention of enzymic activity at temperatures at which unmodified aqueous enzyme solutions are deactivated.

This effect of polyhydroxyl compounds may not be quite as simple as it has been described, as the structure of the polyhydroxyl compound may play some part in effective stabilization of enzymes in "wet" systems. Thus Fujita et al, (20) reported that inositol was more effective than sorbitol in stabilizing lysozyme in aqueous solutions. Both compounds contain six hydroxyl groups, but inositol is cyclic in structure whereas sorbitol is linear, Fig 10. The interaction of polyhydroxyl compounds with water promotes a change in the molecular structure of water. Inositol was reported to have a larger structure-making effect than sorbitol, which accounted for the greater stabilization effect of this compound.

Similarly Ye et al, (21) reported that glucose oxidase was established in solution by a variety of compounds which included polyhydroxyls, (xylitol being the most effective), polyethylene glycols, (PEGs) and inorganic salts, all of which showed the ability to affect the structure of water by making it more "organized". This effect was not purely dependent on the hydroxyl content of the additive used, as PEG molecules

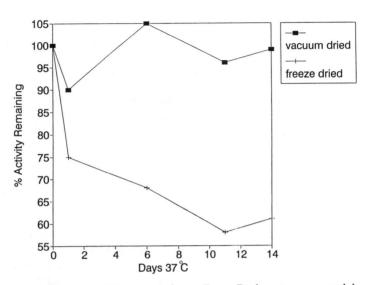

Figure 8.   Effect of Vacuum drying vs Freeze Drying on enzyme activity.

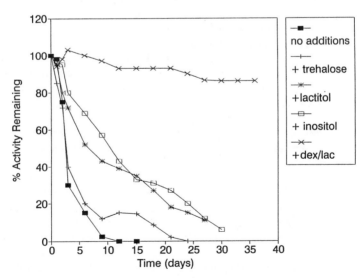

Figure 9.   Effect of Stabilizers on Alcohol Oxidase Electrode Stability.

only have two free hydroxyls and inorganic salts have none, yet both gave stabilization effects. Also solutions of glucose oxidase in D_2O, which has been reported to be a more structured solvent than water, showed enhanced thermal stability when compared to aqueous solutions.

The conclusion that was drawn for the stabilization of protein structure in aqueous solvent systems was as follows: the type of protein, the hydrophilic or hydrophobic character and thus the subsequent interaction of the protein with, (i) water and, (ii) the additives themselves, all play an important part of the stabilization process.

Possible Role of Polyhydroxyl Compounds in dry systems. Polyhydroxyl compounds when dried with enzymes tend to stabilize activity. This is thought to be due to the hydroxyl groups holding or substituting for the "bound" water which is necessary for the retention of the tertiary structure of the protein and the subsequent activity of the molecule, (22). Similarly removal of the carbohydrate, (polyhydroxyl) side chains of certain fungal enzymes rendered them susceptible to the effects of dehydration, in that enzyme activity was lost compared to untreated controls with intact carbohydrate side chains. The water - carbohydrate interaction was thought to be necessary for stability of the enzymes, (23). When drying a protein in the presence of polyhydroxyl compounds which tend to interact with and order the structure of water, it may be envisaged that the molecules may "coat" the surface of the protein with a layer or layers of polyhydroxyl - water complexes. The polyhydroxyl compounds used are usually small molecules and as such, will probably penetrate into the protein structure.

Such infiltration of the protein structure and subsequent drying may account for the increase in stability with certain small polyhydroxyl molecules such as inositol. The efficiency of this proposed process may depend on molecular size, molecular shape, charge and chemical properties of the molecule. This may account for the fact that linear polyhydroxyl compounds, (e.g. sorbitol, mannitol) do not stabilize to the same extent as the cyclic molecule inositol.

The subtle differences between sugar molecules which cause such dramatic effects (i.e. Dulcitol verses sorbitol) are possibly caused by adverse interactions either with water or the amino acids of the protein. It is possible that under certain conditions the position of hydroxyl groups causes strong binding of water and leads to confirmational distortion of the protein rather than stabilization. Monsan and Combes (13) have suggested that this is due to excessive binding to the stabilizer thus disrupting the protein surface/water interactions. As yet we have no evidence to further elucidate this theory.

Polyelectrolyte/sugar stabilization, possible mechanism of action. The association of the polyelectrolyte with the enzyme probably results in the formulation of a "cage" around the enzyme (Fig 11.). If this theory is correct then it may be reasonable to assume that the protein molecule is held by a fairly rigid electrostatic interaction, allowing a greater degree of infiltration of the polyhydroxyl compounds present in solution, (Fig. 11). Drying such a mixture produces a high stabilization effect on the enzyme. This may reflect on a more efficient "layering" of the polyhydroxyl compounds within the three dimensional structure of the protein molecule, which is able to take place as a result of the molecule being "anchored" to some extent by the polyelectrolyte. (This effectively replaces the "free water

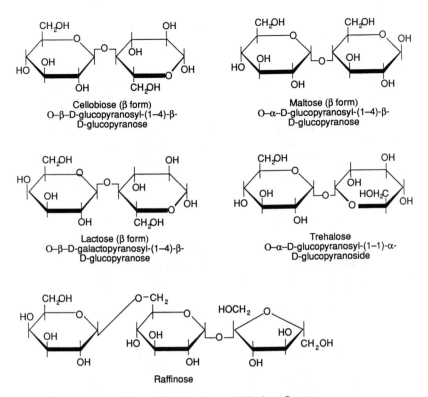

Cellobiose (β form)
O–β–D-glucopyranosyl-(1–4)-β-
D-glucopyranose

Maltose (β form)
O–α–D-glucopyranosyl-(1–4)-β-
D-glucopyranose

Lactose (β form)
O–β–D-galactopyranosyl-(1–4)-β-
D-glucopyranose

Trehalose
O–α–D-glucopyranosyl-(1–1)-α-
D-glucopyranoside

Raffinose

Figure 10.   Structures of Various Sugars.

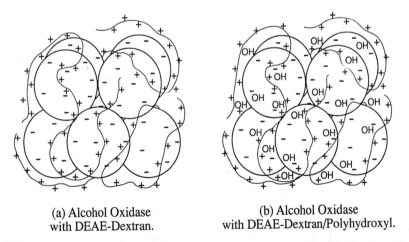

(a) Alcohol Oxidase
with DEAE-Dextran.

(b) Alcohol Oxidase
with DEAE-Dextran/Polyhydroxyl.

Figure 11.   Alcohol oxidase. Interaction of polyelectrolytes/polyhydroxyls. The diagrams above represent the postulated interaction of alcohol oxidase with (a) DEAE–Dextran and (b) the same interaction in the presence of polyhydroxyl compounds. The structure of the enzyme was taken from Woodward 1990 (2).

within the protein structure). Dehydration of the protein-polyelectrolyte-polyhydroxyl-water complex so formed, effectively removes the electrostatic interaction between the protein and polyelectrolyte as described before, but now the protein is packed internally with polyhydroxyl compounds and retains a stable configuration.

If this mechanistic theory is correct, at least in part, then other polyelectrolyte/polyhydroxyl compound combinations should also stabilize alcohol oxidase, this appears to be the case. (17)

The elucidation of the mechanism of stabilization of this generic polyelectrolyte/disaccharide enzyme stabilizing system will be further investigated by determining the interaction between the polyelectrolyte and the protein and the interaction of water molecules with the subtly different structures of sugar in combination with the polyelectrolyte. We have already observed that it is possible to formulate a predictive model which dictates that the charge on a polymer must complement that on the enzyme. It is possible to predict the polyelectrolyte required to stabilize an enzyme by establishing the electric point of the enzyme.

Conclusion. In conclusion then it may be said that through this work we are moving towards a predictive mechanism for enzyme and water/stabilizer interactions which strengthen the structural relationship between the protein and its immediate environment, thus retaining its 3 dimensional structure. It is hoped that further research will lead to a full predictive model of the mechanism of enzyme stabilization and thus provide a completely generic protein stabilization system.

Acknowledgements. The work on alcohol oxidase/peroxidase stabilization on graphite electrodes was carried out by Simon Parker as part of his PhD Thesis work. The authors also acknowledge the work of Ruth Schewior who carried out the experiments on controlled pore glass immobilization of alcohol oxidase. Funds to support the work were provided by the Science and Engineering Research Council (graduate studentship), Cranfield Biotechnology Ltd and Yellow Springs Instrument Ltd.

References
1. Clark, L. C. Jr. and Lyons, C. Ann. N.Y. Acad. Sci.,1962; 105, pp 20-45.
2. Woodward, J.R. In Advances in Autotrophic Microbiology and One Carbon Metabolism. Editors Codd, G.A., Dijkhuizen, L. and Tabita, F.R. Kluwer Academic Publishers, Dordrecht, Netherlands,1990; Vol. 1, pp 193-225.
3. Winter, G. and Fersht, A.R. Trends in Biotechnol,1984; 2, pp 115-119.
4. Arnold, F. Trends in Biotechnol,1990; 8, pp 244-249.
5. Tuengler P. and Pfleiderer G. Biochim BioPhys Res. Comm. 1977; 40, pp 110-116.
6. Pazur J. H., Knull, H.R. and Simpson P.L. Biochim Biophys Res. Comm. 1970; 40, pp 110-116.
7. Wriston, J.C. Jr. Febs. Lett.,1973; 33, pp 93-96
8. Marshal J.J. Trends in Biochem,1978; 3, pp 79-83.
9. Barker S.A. Biosensors, Fundamental and Applications. Editors Turner, A.P.R., Karube I and Wilson G. Oxford Science Publications, Oxford, UK, 1987; pp 85-99.
10. Coughlan, M.P.; Kierstarr, M.P.J.; Border, P.M. and Turner, A.P.F. J. of Microbiol Methods, 1988; Vol. 8, pp 1-50
11. Gibson, T.D. and Woodward J.R. Analyt. Proc. 1986; Vol. 23, pp 360-362.

12. Weetall H. H. Methods in Enzymol. 1976; 44, pp 134-148

13. Monsan P. and Combes D. Ann. N.Y. Acad. Sci. 1983; 434, pp 48-63.

14. Larreta Garde, V.; Zu Feng Xu and Thomas D. Ann. N.Y. Acad. Sci. 1984; 542. pp 294-299

15. Crow J.H.; Crowe L.M.; Carpartner J.F. and Wistrom C.A., Biochemist. 1987; 242, pp 1-10

16. Back, J.F.; Oakenfull, D. and Smith. M.B. Biochemistry 1987; 18, pp 5191-5796.

17. Gibson, T.D., PhD Thesis 1991 Leeds University, Leeds UK.

18. Gibson, T.D. and Woodward, J.R., 1989 Patent Application number PCT/GB89/01346.

19. Arakawa, T. and Masheff S.N. Biochemistry 1982; 21, pp 6536 - 6544.

20. Fujita, Y.; Iwada, Y. and Noda, Y. Bull. Chem. Soc. Jpn 1982; 55, 124, pp 793-800.

21. Ye, W.N.; Combes, D. and Monsan, P. Eng. Microb Tehcnol 1988; 10, pp 498-502.

22. Roser, B.J. 1986, Int Patent Application WO87/00196.

RECEIVED November 26, 1991

Chapter 6

Electrocatalytic Oxidation of Nicotinamide Adenine Dinucleotide Cofactor at Chemically Modified Electrodes

L. Gorton[1], B. Persson[1], P. D. Hale[2], L. I. Boguslavsky[2], H. I. Karan[3], H. S. Lee[4], T. A. Skotheim[2], H. L. Lan[5], and Y. Okamoto[5]

[1]Department of Analytical Chemistry, University of Lund, P.O. Box 124, S–221 00 Lund, Sweden
[2]Moltech Corporation, Chemistry Building, State University of New York, Stony Brook, NY 11794–3400
[3]Division of Natural Science and Mathematics, Chemistry Department, Medgar Evers College, City University of New York, Brooklyn, NY 11225
[4]Materials Science Division, Department of Applied Science, Brookhaven National Laboratory, Upton, NY 11973
[5]Department of Chemistry, Polytechnic University, Brooklyn, NY 11201

Chemically modified electrodes (CMEs) for electrocatalytic oxidation of the reduced form of the nicotinamide adenine dinucleotide cofactor (NADH) are discussed. The work of the authors in the field is reviewed. CMEs based on adsorbed polyaromatic redox mediators (phenoxazines and phenothiazines) and the deposition of aqueous insoluble redox polymers are described.

The nicotinamide adenine dinucleotide (NAD$^+$/NADH) dependent dehydrogenases constitute the largest group of redox enzymes known today. This is not reflected by the number of publications of amperometric biosensors making use of dehydrogenases. On the contrary, most amperometric biosensors are based on other enzyme systems, most often an oxidase (1-11). Unlike other redox enzymes the dehydrogenases rely for activity on the presence of a *soluble* cofactor. The construction of practical amperometric biosensor devices thus becomes more complicated (12) as the necessary cofactor has to be added to any sensor configuration. However, several approaches have been tried to overcome this limitation such as development of coenzymes with enlarged molecular size, which can be retained within the enzyme electrode or binding of the coenzyme to the enzyme molecule, yielding what can be defined as an "artificial prosthetic group". The electrochemistry of both the oxidized (NAD$^+$) and the reduced (NADH) forms of the cofactor is very irreversible and further complicates the use of these enzymes in biosensor research and construction. The basic chemical and electrochemical characteristics and properties of the nicotinamide cofactors have been excellently reviewed by Chenault and Whitesides (13).

When the dehydrogenases are used in analysis the method relies on measuring the change in the redox state of the cofactor, i.e. the change in the concentration of NAD^+ or NADH. NADH is inherently more easily detected photometrically and electrochemically (see below) than its oxidized counterpart, NAD^+ (*13*). When catalyzed by a dehydrogenase, the redox reaction of the nicotinamide adenine dinucleotides $(NAD(P)^+/NAD(P)H)$ is reversible, see Figure 1. A reaction catalyzed by a dehydrogenase can be schematically written as follows:

$$\text{substrate} + NAD^+ <==> \text{product} + NADH + H^+ \tag{1}$$

Reaction (1) involves a stereospecific net transfer of a hydride ion between a substrate and the C(4) of the pyridine ring of the coenzyme and an exchange of a proton with the medium (*14*). The generally accepted formal potential, E°', of the NAD^+/NADH redox couple is -560 mV vs SCE (pH 7, 25°C) (*15*). This value is obtained from equilibrium constants of dehydrogenase catalyzed reactions and thermal data. For most systems the equilibrium of reaction (1) favors the substrate rather than the product side. The reason for this is the low oxidizing power of NAD^+, which is reflected by the low value of E°'.

For a restricted number of reactions the product formed will spontaneously and rapidly react with water, driving the reaction to the product side, thus favoring NADH production. Examples of such systems are the enzymatic oxidation of aldoses e.g. the oxidation of β-D-glucose with glucose dehydrogenase (GDH). In a first reversible reaction the aldose is oxidized to form a lactone:

$$\text{β-D-glucose} + NAD^+ \overset{\text{GDH}}{<=====>} \text{δ-gluconolactone} + NADH \tag{2}$$

The lactone will add water in a second irreversible reaction:

$$\text{δ-gluconolactone} + H_2O ----> \text{D-gluconate} + H^+ \tag{3}$$

The net result will be an irreversible oxidation of β-D-glucose with the production of a stoichiometric amount of NADH. A similar addition of water to the product first formed is also valid for the enzymatic oxidation of aldehydes by dehydrogenases. However, a deliberate addition of a second enzymatic or a "purely" chemical step alternatively to an electrochemical step can be made that will drive the reaction to the product side (*1,9,16-18*). This makes it possible to follow the production of NADH and to correlate it to the concentration of the analyte.

Electrochemistry of NAD^+ and NADH.

The electrochemistry of both the oxidized (NAD^+) and the reduced (NADH) forms of the cofactor at naked electrodes is complicated by large overvoltages and side reactions (*13*). To electrochemically reduce NAD^+, appropriate conditions have to be chosen in order to diminish adsorption effects, ascribed to be mainly caused by the adenine moiety of the cofactor (*19-23*). The polarographic reduction of NAD^+ in neutral aqueous solutions is observed as a single wave at $E_{1/2} \approx$ -0.9 V vs SCE (*24*), whereas in alkaline buffers based on tetra-alkylammonium salts, this single cathodic wave is resolved into two waves, a first one at $E_{1/2} \approx$ -1.1 V, and a second one at $E_{1/2} \approx$ -1.7 V vs SCE (*25*). The first wave at $E_{1/2} \approx$ -1.1 V corresponds to the formation of $NAD\bullet$ radicals (*20,21*) and no dependence of the reaction on pH is observed above pH 5. The re-oxidation of the radicals is observed by cyclic voltammetry at moderatly fast scan rates and thus this step appears reversible (*20,21*) with an $E^{o'}$ at pH 9.1 (25°C) of -1.16 V vs SCE and with an electron transfer rate estimated to exceed 1 cm s^{-1} (*26*). From two independent studies by pulse radiolysis (*27,28*), an average one-electron reduction potential, E^1, of -1.17 V vs SCE at pH 7 is obtained. The reduction of NAD^+ is coupled to a fast dimerization reaction of the $NAD\bullet$ first formed, k_d = 8 10^7 M^{-1} s^{-1}, (*26,29*), where stereoisomers of NAD_2 are produced (*19,20,30*). The dimers are not further reduced in the potential range down to -1.8 V, but can be oxidized to NAD^+ at a potential of $\approx$ - 0.1 V vs SCE (*20,30*). In the continued reduction of the $NAD\bullet$ radicals in the second cathodic polarographic wave both 1,4-NADH and 1,6-NADH are produced, the latter being enzymatically inactive (*20,30*).

Some attempts have been made to find mediators that can catalyze the electrochemical reduction of NAD^+ to the enzymatically active 1,4 NADH. An electrochemical reaction path via a mediator may be an interesting and an economically favourable way to regenerate enzymatically active NADH, stressed to be an economically very important reaction in biotechnology and synthesis (if not in sensors) (*13*). Steckhan and coworkers (*31-34*) have in some papers investigated the possibilty of using various rhodium(III)-bipyridyl complexes with positive results. Preliminary experiments from our laboratory with adsorbed Alizarin Green (a phenoxazine, for structure see Figure 2) on graphite having an $E^{o'}_{pH7}$ slightly more negative than that of $NAD^+/NADH$, have shown some promising results in this direction. In the presence of NAD^+, the reduction peak of the cyclic voltammogram of adsorbed Alizarin Green (not shown) is either increased for a low sweep rate (5 mV s^{-1}) or split into two peaks for a higher sweep rate (50 mV s^{-1}). This is in accordance with the theory of a mediated reaction at a CME (*36*) if the k_{obs} is low. No identification of any products have been made, however.

The electrochemical oxidation of NADH in aqueous solutions is seen as a single peak by cyclic voltammetry and takes place at $\approx$ 0.4, $\approx$ 0.7, and $\approx$ 1V vs SCE at carbon, Pt, and Au electrodes, respectively (*37,38*). No re-reduction of NADH related intermediates is observed in cyclic voltammetry even at fast scan rates (30 V/s) (*39*), reflecting the high chemical irreversibility of the reaction. It was early recognized that the oxidation of NADH resulted in electrode fouling, necessitating

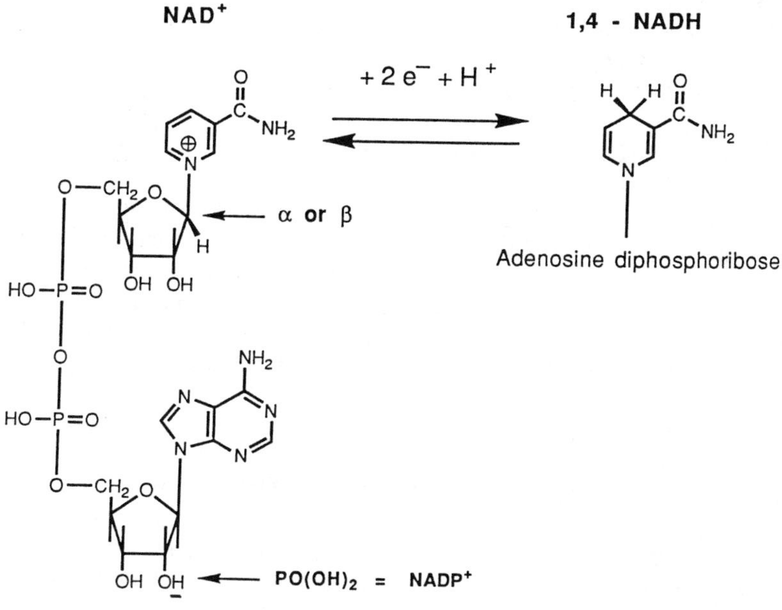

Figure 1. Structural formulae (β-form shown) and overall redox reaction of nicotinamide adenine dinucleotide, NAD^+, and its reduced form, the enzymatically active 1,4-NADH. Reproduced with permission from ref. 35. Copyright 1991 Elsevier Science Publishers.

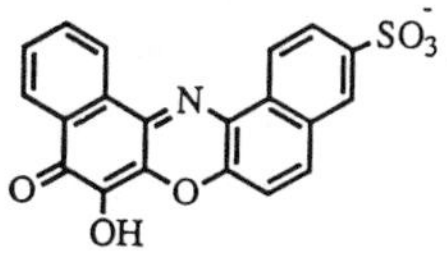

Figure 2. Structural formula of Alizarin Green.

careful pretreatment and conditioning of the electrodes to obtain reproducible results between runs *(37,40-42)*. Furthermore, the major product formed, NAD^+, is an inhibitor of the direct electrochemical oxidation process *(43)*.

The electrochemical oxidation of model compounds of NADH in acetonitrile gives two main oxidation peaks in the absence of a base, clearly demonstrating a stepwise oxidation of NADH *(41)*. An ECE (electrochemical-chemical-electrochemical) mechanism for the electrochemical oxidation of NADH has been proposed in several studies *(41-47)* most often depicted:

$$NADH \xrightarrow{-e^-} NADH\bullet^+ \xleftrightarrow{-H^+} NAD\bullet \xleftrightarrow{-e^-} NAD^+ \qquad (4)$$

although with different views given on the rate and reversibility of the individual steps in the reaction mechanism as well as influences of concurrent reactions. Results from other studies involving both chemical and photochemical redox reactions *(48-52)* give additional credence to an ECE mechanism for the electrochemical oxidation of NADH from which conclusions can be drawn regarding the individual steps in the reaction mechanism.

The first step in reaction scheme (4), believed to be the rate limiting step *(43)*, produces a cation radical. The $E^{o'}_{pH7}$ of the $NADH\bullet^+/NADH$ couple in aqueous solutions has been estimated to 0.69 V *(48,53)*, 0.72V *(49)*, and 0.78 V vs SCE *(52)* by the use of one electron acceptors such as ferrocenes or ferricyanide. The deprotonation in the second step should be considered as irreversible due to the estimated low pK_a ($\approx$ -4 *(51)*), of the cation radical, having a fast deprotonation rate, $k_{H^+} > 10^6$ s^{-1} *(50)*. The second electron transfer in scheme (4) is the equivalent of the first step in the NAD^+ reduction discussed above. It is depicted reversible to stress the chemical reversibility of this reaction, although it now takes place at a large positive working potential and therefore from an electrochemical point of view is irreversible.

It has been suggested that the electrochemical oxidation of NADH may be further complicated by oxidation routes other than the one described by scheme (4) *(42,45)*. Possible routes involving the intermediate radicals in scheme (4) are the oxidation of dimers formed from NAD$\bullet$ radicals or the disproportionation reaction:

$$NADH\bullet^+ + NAD\bullet \longrightarrow NADH + NAD^+ \qquad (5)$$

In Figure 3 the reaction paths are summarized for the electrochemical oxidation of NADH.

In conclusion it can be stated that the electrochemical reactions of the nicotinamide cofactors at metallic or carbonaceous electrodes are highly irreversible, taking place at large overpotentials with the occurrence of side-reactions, also complicated by adsorption (fouling) of cofactor related products. The prospects of using the direct electrochemical reactions of either the oxidized or reduced form of the cofactor for amperometric detection of a dehydrogenase catalyzed reaction therefore seem always to be coupled with interfering reactions and other complications. The reduction of NAD^+ will also suffer from the simultaneous reduction of oxygen due to the low reduction potential of NAD^+ as well as from complications by dimer formation. The electrochemical oxidation of NADH would, however, be a suitable detection reaction if only it could be performed through one single reaction route at a substantially decreased overvoltage where the risk for fouling phenomena and for interfering reactions contributing to the response signal do not prevail.

Electrocatalytic Oxidation of NADH

There exist, however, a restricted number of redox mediators that can shuttle the electrons from NADH to an electrode at a substantially decreased overvoltage (54,55). As a general trend it seems as though organic 2 e⁻ acceptors (also proton acceptors) are more efficient than 1 e⁻ acceptors. The basic mediating structures of the organic 2 e⁻ acceptors can in most cases be found to be derivatives of the oxidized form of *ortho-*, *para*-hydroquinones, *ortho-*, *para*-phenylenediamines, or *ortho-*, *para*-aminophenols (all depicted in Figure 4) or very similar structures. Some of these compound are known to have very high reaction rates with NADH in aqueous solution (54,56-58).

Some derivatives with mediating properties are suitable to form chemically modified electrodes (CMEs) with catalytic properties for NADH oxidation (55). Various attempts have been tried with different classes of mediators to immobilize the mediator onto solid electrodes or in carbon paste electrodes since the first deliberately made CME for electrocatalytic oxidation of NADH was described by Tse and Kuwana in 1978 (56), see Table I. They and others (67-72) based their CMEs on immobilized *ortho*-quinone derivatives. However, these CMEs were rapidly inactivated in the presence of NADH, probably because of side reactions in the catalytic process (72). For some other immobilized mediators one major reaction route could be proposed as the CME turned out to be quite stable in the presence of NADH. The catalytic reaction sequence comprizes two steps, one chemical between NADH and the immobilized mediator (reaction (6)) and one electrochemical between the mediator and the electrode (reaction (7)). The sequence is given below for the simplest case:

$$NADH + Med_{ox} \xrightarrow{k_{obs}} NAD^+ + Med_{red} \tag{6}$$

$$Med_{red} \xrightarrow{E_{appl} > E^{o'}} Med_{ox} + H^+ + 2e^- \tag{7}$$

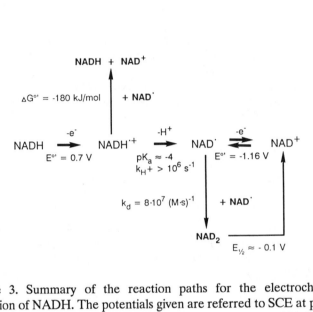

Figure 3. Summary of the reaction paths for the electrochemical oxidation of NADH. The potentials given are referred to SCE at pH 7.0. Reproduced with permission from ref. 35. Copyright 1991 Elsevier Science Publishers.

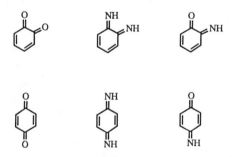

Figure 4. Basic structures of mediators for electrocatalytic oxidation of NADH.

Table I. Chemically modified electrodes for oxidation of NADH

Mediator Group	Immobilization	$E°'_{pH7}$ / mV vs SCE	References
oxidized carbon	covalent	0 to +200	[59-66]
1,2-Benzoquinone	polymeric	+170 a)	[56,67]
1,2-Benzoquinone	polymeric	+160 a)	[68]
1,2-Benzoquinone	polymeric	+170	[69,70]
1,2-Benzoquinone	adsorbed	+170	[71]
1,2-Benzoquinone	adsorbed	+100	[72]
1,4-Benzoquinone	mixed with carbon	+55	[73]
1,4-Benzoquinone, tetrachloro-	mixed with carbon	+50	[74]
Mercaptoquinone	polymeric		[75]
1,2-Naphthoquinone	mixed with carbon	-150 b)	[76]
1,4-Naphthoquinone, 2-metyl- and diaphorase	polymeric	-380 b)	[77]
Anthraquinone-2-sulphonate and NADH dehydrogenase	polymeric		[78]
p-(4-vinylpyridyl)-N-(3,4-dihydroxybenzylidene)aminobenzene	polymeric		[79]
1,4-Phenylenediamine and N,N,N',N'-tetramethyl-1,4-phenylenediamine	mixed with carbon	+100	[80]
Phenazin-5-ium, N-methyl-	adsorbed	-160	[81,82]
Phenazin-5-ium, N-ethyl-	adsorbed	-210	[81]
Phenothiazin-5-ium, 7-(dimethylamino)-2-methyl-3-[(2-naphthalenylcarbonyl)amino]-	adsorbed	-135	[83,84]
Phenothiazin-5-ium, 7-(dimethylamino)-2-methyl-3-[(3-thienylcarbonyl)amino]-	adsorbed	-115	[85]
Phenoxazin-5-ium, 7-(dimethylamino)-2-methyl-3-[(2-naphthalenylcarbonyl)amino]-	adsorbed	-180	[84]
Phenoxazin-5-ium, 7-(dimethylamino)-2-methyl-3-[(2-naphthalenylcarbonyl)amino]-	mixed with carbon	-180, -75	[16]
Phenoxazin-5-ium, 7-(dimethylamino)-2-methyl-3-[(3-thienylcarbonyl)amino]-	adsorbed	-160	[85]
Phenoxazin-5-ium, 7-(dimethylamino)-2-methyl-3-[(3-thienylcarbonyl)amino]-	adsorbed	-130	[85]
Phenoxazin-5-ium, 7-(dimethylamino)-2-methyl-3-[(5-indolylcarbonyl)amino]-	adsorbed	-135	[85]
Benzo[a]phenoxazin-9-one	adsorbed	-210	[86]
Benzo[a]phenoxazin-7-ium, 9-(dimethylamino)-, (Meldola Blue)	adsorbed	-175	[87]
Benzo[a]phenoxazin-7-ium, 9-(dimethylamino)-, (Meldola Blue)	mixed with carbon	-70	[88]
Benzo[a]phenoxazin-7-ium, 5-(dimethylamino)-5-(phenylamino)-	adsorbed	-360	[84]
Benzo[a]phenoxazin-7-ium, 5-(diethylamino)-5-[(1-pyremylidene)imino]-	adsorbed	-190	[89]
Benzo[a]phenoxazin-7-ium, 5-(diethylamino)-5-[(2-naphthalenylcarbonyl)amino]-	adsorbed	-220	[84,90]
Benzo[a]phenoxazin-7-ium, 5-(diethylamino)-5-[(3-thienylcarbonyl)amino]-	adsorbed	-195	[85]
1,4-bis(5-carbonylamino-9-(diethylamino)-benzo[a]phenoxazin-7-ium]benzene	adsorbed	-220	[91,92]
Flavine adenine dinucleotide	covalent	+200	[93]
Tetrathiafulvalinium-7,7,8,8-tetracyanoquinodimethane radical salt			[94-96]
N-methyl-phenazin-5-ium-7,7,8,8-tetracyanoquinodimethane radical salt			[97,98]
Ferrocene and diaphorase			[12]
Ferrocenyl(methanol or ferrocene carboxylic acid and NADH dehydrogenase	polymeric	+100 or +190 c)	[78]
NaWFe(III)(CN)6	electrodeposited		[99]
Mn(III)meso-tetraphenylporphine	adsorbed		[100]
poly(3-methylthiophene)	polymeric		[101]

a) pH 6.2 b) pH 8.5 c) pH 6.8

Reproduced with permission from ref. 35. Copyright 1991 Elsevier Science Publishers.

NADH reacts in a first step with the oxidized form of the immobilized mediator, Med_{ox}, delivers a hydride equivalent whereby NAD^+ and the fully reduced form of the mediator, Med_{red}, are formed, reaction (6). In the following electrochemical step, reaction (7), the mediator is reoxidized to the mediating active from delivering its charge to the electrode.

Since mainly the $E^{o'}$ of the mediator dictates at what potential the heterogenous electron transfer occurs, the oxidation of NADH can now take place at a much lower potential. The different mediator structures used to produce CMEs for NADH oxidation at a decreased overpotential are summarized in Table I. As is seen in the table, not only chemically modified electrodes based on only immobilized redox mediators have been used for this purpose, but also electrodes based on the combination of redox mediators and NADH oxidizing enzymes (diaphorase and NADH dehydrogenase) as well as electrodes made of the conducting radical salts of tetrathiafulvalinium-7,7,8,8-tetracyanoquinodimethan (TTF-TCNQ) and N-methyl-phenazin-5-ium-7,7,8,8-tetracyanoquinodimethan (NMP-TCNQ).

There are several demands that must be more or less fulfilled by the mediator before a successfull amperometric detection of NADH with CMEs can be realized. Despite having a $E^{o'}$, lower or comparable with the optimal working potential range for amperometric detection, the mediator should exhibit fast reaction rates both with the electrode proper and NADH, and also be chemically stable at any redox state. Furthermore, the redox reaction of the mediator should involve two electrons and at least one proton making possible, at least theoretically, a fast inner sphere hydride transfer in the homogeneous reaction with NADH.

Some of the structures reported from this laboratory for making CMEs for electrocatalytic oxidation of NADH are summarized in Figure 5. We have found that a positively charged paraphenylene-diimine is the best basic catalytic structure (55,83,84-87,89,91,92). When it is incorporated as a part of a phenoxazine or a phenothiazine (see Figure 5) some very beneficial properties are added to the mediator. The $E^{o'}$ values are decreased with some 300 to 400 mV compared with the free paraphenylenediimine structure (57,58,102) and they strongly adsorb on graphite electrodes to form CMEs allowing NADH to be electrocatalytically oxidized well below 0 mV (55,83,84-87,89,91,92).

The close coupling between the phenoxazine or phenothiazine modifier and a graphite electrode (Π-electron overlapping) results in very fast charge transfer rates between the modifier and the electrode seen as small peak separations in cyclic voltammetry (83,86,87,92). The adsorption of the mediators also results in a further decrease in the $E^{o'}$ values with about 50 to 100 mV compared with the bulk values (83,84,92). Drastic changes in the pK_a values of both the reduced and the oxidized forms of the adsorbed mediator have also been noticed (83,84,92).

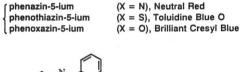

N-methyl-phenazin-5-ium   N-ethyl-phenazin-5-ium

3-amino-7-dimethylamino-2-methyl-

phenazin-5-ium	(X = N), Neutral Red
phenothiazin-5-ium	(X = S), Toluidine Blue O
phenoxazin-5-ium	(X = O), Brilliant Cresyl Blue

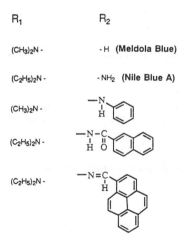

9-R$_1$-5-R$_2$-benzo[a]phenoxazin-7-ium

R$_1$	R$_2$
(CH$_3$)$_2$N -	- H (Meldola Blue)
(C$_2$H$_5$)$_2$N -	- NH$_2$ (Nile Blue A)
(CH$_3$)$_2$N -	
(C$_2$H$_5$)$_2$N -	
(C$_2$H$_5$)$_2$N -	

Figure 5. Some structural formulae of phenazine, phenoxazine, and phenothiazine mediators used at this laboratory for making CMEs used for electrocatalytic oxidation with NADH. Continued on next page.

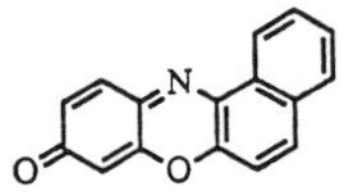

benzo[a]phenoxazin-9-one

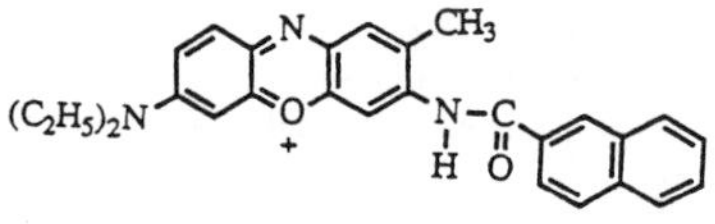

7-diethylamino-2-methyl-3-β-naphthoyl-phenoxazin-5-ium

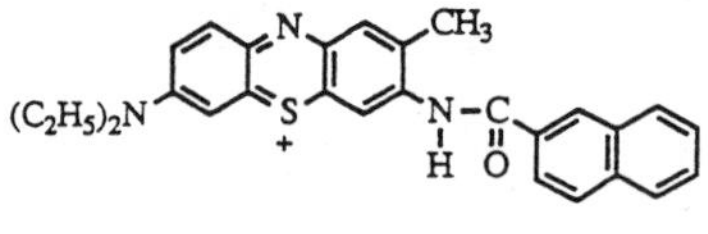

7-diethylamino-2-methyl-3-β-naphthoyl-phenothiazin-5-ium

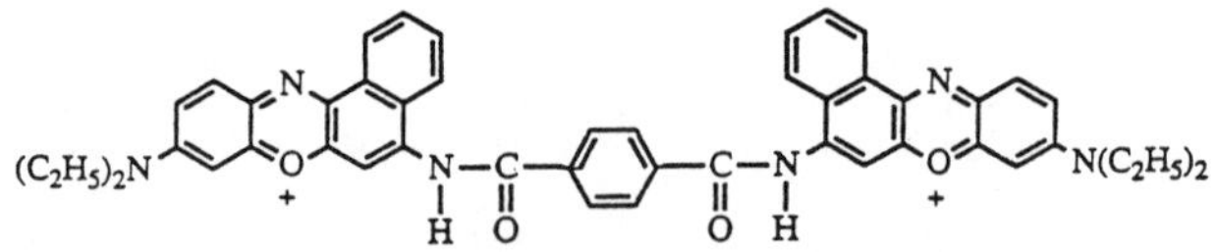

1,4-bis(5-carbonylamino-9-diethylamino-benzo[a]phenoxazin-7-ium)benzene

Figure 5. Continued.

This is illustrated in Figure 6A, where it is clear that different $E^{\circ\prime}$ values are found depending on whether the redox compound is in a soluble or in an adsorbed state and also depending on the nature of the surface on which the compounds is adsorbed. That these observations are not the results of local pH effects at the electrode surface is revealed by comparing the variation of the $E^{\circ\prime}$ of an adsorbed flavin with that of the soluble form (*103*). As is clear in Figure 6B, the $E^{\circ\prime}$ values for both the adsorbed and soluble formes of the flavin are very close up to about pH 6 where they start to differ because of a pK_a-shift of the adsorbed flavin. These observations indicate the close interaction between the electrode surface and the adsorbed modifier.

The reaction rate with NADH is largely a function of the $E^{\circ\prime}$ value of the adsorbed phenoxazine or phenothiazine being higher for mediators with more positive $E^{\circ\prime}$ values (*55,83,84,89-92*). For some adsorbed phenoxazine derivatives a linear correlation was found between the logarithm of the observed second order rate constant (k_{obs}) and the $E^{\circ\prime}$ value (*55*). The reaction rate between the adsorbed phenoxazines and NADH was also found to be a function of the concentration of NADH so that higher concentrations of NADH resulted in a lower reaction rate (*55,83,84,86,87,89,92*). This observation and other results from rotating disk electrode experiments gave credence to a belief that this behavior could be explained by the formation of an intermediate charge transfer (CT) complex in the reaction mechanism (*55,83,84,86,87,89*), c.f. reaction (6) above:

$$\text{NADH} + \text{Med}_{ox} \underset{k_{-1}}{\overset{k_{+1}}{\xleftarrow{\hspace{1.5em}}\rightarrow}} \text{CT-complex} \overset{k_{+2}}{\xrightarrow{\hspace{1.5em}}} \text{NAD}^+ + \text{Med}_{red} \qquad (8)$$

$$\xrightarrow{\hspace{3em}k_{obs}\hspace{3em}}$$

Two constants, k_{+2} and K_M ($=(k_{-1} + k_{+2})/k_{+1}$), describe the reaction and the overall second order rate constant, k_{obs}, is then expressed as; $k_{obs} = k_{+2}/(K_M + [\text{NADH}])$.

For adsorbed mediators with $E^{\circ\prime}_{pH7}$ values between -200 and -150 mV *vs* SCE, rotating disk electrode studies gave k_{obs} values in the order of 10^4 and 10^5 M^{-1} s^{-1} and K_M values between 1 and 5 mM. The K_M was found to depend somewhat on the surface coverage (*84*). When using these CMEs as amperometric sensors in flow systems, they responded linearly to sample concentrations of NADH up to around 2 mM. Due to a less efficient mass transport of NADH to the electrode and to the dilution in the flow system the expected upper linear response limit to NADH concentrations equal to 0.1 K_M or more is thus observed. The real working range can furthermore be adjusted according to the dilution that occurs in the flow (*104-106*). From the reaction mechanism (scheme (8)) it can be concluded that the response to NADH will be independent of the mediator coverage once it has reached a certain minimum value (*55,83,84,87*). If the initial

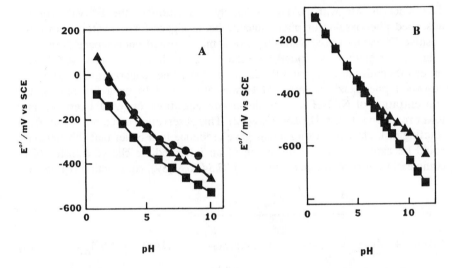

Figure 6. Variation of E°' with pH for (A) Nile Blue A adsorbed on graphite (■), Nile Blue A adsorbed on Ag (●), and Nile Blue A dissolved in aqueous solution (▲) and (B) flavine adenine dinucleotide, FAD, adsorbed on graphite (■) and FAD dissolved in aqueous solution (▲). Reproduced with permission from ref. 35. Copyright 1991 Elsevier Science Publishers.

coverage is large compared with this critical coverage, a slight desorption or deactivation of the immobilized mediator will pass unnoticed in the response to NADH. Phenoxazines derivatized in only position 3 or 7 are alkali unstable (c.f. Meldola Blue in Figure 5) (*107,108*). As many dehydrogenase catalyzed reactions have pH optima in the alkaline region, the mediator should be stable also in this pH interval. Several commercially available phenoxazines and phenothiazines are derivatized both in position 3 and 7, in one position with a dialkylamino group and in the other with a primary amino group (*84,102*), e.g. Nile Blue A, NBA, and Brilliant Cresyl Blue, BCB, (Figure 5). By reacting the primary amino group with an aromatic acid chloride several improvements of the mediator characteristics can be obtained (*55,83,84,89-92*). The $E^{\circ\prime}_{pH7}$ value is drastically increased with about 200 mV to a final value of about -200 to -150 mV *vs* SCE, also resulting in an increased reaction rate with NADH (*55,83,84,89-92*). The mediator becomes alkali stable and the number of aromatic rings can be increased further stabilizing the electrode modification.

Many of our observations with the CMEs based on adsorbed phenoxazine and phenothiazine derivatives cannot be fully explained by electrochemical experiments. The facts that the $E^{\circ\prime}$ value and the pK_a values of the adsorbed mediator are shifted compared with their solution values can be rationalized by specific interactions of some parts of the adsorbed molecule with the electrode surface. A full explanation can, however, never be realized with only electrochemical experiments. In the process of making CMEs aimed for catalytic reactions with biological molecules it is therefore most essential to increase the understanding of the interactions between the mediating molecules and the electrode surface so that new molecules can be synthesized with anticipated (and increased) catalytic properties. With electrochemical experiments it was strongly suggested that a charge transfer intermediate exists in the reaction mechanism between the adsorbed mediator and NADH, scheme (8) (*55,83,84,86,87,89*). This observation might indicate that the transfer of charge from NADH includes several steps opposing a single hydride transfer very much debated in the literature (*109*). It was therefore concluded that additional information on the nature of the chemically modified electrode surface with a spectrophotometric technique should give valuable information both on the nature of the specific interactions between the adsorbed mediator and the electrode surface as well as on the existence of a complex formation in the reaction sequence between the mediator and NADH. A collaboration was therefore entered in 1988 with Prof. T. M. Cotton (Iowa State University, Ames, Iowa, USA) who has specialized in the combination of electrochemistry and spectroscopy (especially surface enhanced Raman scattering, SERS, (*110,111*)). Further support to the belief of the existence of an intermediate complex was revealed by SERS of adsorbed Nile Blue A on silver and glassy carbon electrodes in the presence of NADH (*112*). NBA was chosen as the mediator in this investigation because of its known low reaction rate with NADH (*55*), making a possible intermediate complex long lived. A new species on the electrode surface was traced in the presence of

NADH, whose Raman spectrum could not be interpreted as originating from NAD^+, NADH, oxidized or reduced NBA but rather from the proposed intermediate CT-complex. The formation of an intermediate complex may also be the cause why up to now no other compounds have been found, which are catalytically oxidized at these CMEs (104-106) that thus further favors a selective detection of NADH. Specific interactions of some parts of the adsorbed mediator could be evaluated and a proposed orientation of the mediator on the electrode surface could be given explaining parts of the pK_a and $E^{o'}$ shifts. Moreover, an explanation to the pH dependence of the reaction rate with NADH could be given by a reorientation of the adsorbed phenoxazine with pH resulting in a less accessible catalytic functionality at higher pHs, see below. It was recently noticed that experimental results for very low concentrations of NADH with mediators with high K_M values did not fit the proposed overall reaction scheme given by scheme (8) (84). Obviously the charge transfer reaction is more complex and includes further steps. Future plans also include an investigation on the effects of metal ions on the charge transfer reaction. In some analytical applications where the immobilized enzymes depend on a metal ion (e.g. Co^{2+}, Mg^{2+}, Ca^{2+}) a slow decrease in the response factor to NADH was noticed without a decrease in the amount of electrochemically active mediator as seen with cyclic voltammetry.

As has been observed in many of our investigations the observed rate constant, k_{obs}, between the mediator and NADH varies with the pH of the contacting buffer in a manner so that a higher pH results in a lower reaction rate (55,83,84,86,89,91,92,112). There may be several explanations to this behavior. We have up till now indentified the following possible reasons why the reaction rate could vary:

The orientation of the adsorbed mediator on the electrode surface may change with a change in pH. That this is the case for adsorbed Nile Blue A was found using surface enhanced Raman scattering (112). From experimental data it was proposed that Nile Blue A adsorbed at pH 6 in a perpendicular fashion resulting in that the catalytic site of the mediator is available for reaction with NADH diffusing in from the buffer. At a higher pH Nile Blue A adsorbs in a more perpendicular fashion resulting in that the catalytic site is less available for NADH resulting in a lower reaction rate.

The variation of the reaction rate with pH may also be the result of an inherent reaction mechanism according to the following, proposed from several investigations in solution using 1 electron as well as 2 electron acceptors (quinoids) (113-116). In a first reaction between the mediator and NADH the complex is formed followed by the first electron donation from NADH to the mediator part within the complex:

$$\text{NADH} + \text{Med} \overset{K_{CT}}{<====>} [\text{NADHMed}] \overset{k_{+1}}{\underset{k_{-1}}{<====>}} [\text{NADH}\bullet^+\text{Med}\bullet^-] \quad (9)$$

The proton within the $[\text{NADH}\bullet^+\text{Med}\bullet^-]$ complex is very acid, c.f. above on $\text{NADH}\bullet^+$ in reaction (4). Two competing reactions may follow reaction (9). In one possible reaction route the proton is donated to the mediator part within the complex, reaction (10):

$$[\text{NADH}\bullet^+\text{Med}\bullet^-] \overset{k_H}{----->} [\text{NAD}\bullet\text{MedH}\bullet] \overset{\text{fast}}{------>} \text{NAD}^+ + \text{MedH}^- \quad (10)$$

In another, reaction (11), the proton is donated to a competing base, which in an aqueous environment very well could be hydroxide ions:

$$[\text{NADH}\bullet^+\text{Med}\bullet^-] + \text{OH}^- \overset{k_B}{----->} \text{NAD}\bullet + \text{Med}\bullet^- + \text{H}_2\text{O} \quad (11)$$

The competition between reactions (10) and (11) should then result in different reaction rates at different pHs. This scheme may be even further complicated if any or more than one of the forms of the complex ([NADHMed], [NADH•⁺Med•⁻], or [NAD•MedH•]) is electrochemically active at the applied potential given to the electrode, see below.

For adsorbed mediators with high reaction rates with NADH ($\approx 10^4$ - 10^5 M^{-1} s^{-1}), the oxidation peak potential seen in cyclic voltammetry in plain buffer and in the presence of NADH are very close. A contrasting picture is obtained for adsorbed mediators with very low reaction rates (≈ 10 M^{-1} s^{-1}), e.g. Nile Blue A ($E^{\circ\prime}_{pH7}$ = -430 mV vs SCE) (*55,112*) and FAD ($E^{\circ\prime}_{pH7}$ = -460 mV vs SCE) (*93*). The decrease in the overvoltage for electrochemical oxidation of NADH is observed in a potential range much more positive than the $E^{\circ\prime}$ of the adsorbed mediator, seen as a peak in cyclic voltammetry at $\approx$ -100 to +100 mV. That the catalytic peak appears that far away from the $E^{\circ\prime}$ of the mediator cannot be explained from the low reaction rates and should not be seen as a peak (*36*). The most usual explanation given to the reaction with faster mediators ($E^{\circ\prime}_{pH7}$ around -250 to -150 mV vs SCE), where the catalytic current starts to appear very close to the $E^{\circ\prime}$ of the mediator, is that the electrochemically active part is the fully reduced form of the mediator, Med$_{red}$ (*55,86,87*), see reactions (6) and (7). However, one possible reason could be that for all mediators one of the complex forms ([NADHMed], [NADH•⁺Med•⁻], or [NAD•MedH•]) is electrochemically active in the potential range between -200 and +100 mV. For the mediators with a low k_{obs}, the peak potential in cyclic voltammetry will appear far away from the $E^{\circ\prime}$ value, whereas for the fast reacting mediators the competion will be such as to whether it will be the complex or the fully reduced form of the mediator, which will be the predominantly electrochemically active component. For both alternatives the actual reaction will occur in the same potential range.

A fourth explanation to the variation of the reaction rate is valid for the mediators being derivatized with two amine groups both in para-position to the nitrogen heteroatom within the ring (84), shown in Figure 7 for 3-β-naphthoyl-Nile Blue A and its iminoform. At a low pH the bridging group between the benzo[a]phenoxazine part and the naphthalene, -NH-(C=O)-, is protonated. This results in that the the optimal catalytic functionality, a charged paraphenylenediimine, is located around position 9 of the benzo[a]phenoxazine part. At a higher pH when the bridging group is deprotonated, =N-(C=O)-, the mediator will lose its preferred catalytic structure into its less active counterpart, the uncharged iminoform (83,84).

By comparing analogues mediators, one prepared by reacting BCB (a *phenoxazine*) with 2-naphthoyl chloride (Figure 5) and another one from NBA (a *benzo[a]phenoxazine*) (Figure 5) it became clear that the oxidized form of the mediator prepared from BCB had a much higher pK_a value than the one prepared from NBA (84). It was also shown that k_{obs} for the adsorbed mediator based on BCB was much less pH sensitive than that with the mediator prepared from NBA. Some minor pH dependence on k_{obs} of this mediator is still noticed, however. Further studies indicated that the reaction with NADH with the uncharged paraphenylene-diimine formed when deprotonating the mediator might be complicated by side reactions (84). A similar observed deactivation in the presence of NADH was found for electrodes modified with *ortho*-quinone derivatives (72).

A possible fifth explanation would be that the variation of the $E^{o'}$ of $NAD^+/NADH$ with pH moves with 30 mV per pH unit ($NAD^+ + H^+ + 2e^-$ <==> NADH) whereas the variation for the mediator may vary with 60 mV per pH ($2H^+$ and $2e^-$ take part in the redox reaction per mediator molecule). The resulting difference between the $E^{o'}$ values, $\Delta E^{o'}$, constituting the thermodynamic driving force for the oxidation, is then a function of pH and will be less at a higher pH. As shown in earlier investigations (55,83,84) k_{obs} is strongly dependent on the $E^{o'}$ of the mediator. As the $\Delta E^{o'}$ may vary with pH this may result in that k_{obs} also varies with pH.

Our previous reports on the synthesis of new mediators have mainly focused on the phenoxazine group (55,84,89,91,92) but analogues phenothiazines were shown to be equally good mediators (Figure 5) (83,84). The basic aims of the work have previously been 1/ to increase the adsorption strength by increasing the number of aromatic rings of the mediator as the immobilization has been based on irreversible adsorption to carbon electrodes, 2/ to make the phenoxazine mediator alkaline stable (derivatized both in position 3 and 7), 3/ to increase k_{obs} by increasing the $E^{o'}$ value and thus to make the current for NADH mass transfer

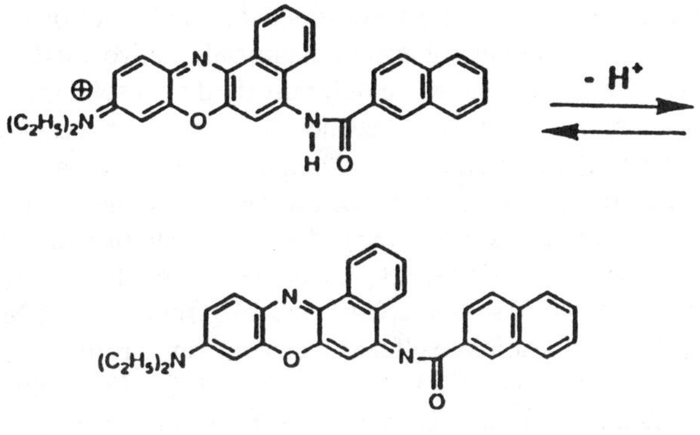

Figure 7. Structural formulae of 3-β-naphthoyl-Nile Blue A and its imino form.

rather than kinetically controlled at the CME and 4/ to make the response current to NADH pH independent. The basic coupling of the parent (benzo[a])phenoxazine (or phenothiazine) has been the reaction of an acid chloride with the primary amino group of e.g. NBA and BCB to form a -NH-CO- linkage containing an acid proton on the nitrogen. If deprotonated the mediator will lose its preferred catalytic structure (Figure 7). The pK_a of the proton is, however, affected by functionalities in close connection to the bonding as described above.

Almost all CMEs for catalytic NADH oxidation reported from this laboratory have been based on adsorbed mediators on graphite electrodes, see Table I. Chemically modified carbon paste (CMCP) electrodes have also been studied, where the mediator was added to the carbon as a dry powder or dissolved in an organic solvent evaporated before addition of the pasting liquid (16,88). The existence of two forms of the mediator in the paste could be seen with cyclic voltammetry. This could be explained by one form strongly adsorbed on the graphite in the paste and another form more loosely bound or dissolved in the paste (16). The total amount of electrochemically active mediator was much higher in the CMCP than found when only adsorbed on a solid graphite electrode (88). This is advantageous in an enzyme electrode to prevent the mediated reaction from being the rate limiting step and also to drive unfavorable equilibria to the product side (16). It was noticed, however, that the loosely bound form slowly dissolved from the CMCP (88). The research program has therefore included studies on the possibilities to introduce the mediator into an insoluble polymer to prevent its dissolution from the electrode. The CMEs based on adsorbed mediators on graphite suffer somewhat from a rather high background current (even at 0 mV vs SCE). Polymer based CMEs could be made on alternative electrode materials such as glassy carbon or Pt for which lower background currents are expected. This should make it possible to measure even lower concentrations of NADH than 0.2 μM, which is the detection limit in a flow system obtained with graphite modified with an adsorbed aromatic mediator (83). Previous attempts by other research groups to make stable polymer based CMEs for catalytic NADH oxidation have, however, failed probably because of choosing a mediating structure that is not long term stable in the presence of NADH (68-70).

Figure 8 shows cyclic voltammograms of a naked graphite electrode and of electrodes modified with two different and very recently synthesized aqueous insoluble polymers incorporating a phenothiazine mediator registered in plain buffer (8Aa, 8Ba, and 8Ca, respectively) and in buffer also containing 7.5 mM NADH (four consecutive voltammograms of each electrode are shown; 8Ab1-4, 8Bb1-4, and 8Cb1-4, respectively). The figure also shows the structure of the polymers used for electrode modification. The phenothiazine compound used for covalent incorporation into the polymer was Toluidine Blue O (TBO, see Figure 5). The synthesis and purification of the polymers will be described elsewhere

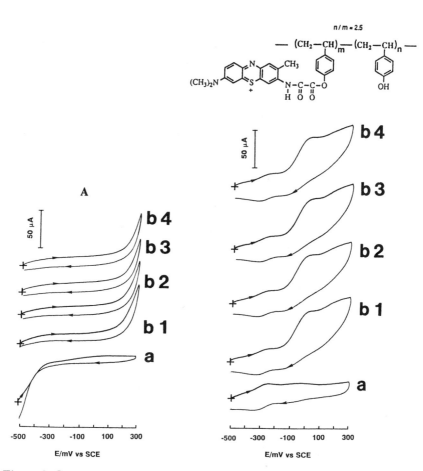

Figure 8. Cyclic voltammograms of (A) a naked solid graphite electrode (surface area 0.0731 cm^2) and (B and C) two equivalent graphite electrodes also chemically modified with the structures included in the figure.

Voltammograms indicated (a) are registered in plain 0.25 M phosphate buffer pH 7.0 and voltammograms indicated (b1 to 4) are registered in the same buffer with the addition of 7.5 mM NADH. The sweep rate was 50 mV s^{-1}. + indicates start potential and the arrows the sweep direction. Continued on next page.

C

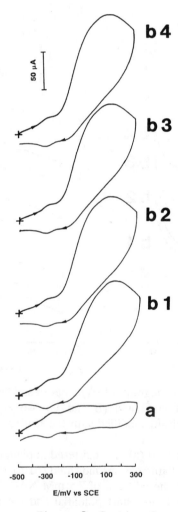

Figure 8. Continued.

(117). The modification was performed by dissolving the polymer in dimethylformamide, applying a small droplet on the electrode surface and allowing the solvent to evaporate before dipping the electrode into the buffer. As is clear from the cyclic voltammograms there seem to be at least two electrochemically active forms of the mediator, one with a lower $E^{o'}$ value and one with a $E^{o'}$ value about 200 to 250 mV more positive. The reason for having two forms is not clear. At pH 7 the $E^{o'}$ value of adsorbed TBO is -285 mV vs SCE *(83,84)*. Both polymers show at this pH a wave close to this value, which might indicate traces of TBO left in the purified polymer. As mentioned above, previous results with CPEs chemically modified with phenoxazine and phenothiazine mediators have shown that there exist two forms of the mediator, one that could be explained to be a form where the mediator is strongly bound to the graphite surface and one form dissolved in the pasting liquid. The results shown in the figure (8Ba and 8Ca) may therefore indicate that the mediator in the polymer will take different $E^{o'}$ values depending on whether the mediator moiety is in close contact with the graphite electrode surface or not. The wave with the lower $E^{o'}$ may therefore indicate a direct adsorption of the TBO moiety on the electrode rather than traces of unreacted TBO. This is also in accordance with the careful purification of the polymer formed in the synthetic reaction *(117)*. In the presence of NADH in the contacting buffer a drastic increase in the oxidation current is observed with a peak potential very close to the $E^{o'}$ value of the more positive wave clearly indicating the strong catalytic properties of the mediator containing polymer (Figure 8Bb1-4 and 8Cb1-4). The reason for obtaining the catalytic peak at the more positive potential form of the mediator may be that this form is the more active (the higher the $E^{o'}$ value the more active is the the mediator) or that the form with the lower $E^{o'}$ value is the innermost layer of the polymer at the electrode surface where NADH has difficulties in reaching from the contacting solution. Four consequtive voltammograms are shown to illustrate the reproducibility of the catalytic effect. After each cyclic scan, the solution was carefully stirred ($\approx$ 20s) before the next scan was registered. As can be seen, these voltammograms appear very similar. For comparison the voltammograms obtained with a naked graphite electrode in plain buffer and in the NADH containing buffer are included. The four consecutive voltammograms registered in the NADH containing buffer show clear indications of electrode fouling. The overvoltage is increased after each scan. This is not observed for the mediator-polymer based CMEs. A rather high NADH concentration (7.5 mM) was chosen for these experiments as any fouling effects would be more clearly noticed compared with a lower concentration.

From the fact that the dehydrogenases depend on a soluble cofactor follows that the enzyme need not be immobilized in close proximity to the detection device. There are several examples in flow systems (FIA and LC) where the enzymatic reaction and NADH production occur in an immobilized enzyme reactor *(18,66, 104-106,118,119)*. The NADH is then transported downstream to the electrode situated in an amperometric flow through cell. As we have reported, the enzyme can also be immobilized directly on the surface of the CME. GDH was shown to retain its catalytic properties when just adsorbed on a phenoxazine modified graphite electrode (120). The stability of this kind of electrode was greatly improved when glutaraldehyde was added to the enzyme solution as a cross-linking agent just before application to the electrode surface (91).

Several attempts have been made over the years to overcome the need for the continuous addition of the costly cofactor. Derivatives of NAD^+ have been immobilized together with the enzyme directly on electrode surfaces (120). The cofactor has also been covalently bound to the enzyme (122,123) or been coupled to macromolecules to prevent its dissolution and escape through protecting membranes (124). The cofactor has been deposited together with mediators in carbon pastes with the enzyme held in close proximity to the surface with a membrane coating (125). A silk screen printed electrode was reported (126) containing both a dehydrogenase and NAD^+ as well as a mediator. Some very interesting recent results have been reported from Aizawa's group (127,128). They constructed an ethanol and a L-lactate sensor by co-immobilizing either alcohol or L-lactate dehydrogenase with Meldola Blue (for structure see above Figure 5) and NAD^+ into a polypyrrole membrane electropolymerized on the electrode surface.

Recently we were able to show for the very first time that all the necessary chemicals can be mixed into a carbon paste electrode, CPE, (graphite-paraffin oil) with retained activity (a dehydrogenase, NAD^+, and a phenoxazine mediator) so that a response current to the substrate could be obtained at 0 mV vs SCE and below (16,88). A glucose and an ethanol sensor were described, based on GDH and alcohol dehydrogenase, ADH, respectively (16,88). Contrary to what has been previously believed, the organic nature of the carbon paste was shown not to be detrimental to the enzyme. If the enzymatic, the mediated reoxidation of the cofactor, and the electrochemical reactions were to take place very close to one another, as should be the case in a carbon paste chemically modified with both the enzyme, the cofactor, and the mediator, the close coupling of these systems should drive any unfavorable equilibrium to the product side. It was thus observed that with ADH in a cofactor and mediator modified electrode at pH 7, a true shift of the overall equilibrium really took place and calibration curves (if not fully linear) for ethanol could be obtained between the μM and mM levels (16). However, ADH was found to be unstable in the paste and a decrease in the response with time was observed. We are currently studying how to stabilize ADH and other related dehydrogenases (129,130). Further studies will therefore include other dehydrogenases, whose reactions suffer from unfavorable equilibria. Naturally the newly synthesized polymers containing mediators will be used instead of the previously used adsorbed mediators.

Financial support from the Swedish Natural Science Research Council (NFR), the Swedish Board for Technical Development (STU and STUF), and the Swedish Energy Administration (STEV) is gratefully acknowledged.

Literature Cited

1. Wise, D. L., *Bioinstrumentation: Research, Development and Applications;* Butterworths, Boston, MA, 1990.
2. *Biosensors International Workshop;* Schmid, R. D.; Guilbault, G. G.; Karube, I.; Schmidt, H.-L.; Wingard, L. B., Jr., Eds.; GBF Monographs, VCH, Weinheim, 1987; Vol. 10.
3. *Flow Injection Analysis (FIA) Based on Enzymes or Antibodies;* Schmid, R. D., Ed.; GBF Monographs, VCH, Weinheim, 1990; Vol. 14.
4. *New Electrochemical Sensors;* special issue of *J. Chem. Soc., Faraday Trans., I* **1986**, *82,*.
5. *Biosensors;* Akhtar, M.; Lowe, C. R.; Higgins, I. J.; Eds.; Proceedings of a Royal Society Discussion Meeting held on 28 and 29 May 1986, The Royal Society, London, 1987.
6. *Biosensors, Fundamentals and Applications;* Turner, A. P. F.; Karube, I.; Wilson, G. S., Eds.; Oxford University Press, Oxford, 1987.
7. Scheller, F.; Schubert, F. In *Biosensoren;* Döring, D.; Etzhold, G.; Köster, T.; Schulze, D., Eds.; Akademie-Verlag Berlin, Berlin, 1989.
8. *Applied Biosensors;* Wise, D. L., Ed.; Butterworths, Boston, 1989.
9. Guilbault, G. G. *Analytical Uses of Immobilized Enzymes;* Dekker, New York, NY, 1984.
10. Hendry, S. P.; Higgins, I. J.; Bannister, J. V. *J. Biotechnol.* **1990**, *15*, 229-237.
11. Carr, P. W.; Bowers, L. D. In *Immobilized Enzymes in Analytical and Clinical Chemistry, Fundamentals and Applications;* Elving, P. J.; Winefordner, J. D., Eds.; Wiley, New York, NY, 1980.
12. Turner, A. P. F. *The World Biotech Report;* Online Publications, Pinner: London, 1985; Vol. 1, pp 181-192.
13. Chenault, H. K.; Whitesides, G. M. *Appl. Biochem. Biotechnol.* **1987**, *14*, 147-197.
14. *The Enzymes;* Boyer, P. D., Ed.; Academic Press, New York, NY, 1975-1976. Vols. XII and XIII.
15. Clark, W. M. *Oxidation-Reduction Potentials of Organic Systems;* William and Wilkins: Baltimore, ML, 1960.
16. Gorton, L.; Bremle, G.; Csöregi, E.; Jönsson-Pettersson, G.; Persson, B. *Anal. Chim. Acta* **1991**, *249*, 43-54.
17. Olsson, B.; Marko-Varga G.; Gorton, L.; Appelqvist, R.; Johansson, G. *Anal. Chim. Acta* **1988**, *206*, 49-55.
18. Gorton, L.; Hedlund, A. *Anal. Chim. Acta* **1988**, *213*, 91-100.
19. Underwood, A. L.; Burnett, R. W. In *Electroanalytical Chemistry;* Bard, A. J., Ed.; Dekker: New York, NY, 1973, Vol. 6; pp. 1-85.

20. Elving, P. J.; Schmakel, C. O.; Santhanam, K. S. V. *Crit. Rev. Anal. Chem.* **1976**, *6,* 1-67.
21. Bresnahan, W. T.; Elving P. J. *J. Am. Chem. Soc.* **1981**, *103,* 2379-2386.
22. Jensen, M. A.; Bresnahan W. T.; Elving, P. J. *Bioelectrochem. Bioenerg.* **1983**, *11,* 299-306.
23. Moiroux, J.; Deycard, S.; Malinski, T. *J. Electroanal. Chem.* **1985**, *194,* 99-108.
24. Kaye, R. C.; Stonehill, H. I. *J. Chem. Soc.* **1952**, *56,* 3244-3247.
25. Burnett, J. N.; Underwood; A. L. *Biochemistry* **1965**, *4,* 2060-2064.
26. Jensen, M. A.; Elving, P. J. *Biochim. Biophys. Acta* **1984**, *764,* 310-315.
27. Farrington, J. A.; Land, E. J.; Swallow, A. J. *Biochim. Biophys. Acta* **1980**, *590,* 273-276.
28. Andersson, R. F. *Biochim. Biophys. Acta* **1980**, *590,* 277-281.
29. Bielski, B. H. J.; Chan, P. C. *J. Am. Chem. Soc.* **1980**, *102,* 1713-1716.
30. Jaegfeldt, H. *Bioelectrochem. Bioenerg.* **1981**, *8,* 355-370.
31. Weinkamp, R.; Steckhan, E. *Angew. Chem. Int. Ed. Engl.* **1982**, *21,* 782-783.
32. Weinkamp, R.; Steckhan, E. *Angew. Chem. Int. Ed. Engl.* **1983**, *22,* 497.
33. Ruppert, R.; Herrmann, S.; Steckhan, E. *Tetrahedron Lett.* **1987**, *28,* 6583-6586.
34. Grammenudi, S.; Franke, M.; Vögtle, F.; Steckhan, E. *J. Incl. Phenomena* **1987**, *5,* 695-707.
35. Gorton, L.; Csöregi, E.; Domínguez, E.; Emnéus, J.; Jönsson-Pettersson, G.; Marko-Varga, G.; Persson, B. *Anal. Chim. Acta* **1991**, *250,* 203-248.
36. Andrieux, C. P.; Savéant, J. M. *J. Electroanal. Chem.* **1978**, *93,* 163-168.
37. Blaedel, W. J.; Jenkins, R. A. *Anal. Chem.* **1975**, *47,* 1337-1343.
38. Samec, Z.; Elving, P. J. *J. Electroanal. Chem.* **1983**, *144,* 217-234.
39. Braun, R. D.; Santhanam, K. S. V.; Elving, P. J. *J. Am. Chem. Soc.* **1975**, *97,* 2591-2598.
40. Moiroux, J.; Elving, P. J. *Anal. Chem.* **1979**, *51,* 346-350.
41. Blaedel, W. J., Haas, R. G. *Anal. Chem.* **1970**, *42,* 918-927.
42. Jaegfeldt, H. *J. Electroanal. Chem.* **1980**, *110,* 295-302.
43. Blankespoor, R. L.; Miller, L. L. *J. Electroanal. Chem.* **1984**, *171,* 231-241.
44. Elving, P. J.; Bresnahan, W. T.; Moiroux, J.; Samec, Z. *Bioelectrochem. Bioenerg.* **1982**, *9,* 365-378.
45. Moiroux, J.; Elving, P. J. *J. Am. Chem. Soc.* **1980**, *102,* 6533-6538.
46. Leduc, P.; Thevénot, D. *Bioelectrochem. Bioenerg.* **1974**, *1,* 96-107.
47. Ludvík, J.; Volke, J. *Anal. Chim. Acta* **1988**, *209,* 69-78.
48. Carlson, B. W.; Miller, L. L. *J. Am. Chem. Soc.* **1983**, *105,* 7453-7454.
49. Cénas, N. K.; Kanapieniené, J. J.; Kulys, J. J. *Org. React. (N. Y. Engl. Transl.)* **1987**, *24,* 70-76.
50. Grodkowski, J.; Neta, P.; Carlson, B. W.; Miller, L. *J. Phys. Chem.* **1983**, *87,* 3135-3138.
51. Martens, F. M.; Verhoeven, J. W. *Recl. Trav. Chim. Pays-Bas,* **1981**, *100,* 228-236.

52. Matsue, T.; Suda, M.; Uchida, I.; Kato, T.; Akiba U.; Osa, T. *J. Electroanal. Chem.* **1987**, *243*, 163-173.
53. Carlson, B. W.; Miller, L. L.; Neta, P.; Grodkowski, J. *J. Am. Chem. Soc.* **1984**, *106*, 7233-7239.
54. Čėnas, N. K.; Kanapieniené, J. J.; Kulys, J. J. *Biochim. Biophys. Acta* **1984**, *767*, 108-112.
55. Gorton, L. *J. Chem. Soc., Faraday Trans., I* **1986**, *82*, 1245-1258.
56. Tse, D. C.-S.; Kuwana, T. *Anal. Chem.* **1978**, *50*, 1315-1318.
57. Kitani, A.; Miller, L. L. *J. Am. Chem. Soc.* **1981**, *103*, 3595-3597.
58. Kitani, A.; So, Y.-H.; Miller, L. L. *J. Am. Chem. Soc.* **1981**, *103*, 7636-7641.
59. Evans, J. F.; Kuwana, T. *Anal. Chem.* **1977**, *49*, 1632-1635.
60. Scheurs, J.; van den Berg, J.; Wonders, A.; Barendrecht, E. *Recl. Trav. Chim. Pays-Bas* **1984**, *103*, 251-259.
61. Čėnas, N.; Rozgaitė, J.; Pocius, A.; Kulys, J. *J. Electroanal. Chem.* **1983**, *154*, 121-128.
62. Falat, L.; Cheng, H.-Y. *J. Electroanal. Chem.* **1983**, *157*, 393-397.
63. Ravichandran, K.; Baldwin, R. P. *Anal. Chem.* **1984**, *56*, 1744-1747.
64. Bartalits, L.; Nagy, G.; Pungor, E. *Anal. Lett.* **1984**, *17*, 13-41.
65. Čėnas, N. K.; Kanapieniené, J. J.; Kulys, J.J. *J. Electroanal. Chem.* **1985**, *189*, 163-169.
66. Eisenberg, E. J.; Cundy, K. C. *Anal. Chem.* **1991**, *63*, 845-847.
67. Ueda, C.; Tse, D. C.-S.; Kuwana, T. *Anal. Chem.* **1982**, *54*, 850-856.
68. Degrand, C.; Miller, L. L. *J. Am. Chem. Soc.* **1980**, *102*, 5728-5732.
69. Fukui, M.; Kitani, A.; Degrand, C.; Miller, L. L. *J. Am. Chem. Soc.* **1982**, *104*, 28-33.
70. Lau, A. N. K.; Miller, L. L. *J. Am. Chem. Soc.* **1983**, *105*, 5271-5277.
71. Jaegfeldt, H.; Torstensson, A. B. C.; Gorton, L. G. O.; Johansson, G. *Anal. Chem.* **1981**, *53*, 1979-1982.
72. Jaegfeldt, H.; Kuwana, T.; Johansson, G. *J. Am. Chem. Soc.* **1983**, *105*, 1805-1814.
73. Ikeda, T.; Shibata, T.; Senda, M. *J. Electroanal. Chem.* **1989**, *261*, 351-362.
74. Huck, H.; Schmidt, H.-L. *Angew. Chem. Int. Ed. Engl.* **1981**, *20*, 402-403.
75. Arai, G.; Matsushita, M.; Yasumori, I. *Nippon Kagaku Kaishi* **1985**, *5*, 894-897.
76. Ravichandran, K.; Baldwin, R. P. *J. Electroanal. Chem.* **1981**, *126*, 293-300.
77. Miki, K.; Ikeda, T.; Todoriki, S.; Senda, M. *Anal. Sci.* **1989**, *5*, 269-274.
78. Matsue, T.; Kasai, N.; Narumi, M.; Nishizawa, M.; Yamada, H.; Uchida, I. *J. Electroanal. Chem.* **1991**, *300*, 111-118.
79. Kunitake, M.; Akiyoshi, K.; Kawatana, K.; Nakashima, N.; Manabe, O. *J. Electroanal. Chem.* **1990**, *292*, 277-280.
80. Ravichandran, K.; Baldwin, R. P. *Anal. Chem.* **1983**, *55*, 1586-1591.
81. Torstensson, A.; Gorton, L. *J. Electroanal. Chem.* **1981**, *130*, 199-207.
82. Kimura, Y.; Niki, K. *Anal. Sci.* **1985**, *1*, 271-274.
83. Persson, B. *J. Electroanal. Chem.* **1990**, *287*, 61-80.
84. Persson, B.; Gorton, L. *J. Electroanal. Chem.* **1990**, *292*, 115-138.

85. Persson, B.; Gorton, L. unpublished results, 1991.
86. Gorton, L.; Johansson, G.; Torstensson, A. *J. Electroanal. Chem.* **1985**, *196*, 81-92.
87. Gorton, L.; Torstensson, A.; Jaegfeldt, H.; Johansson, G. *J. Electroanal. Chem.* **1984**, *161*, 103-120.
88. Bremle, G.; Persson, B.; Gorton, L. *Electroanalysis* **1991**, *3*, 77-86.
89. Persson, B.; Gorton, L.; Johansson, G. In *Proceedings of the 2nd International Meeting on Chemical Sensors;* Aucouturier, J.-L.; Cauhapé, J.-S.; Destriau, M.; Hagenmuller, P.; Lucat, C.; Ménil, F.; Portier, J.; Salardenne, J., Eds.; Imprimerie Biscaye: Bordeaux, 1986, 584-587.
90. Huck, H. *Fresenius' Z. Anal. Chem.* **1982**, *313*, 548-552.
91. Polášek, M.; Gorton, L.; Appelqvist, R.; Marko-Varga, G.; Johansson, G. *Anal. Chim. Acta* **1991**, *246*, 283-292.
92. Gorton, L.; Persson, B.; Polasek, M.; Johansson, G. In *Contemporary Electroanalytical Chemistry, Proceedings of the ElectroFinnAnalysis Conference (1988);* Ivaska, A.; Lewenstam, A.; Sara, R., Eds.; Plenum Press: New York, NY, 1991, 183-189.
93. Miyawaki, O.; Wingard, Jr., L. B. *Biochim. Biophys. Acta* **1985**, *838*, 60-68.
94. Kulys, J. *Enzyme Microb. Technol.* **1981**, *3*, 344-352.
95. Albery, W. J.; Bartlett, P. N. *J. Chem. Soc., Chem. Commun.* **1984**, *4*, 234-236.
96. Todoriki, S.; Ikeda, T.; Senda, M.; Wilson, G. S. *Agric. Biol. Chem.* **1989**, *53*, 3055-3056.
97. Kulys, J. J. *Biosensors* **1986**, *2*, 3-13.
98. McKenna, K.; Boyette, S. E.; Brajter-Toth, A. *Anal. Chim. Acta* **1988**, *206*, 75-84.
99. Yon Hin, B. F. Y.; Lowe, C. R. *Anal. Chem.* **1987**, *59*, 2111-2115.
100. Wang, J.; Golden, T. *Anal. Chim. Acta* **1989**, *217*, 343-351.
101. Atta, N. F.; Galal, A.; Karagözler, A. E.; Zimmer, H.; Rubinson, J. F.; Mark, Jr., H. B. *J. Chem. Soc., Chem. Commun.* **1990**, *19*, 1347-1349.
102. Ottaway, J. M. In *The Indicators;* Bishop, E., Ed.; Pergamon Press: Oxford, 1972, pp. 498-503.
103. Gorton, L.; Johansson, G. *J. Electroanal. Chem.* **1980**, *113*, 151-158.
104. Appelqvist, R.; Marko-Varga, G.; Gorton, L.; Torstensson, A.; Johansson, G. *Anal. Chim. Acta* **1985**, *169*, 237-247.
105. Marko-Varga, G. *Anal. Chem.* **1989**, *61*, 831-838.
106. Marko-Varga, G.; Domínguez, E.; Hahn-Hägerdal, B.; Gorton, L. *J. Chromatogr.* **1990**, *506*, 423-441.
107. Kotouček, M.; Tomášová, J.; Durčáková, S. *Collect. Czech. Chem. Commun.* **1969**, *34*, 212-220.
108. Kotouček, M.; Zavadilová, J. *Collect. Czech. Chem. Commun.* **1972**, *37*, 3212-3218.
109. Powell, M. F.; Bruice, T. C. *J. Am. Chem. Soc.* **1983**, *105*, 7139-7149.
110. Ni, F.; Thomas, L.; Cotton, T. M. *Anal. Chem.* **1989**, *61*, 888-894.

111. Ni, F.; Cotton, T. M. *J. Raman Spectrosc.* **1988**, *19*, 429-438.
112. Ni, F.; Feng, H.; Gorton, L.; Cotton, T. M. *Langmuir* **1990**, *6*, 66-73.
113. Carlson, B. W.; Miller, L. L. *J. Am. Chem. Soc.* **1985**, *107*, 479-485.
114. Miller, L. L.; Valentine, J. R. *J. Am. Chem. Soc.* **1988**, *110*, 3982-3989.
115. Fukuzumi, S.; Nishiazawa, N.; Tanaka, T. *J. Org. Chem.* **1984**, *49*, 3571-3578.
116. Fukuzumi, S.; Koumitsu, S.; Hironaka, K.; Tanaka, T. *J. Am. Chem. Soc.* **1987**, *109*, 305-316.
117. Skotheim, T. A.; Okamoto, Y.; Gorton, L. G.; Lee, H. S.; Hale, P. D. *US Patent Appl.;* 1991.
118. Marko-Varga, G.; Gorton, L. *Anal. Chim. Acta* **1990**, *234*, 13-29.
119. Marko-Varga, G.; Domínguez, E.; Hahn-Hägerdal, B.; Gorton, L. *J. Pharm. Biomed. Anal.* **1990**, *8*, 817-823.
120. Marko-Varga, G.; Appelqvist, R.; Gorton, L. *Anal. Chim. Acta* **1986**, *179*, 371-379.
121. Blaedel, W. J.; Jenkins, R. A. *Anal. Chem.* **1976**, *48*, 1240-1247.
122. Torstensson, A.; Johansson, G.; Månsson, M.-O.; Larsson, P.-O.; Mosbach, K. *Anal. Lett.* **1980**, *13*, 837-849.
123. Persson, M.; Månsson, M.-O.; Bülow, L.; Mosbach, K. *Bio/Technology* **1991**, *9*, 280-284.
124. Månsson, M.-O.; Mosbach, K. In *Pyridine Nucleotide Coenzymes, Chemical, Biochemical, and Medical Aspects, Part B;* Dolphin, D.; Poulson, R.; Avramović, O., Eds.; Coenzymes and Cofactors, Wiley: New York, NY, 1987, Vol. 2; pp. 217-273.
125. Miki, K.; Ikeda, T.; Todoriki, S.; Senda, M. *Anal. Sci.* **1989**, *5*, 269-274.
126. Batchelor, M. J.; Green, M. J.; Sketch, C. L. *Anal. Chim. Acta* **1989**, *221*, 289-294.
127. Yabuki, S.; Shinohara, H.; Ikariyama, Y.; Aizawa, M. *J. Electroanal. Chem.* **1990**, *277*, 179-188.
128. Ikariyama, Y.; Ishizuka, T.; Sinohara, H.; Aizawa, M. *Denki Kagaku* **1990**, *58*, 1097-1102.
129. Domínguez, E.; Marko-Varga, G.; Hahn-Hägerdal, B.; Gorton, L. unpublished data.
130. Domínguez, E.; Marko-Varga, G.; Hahn-Hägerdal, B.; Gorton, L. *Anal. Chim. Acta*, **1991**, *249*, 145-154.

RECEIVED November 5, 1991

Chapter 7

Preparation and Characterization of Active Glucose Oxidase Immobilized to a Plasma-Polymerized Film

Michael J. Danilich[1], Dominic Gervasio[2], and Roger E. Marchant[1,3]

Departments of [1]Biomedical Engineering and [2]Chemistry, Case Western Reserve University, Cleveland, OH 44106

Plasma polymerized N-vinyl-2-pyrrolidone films were deposited onto a poly(etherurethaneurea). Active sites for the immobilization were obtained via reduction with sodium borohydride followed by activation with 1-cyano-4-dimethyl-aminopyridinium tetrafluoroborate. A colorometric activity determination indicated that 2.4 cm^2 of modified poly(etherurethaneurea) film had an activity approximately equal to that of 13.4 nM glucose oxidase in 50 mM sodium acetate with a specific activity of 32.0 U/mg at pH 5.1 and room temperature. Using cyclic voltammetry of gold in thin-layer electrochemical cells, the specific activity of 13.4 nM glucose oxidase in 0.2 M aqueous sodium phosphate, pH 5.2, was calculated to be 4.34 U/mg at room temperature. Under the same experimental conditions, qualitative detection of the activity of a modified film was demonstrated by placing it inside the thin-layer cell.

Radiofrequency (RF) plasma methods have become popular for the surface-selective modification of materials surfaces (1). In the case of plasma treatment, the surface chemistry of the treated material is altered, generally by the introduction of oxygen- or nitrogen-containing moieties. In the case of plasma polymerization, a thin film derived from an organic monomer plasma is deposited onto the substrate material surface. Plasma treatment followed by plasma polymerization generally results in excellent film adhesion to inorganic substrates and film grafting to polymeric substrates.

[3]Corresponding author

0097–6156/92/0487–0084$06.00/0

A great deal of research has been focused on the evaluation of plasma polymers and plasma treated materials for blood and soft tissue contacting-applications (*2, 3*). A number of studies have involved the physical adsorption or covalent attachment of a variety of biomolecules to various gas plasma-treated polymer surfaces (*4, 5*). In such studies, however, the covalent immobilization is often assumed to take place through precursor groups formed at the biomaterial surface from ill-defined oxygen and nitrogen functionalities obtained directly from the plasma.

Previously, we have reported the preparation of RF plasma polymerized N-vinyl-2-pyrrolidone with a hydroxylated surface, prepared by chemical derivatization of the plasma product (*6*). The preparation of this new interface material has provided a well-characterized functional group suitable for the covalent attachment of bioactive molecules. Glucose oxidase from *Aspergillus niger* was selected for immobilization onto a plasma-modified poly(etherurethaneurea) on the basis of its functional importance in glucose biosensor devices (*7*), its well-characterized structure and properties (*8*), its well-established stability to denaturing agents (*9*), and its retention of properties after immobilization (*10*). These valuable properties permitted a rigorous evaluation of the binding and activity of the immobilized enzyme (*11*).

Experimental

Materials. Unless indicated otherwise, all water was deionized and distilled or prepared by reverse osmosis and distillation under N_2. Acetone was glass distilled HPLC grade (Fisher Scientific Co.). All glassware was washed in a sulfuric acid (H_2SO_4) / Nochromix (Godax Laboratories, Inc.) solution, followed by a Liquinox (Alconox, Inc.) soap solution, rinsed with tap water, water, and HPLC grade methanol, and dried at 100°C. Stainless steel monomer inlet lines were immersed overnight in 100% ethanol and dried at 100°C prior to each plasma reaction. Glass microslides were washed (1-2 h) in H_2SO_4 / Nochromix solution, thoroughly rinsed with tap water, washed with refluxing acetone in a Soxhlet extraction system for at least 24 h, dried overnight at 80°C under vacuum, and stored in a dessicator over P_2O_5. Poly(etherurethaneurea) (PEUU, Mercor, approximately 0.1 mm thick on Mylar backing) samples (2 cm x 5 cm) and silicon wafers (approximately 0.75 cm²) were cleaned by sonication (15-30 min) in 100% ethanol, dried overnight at 60°C under vacuum and stored in a dessicator over P_2O_5.

Plasma polymerization and surface modification. Thin films of plasma polymerized N-vinyl-2-pyrrolidone (NVP, 98%, Aldrich) (PPNVP) were prepared by radiofrequency (13.56 MHz) glow discharge using an inductively coupled flow-through Pyrex reactor system which has been described previously (*12*). PEUU substrate films and silicon wafers were mounted on clean glass microslides suspended on a glass-rod tray. Two PEUU film substrates were placed with their leading edges beneath the trailing edge of the last turn of the induction coil (seven turns of 1.5 mm diameter copper wire). The two PEUU substrates were placed adjacent to each other, resulting in an equivalent coating thickness and composition on each, as described in previous studies (*12*). Two silicon wafers were placed immediately behind the PEUU samples at approximately 6 cm from the end of the coil, and a clean glass microslide was placed with its leading edge approximately 7.5 cm from the end of the coil. No substrates were placed beneath the induction coil. Prior to plasma polymerization, NVP was degassed three or more times by the freeze (dry ice / acetone bath)-thaw method under reduced pressure. After plasma polymerization reactions, NVP monomer was stored under argon at -4°C. An initial argon plasma treatment of the PEUU films was carried out for 5 min. at 35-40 W net discharge power, 0.03 Torr, and approximately 2.0 cc(STP)/min. argon flow rate prior to the deposition reaction. The NVP plasma polymerizations were carried out for 5 min. at 25 W net power and 0.03 Torr with a monomer flow rate of approximately 0.23 cc (STP)/min. The resulting PPNVP/PEUU films and the PPNVP coated silicon wafers and glass slide were removed from the plasma reactor and placed in sterile plastic petri dishes and stored in a dessicator over P_2O_5.

Reduction reactions were carried out in freshly prepared, continuously stirred 0.26 M aqueous sodium borohydride ($NaBH_4$, powder, 98%, Aldrich Chemical Co.) at room temperature under nitrogen for 30 h according to the methods of Yu and Marchant (*6*). Two PPNVP/PEUU films, generally from different PPNVP deposition reactions were placed in the reduction reaction medium. After reduction, the resulting hydroxylated PPNVP/PEUU films were washed three times with water and once with acetone, dried overnight at 60°C under vacuum, and stored in a dessicator over P_2O_5.

Activation of the hydroxylated PPNVP/PEUU films was accomplished according to a modification of the cyano-transfer methods of Kohn and Wilchek (*13*). A single hydroxylated PPNVP/PEUU film was washed with water, acetone/water (35:65), and acetone/water (60:40). The film was transferred to a glass bowl containing approximately 50 mL continuously stirred acetone/water (60:40) cooled to 0°C with an ice-water bath. 1-cyano-4-dimethylamino pyridinium tetrafluoroborate solution (CDAP, Sigma Chemical Co., 0.1 g/mL acetonitrile, anhydrous 99+%, Aldrich) (0-0.5 mL) was added rapidly to the solution and stirred for 60 sec. Aqueous triethylamine solution (TEA, 99+%, Aldrich, 0.2 M) (0-0.4 mL) was then added over a 1-2 min. period, and stirring was continued for 60 sec. Cold (4-5°C) 0.05 N HCl (approximately 180 mL) was added and stirring was stopped. After 15 min., the activated PPNVP/PEUU film was removed from the reaction mixture and washed with cold water and cold 20 mM sodium phosphate (pH 7.0). The film was transferred to a small glass vial containing approximately 25 mL of glucose oxidase (β-D-glucose:1-oxygen oxidoreductase, EC 1.1.3.4, Type X-S from *Aspergillus niger*, Sigma, 2.0 mg/mL 20 mM sodium phosphate, pH 7.0) for immobilization. The vial was rocked gently (18 cycles/min.) on an aliquot mixer (American Dade) at 4°C for up to 24 h.

The sample (GOx-PPNVP/PEUU) was removed from the GOx solution and washed exhaustively to remove adsorbed enzyme by continuous stirring in 900 mL of sodium dodecyl sulfate (SDS, Sigma, 2% v/v in H2O), Triton X-100 (octyl phenoxy polyethoxyethanol, Sigma, 2% v/v in H2O), and 20 mM sodium phosphate (pH 7.0) for 24 h each at 5°C and stored in 20 mM sodium phosphate (pH 7.0) at 4°C. It has been shown that this wash and storage procedure is effective in removing adsorbed GOx and does not denature the immobilized GOx (*11*).

Characterization. The PPNVP and PPNVP/PEUU films were characterized by attenuated total reflectance Fourier transform infrared spectroscopy (ATR-FTIR), electron spectroscopy for chemical analysis (ESCA), ellipsometry, and water contact angles in air. Two independent methods, a modification of the radioimmunoassay used by Ziats, et. al. (*14*) and an "immunochemical stain" based on the ABC immunohistochemical staining method (*15*) were developed to assay the binding of GOx to the PPNVP/PEUU support.

GOx activity was assayed colorometrically and electrochemically. The colorometric assay was the determination of hydrogen peroxide (H2O2) produced during the coupled reaction with o-dianisidine and peroxidase (POx) (*16*). The formation of oxidized o-dianisidine was monitored continuously at 500 nm with a DMS 200 UV/Visible spectrophotometer (Varian). The reference solution consisted of 1.2 mL of the reduced form of o-dianisidine (dihydrochloride, Sigma, 0.21 mM in 50 mM sodium acetate, pH 5.1), 0.25 mL β-D-glucose (Sigma, 10% w/v in H2O), and 0.05 mL of peroxidase solution, (POx, EC 1.11.1.7, Type II from horseradish, Sigma, 60 Purpurogallin units/ mL H2O). A 0.8 cm x 3 cm GOx-PPNVP/PEUU film sample, from which the Mylar backing had been removed, was placed in a sample cuvette filled with reference solution. The absorbance at 500 nm was monitored until it became constant. Reference solution and unmodified PEUU films treated with GOx and washed were run as controls. The activity of various dilutions of free GOx in solution was also measured spectrophotometrically for comparison with the activity of covalently bound GOx. Each of the components of the reference solution as well as 50 mM sodium acetate (pH 5.1), GOx (0.5 units/mL 50 mM sodium acetate, pH 5.1), and PEUU film samples with and without Mylar backing were scanned individually from 200 nm to 800 nm to check for possible interferences at 500 nm.

The electrochemical assay was the determination of H2O2 production by cyclic voltammetry of gold in a thin-layer electrochemical cell during the GOx-catalyzed oxidation of glucose. The thin-layer cells used for the second electrochemical assay were a modification of the design used by Reilley, et. al. (*17*). Glass microslides, 25 mm x 75 mm, were cleaned by boiling in H2SO4/Nochromix for approximately 2 h, rinsing with tap water, distilled deionized water, and HPLC grade acetone, washing overnight with acetone in a refluxing Soxhlet vessel, rinsing with distilled acetone and 100% ethanol, and drying overnight at 80°C under vacuum. In addition, the slides were subjected to an argon plasma etch (2-3 min) immediately prior to the sputter deposition of approximately 100 nm of gold. The gold deposition was carried out in an argon atmosphere (2×10^{-4} Torr), using a 99.999% pure gold target and a 15 mA/1000 V ion beam in a dual beam ion beam sputtering system (model CSC 330, Commonwealth Scientific Co.) The gold-coated slides were cut into rectangular sections approximately 35 mm x 13 mm, which were rinsed in HPLC grade acetone and air dried. A thin strip of gold, approximately 1 mm wide, was abraded from the gold surface along the narrow (13 mm) edge. To form the electrode, an electrical lead was attached to the gold surface with a silver epoxy. After curing in an oven, the silver epoxy was encapsulated with RTV silicone cement. A Teflon TFE spacer (temp-r-tape, CHR Industries, Inc., HM250, 64 mm thick or HM650, 165 mm thick) with a solvent resistant acrylic adhesive, approximately 1 mm wide, was placed on each parallel edge of the gold surface. Rectangular covers for the thin-layer cells, approximately 17 mm x 13 mm, were cut from microslides, and were placed in boiling sulfuric/formic acid for 1 h, rinsed thoroughly with acetone and ultrapure water, and air dried.

"Glass" thin-layer cells were formed by sandwiching a glass cover and a gold coated glass slide around the Teflon spacers. To form a "GOx-PPNVP/ PEUU" thin-layer cell, a rectangular piece of washed GOx-PPNVP/PEUU film was cut to match a glass cover (i.e., approximately 17 mm x 13 mm). The Mylar backing material was removed and the sample was placed on the glass cover, to which it adhered readily. The cover was placed on the Teflon spacers with the GOx layer facing the gold electrode surface. The cover and the gold coated glass slide electrode were held together by wrapping with HM250 Teflon tape. Several thin-layer cells could be prepared in a one h period. Optimal operating conditions with minimal ohmic polarization and leakage current were obtained by using a spacer thickness of 165 μm, a sweep rate of 1.0 mV/s, and an electrolyte concentration of 0.2-1.0 M. The leakage current was reduced further by the removal of the thin strip of the gold electrode from the portion of the thin-layer cell which comes into contact with the bulk electrolyte.

The experiments were carried out using a Princeton Applied Research Co. PARC 173 potentiostat and a PARC 175 universal programmer, and were recorded on a Yokagawa Technicorder analog x-y recorder. The potential of the working electrode was cycled relative to a reference electrode and the resulting current flow between the working (thin-layer) and counter electrodes was measured. The potential cycle was linear in time with a triangular waveform. The initial potential was selected to be a potential at which only capacitive charging of the working electrode was observed, resulting in zero current at steady-state. Potential scans were carried out with a positive (anodic) initial sweep direction. Experiments were carried out in a three-electrode configuration in a standard three compartment electrochemical cell under a controlled atmosphere. The counter electrode was a gold wire coil. A normal hydrogen electrode (NHE, -329 mV vs saturated calomel electrode, SCE) or a reversible hydrogen electrode (RHE, -554 mV vs SCE) was used as the reference electrode. The external reference was connected to the bulk solution via a salt bridge filled with the supporting buffer. The thin-layer cells were filled at the bottom by capillary action and emptied from the top by applying a mild vacuum at the top edge of the glass cover. Residual analyte was removed between experiments by emptying and refilling the thin-layer cells several times. The GOx-PPNVP/PEUU thin-layer cells were filled with test buffer immediately after fabrication to prevent dehydration of the immobilized enzyme. The purity of the buffer solutions was checked by taking cyclic voltammograms of "gold flags" (i.e., approximately 2 cm^2 of gold, heated in a gas/air flame and annealed in water while red hot).

Results and Discussion

Plasma Polymerization and Surface Modification. The characteristics of the PPNVP films are given in table I. The atomic composition, water contact angle, thickness, and index of refraction values for PPNVP were consistent with those published previously (*12*). The high resolution C$_{1s}$ spectra were resolved into four Gaussian peaks, the positions of which were consistent with the presence of hydrocarbon (285.0 eV), hydroxyl/amine (286.4 eV), ketone (288.0 eV), and acid (289.0 eV) moieties (*18*). However, due to the wide range of species formed in the plasma reaction and the correspondingly broad peaks observed by ESCA, these results do not represent a unique resolution of the spectra into components. In our previous study (*12*), the PPNVP deposition rate (and thus PPNVP film thickness) was shown to decrease with axial distance from the induction coil. Thus, the film thickness values shown in table I are accurate only at the exact position of the silicon wafers (immediately behind the PEUU substrate films), but do provide a reasonable estimate of the minimum thickness of the PPNVP films. The cleaned PEUU films had a mean advancing contact angle with water of 77.4° and a first receding contact angle of 66.9° (n=2).

ATR-FTIR spectra of PPNVP on PEUU in the 1900-900 cm^{-1} range, in which the spectral contributions from the PEUU support were subtracted out confirmed the presence of PPNVP on the surface of the PEUU films. Prominent features of the spectra were the very strong carbonyl absorption (n(C=O), 1666 cm^{-1}) and poor resolution in the 1500-1000 cm^{-1} range attributed to the formation of a wide range of C-H deformations caused by the wide variety of reactions which can occur in the energetic plasma environment. Applying a similar spectral subtraction technique to confirm the presence of hydroxyl in the reduced PPNVP on the surface of the PEUU films was not practicable due to the overlap in the 1110-1100 cm^{-1} region of the very strong C-O-C etheral absorption in PEUU and the hydroxyl C-O stretching absorption in reduced PPNVP. The presence of the hydroxyls at the surface of reduced PPNVP and their availability for further reaction has been demonstrated by ESCA and ATR-FTIR analyses of gas-phase trifluoroacetic acid derivatized-reduced PPNVP (*19*).

TABLE I
CHARACTERISTICS OF PPNVP FILMS

Atomic Composition[a]			H_2O Contact Angle[b]		thickness[c]
(%)			(°)		(A)
C_{1s}	O_{1s}	N_{1s}	θ_A	θ_R[d]	
71.2	18.3	9.0	33.3	25.5	1013
(1.8)	(0.7)	(0.4)	(2.7)	(2.6)	(208)

SOURCE: Adapted from ref. 11.
Values are mean values with the standard deviation shown in parentheses; n=7
in each case. a: ESCA measurements on silicon substrates. b: Sessile drop
measurements on glass slides. c: Ellipsometry measurements on silicon
substrates. d: First receding contact angle.

Binding. In the immunochemical stain studies (*11*), positive stain indicated
the presence of GOx whereas the lack of stain indicated its absence.
Untreated PEUU controls (which had been exposed to GOx) exhibited no
stain after continuous stirring in SDS, Triton, and sodium phosphate. GOx-
PPNVP/PEUU samples washed by the same method as well as PEUU controls
washed by less stringent methods exhibited positive stain. Thus the
immunochemical stain assay demonstrated that the continuous wash in SDS,
Triton, and sodium phosphate removed physically adsorbed GOx from the
surface of GOx-PPNVP/PEUU, leaving covalently bound GOx. Positive
stain was easily observable with the naked eye, making the immunochemical
stain an effective novel technique for the quick screening of wash procedures
for thin film samples.

Activity

Colorometric Studies. None of the solution components of the o-
dianisidine activity assay interfere with the detection of chromophore
formation at 500 nm. The UV/visible spectrum of PEUU (on Mylar backing)
is approximately flat in the vicinity of 500 nm, indicating that PEUU should
not interfere with the assay. Inclusion of PEUU in the reference cuvette
further reduced the possibility of interference from PEUU.
Enzyme activity is generally calculated from the initial rate of enzyme

catalyzed substrate consumption or product generation with time. One unit (U) of glucose oxidase activity will oxidize 1.0 mmole of β-D-glucose to D-gluconic acid and H_2O_2 per minute at pH 5.1 at 35°C (*20*). In the colorometric activity assay, glucose oxidase specific activity, A, was calculated from equation 3,

$$A = \frac{dA_{500}/dt}{\varepsilon l m_{GOx}} \qquad (3)$$

where A_{500} is the absorbance at 500 nm, thus dA_{500}/dt is the initial rate of increase in the absorbance at 500 nm with time, ε is the molar extinction coefficient for the oxidized form of o-dianisidine at 500 nm (7.5 cm^{-1}mM^{-1}, manufacturer's specification), l is the optical path length (0.1 cm), and m_{GOx} is the mass of GOx in the assay solution. The specific activities of three dilutions of soluble GOx (13.4 nM, 26.8 nM, and 53.6 nM) were all found to be approximately 32.0 U/mg. The visible absorption curve obtained with the washed GOx-PPNVP/PEUU sample indicated some initial GOx activity which rapidly dropped off, presumably due to diffusional and partitioning effects of the immobilized enzyme microenvironment (*21*). Such effects, common among immobilized enzymes, results in a local decrease in enzyme substrate concentration at the surface of the support material. The initial response of the GOx-PPNVP/PEUU sample was approximately equivalent to that of the 13.4 nM GOx in solution with a specific activity of 32.0 U/mg at room temperature and at pH 5.1.

The GOx molecule is a rigid prolate ellipsoid with long and short axes of approximately 140 A and 50 A (*22*). If GOx is assumed to be immobilized to PPNVP/PEUU in a close-packed monolayer, and interstitial spaces are accounted for, then 0.91 µg and 2.55 µg of GOx will be present on the surface when the long axis is oriented parallel and perpendicular to the surface, respectively. From the initial response of the GOx-PPNVP/PEUU sample, the specific activity of the immobilized GOx was calculated to be 0.85 U/mg and 0.30 U/mg in the respective monolayer orientations (*11*). Similarly, if immobilization is assumed not to alter the initial activity, then the number of GOx molecules (13.4 nM x 0.3 mL) immobilized in the GOx-PPNVP/PEUU sample is approximately 71% and 25% of the maximum number that would be present on the surface in the parallel and perpendicular orientations.

The visible absorption curves corresponding to the assay reference solution and a PEUU control sample which had been exposed to GOx solution and washed in the same wash media as the active GOx-PPNVP/PEUU sample were essentially identical.This result suggests the absence of active GOx on the surface of the washed PEUU control. Whereas the immunochemical stain indicated the presence of covalently bound GOx on washed GOx-PPNVP/PEUU, the colorometric activity assay indicated that the bound enzyme had retained its active conformation.

Thin-layer Studies. The thin-layer electrochemical system was developed to address the lack of sensitivity of a preliminary bulk amperometric activity assay (*11*). The first set of thin-layer studies was taken to characterize the thin-layer cells in soluble enzyme solutions and to determine if there were any interferences to the detection of hydrogen peroxide. Preliminary thin-layer studies (*23*) indicated that the oxidation of hydrogen peroxide could be detected at approximately 1080 mV with only minimal interference from the oxidation of glucose by gold. The addition of chloride ion to the solution further suppressed the glucose electrooxidation interference.

Figure 1 is the cyclic voltammogram of gold in the 750 to 1750 mV vs RHE region using a glass thin-layer cell as the working electrode in air-saturated 1.0 M sodium phosphate (pH 5.2) with 10^{-5} M sodium chloride, 5 mM glucose, and 13.4 nM soluble GOx in the solution. Throughout the experiment, glucose was continually being oxidized to gluconic acid and H_2O_2 in the bulk solution. The thin-layer cell was emptied and refilled with this bulk solution before each of the cycles indicated in figure 1. The continued generation of H_2O_2 was evidenced by the continual growth in the anodic H_2O_2 oxidation peak (approximately 1080 mV) in figure 1. The inset is a continuation of figure 1 with an expanded current scale. It gives the fifth and sixth sweeps taken after emptying and refilling the thin-layer cell. Subsequent sweeps were essentially identical to the sixth sweep, indicating no further generation of H_2O_2 in the bulk solution, and thus virtual completion of the glucose air oxidation reaction.

The charge passed through the working electrode during the electro-oxidation of H_2O_2 to O_2 on gold was obtained by integrating the current measured during the forward sweeps (1 mV/s) in figure 1 from 750 to 1750 mV with respect to time. The baseline charge value was obtained by integrating the current in the voltammogram of gold using a glass thin-layer cell in air-saturated 1.0 M sodium phosphate (pH 5.2) with 5 mM glucose and 10^{-5} M sodium chloride. The charge values were converted into H_2O_2 concentration values using Farraday's law and the measured geometrical volume of the thin-layer cell (2.48 x 10^{-5} L). Time-zero was the point at which GOx was added to the other solution components in the electrochemical cell. The sampling time, t, of each sweep was defined as the time at which the forward sweep was completed (i.e., the time at which the potential reached 1750 mV).

The resulting data are plotted in figure 2. When the concentration vs time data is linear, the departure from initial conditions is negligible, ensuring that the measured rate of change of substrate or product concentration will yield an accurate measure of enzyme activity. Unfortunately, the slope of the H_2O_2 /time curve (figure 2) is not constant but decreases at each point. Nevertheless, assuming that the first electrochemical data point was taken in the initial linear region of the curve, a determination of a lower limit value for the specific activity was calculated to be 4.34 U/mg. In the colorometric activity study, the generation of H_2O_2

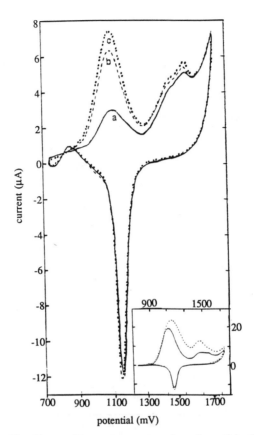

Figure 1.   Cyclic voltammogram of a glass thin-layer cell in 1.0 M sodium phosphate/5 mM glucose/10^{-5} M NaCl/13.4 nM GOx (pH 5.2) under air. Reference: RHE. **a.** First sweep (start time: 7.0 min after the addition of GOx). **b.** Second sweep (start time: 50.0 min after the addition of GOx). **c.** Third sweep (start time: 82.0 min after the addition of GOx). **Inset:** Solid line: Fifth scan (start time: 310.0 min after the addition of GOx). Broken line: Sixth scan (start time: 707.0 min after the addition of GOx).

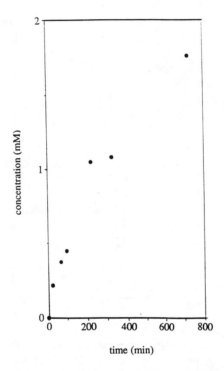

Figure 2. Concentration of H$_2$O$_2$ vs time from figure 1.

was measured in the linear response region, and the GOx specific activity was calculated to be approximately 32.0 U/mg. The experimental variation in results between the UV/Visible and electrochemical techniques can be attributed to the uncertainty associated with the initial rate assumption of the first measured electrochemical datum.

The second set of thin-layer studies was taken with the GOx-PPNVP/PEUU thin-layer cells to confirm the activity of the immobilized GOx in the polymer layer and to establish the validity of the thin-layer approach for measuring immobilized enzyme activity. Preliminary studies with the GOx-PPNVP/PEUU thin-layer cells indicated that inclusion of the GOx-PPNVP/PEUU film into the thin-layer cell did not have any adverse effects on its performance.

Figure 3 is the first two cycles of the cyclic voltammogram of gold in the 750 to 1750 mV vs RHE region, using a GOx-PPNVP/PEUU thin-layer cell as the working electrode in air-saturated 0.2 M sodium phosphate (pH 5.2) with 5 mM β-D(+)-glucose. The thin-layer cell was not emptied and refilled between cycles. The initial voltammogram (solid line) consists of the gold oxidation current (at approximately 1500 mV on the positive sweep), the gold oxide reduction (approximately 1100 mV) and anodic glucose oxidation (approximately 900 mV) currents on the negative sweep, and a fourth anodic current peak occurring in the 900 to 1300 mV region. The forward oxidation peak between 900 and 1300 mV might appear to be exclusively attributable to direct oxidation of glucose by gold. The peak is slightly positive of typical glucose oxidation potentials (*24*) however, indicating a contribution from an overlapping H_2O_2 electrooxidation peak. The second voltammogram (broken line) is similar to that taken during the glass thin-layer cell control experiment under similar conditions. During the second sweep the peak shifted back to its original position in the glucose oxidation potential region, and the H_2O_2 component at more positive potentials was missing. The upward shift of the baseline during the second sweep is attributed to electrode capacitive charging and is otherwise insignificant. The behavior observed during the initial positive sweep is therefore attributed to the combination of the effects of oxidation of the H_2O_2 produced by GOx-PPNVP/PEUU activity and direct glucose electrooxidation. On the negative sweep and on all subsequent cycles, the oxidation current may be attributed exclusively to the direct oxidation of glucose by gold, a process which generates no peroxide. This behavior is similar to, and consistent with, that found in the colorometric assay where the GOx-PPNVP/PEUU had a relatively high initial activity which then decreased, presumably due to diffusional and partitioning effects.

The quantitative characterization of reaction energetics using cyclic voltammetry may be improved by using differential pulse voltammetry to obtain more highly resolved current/potential curves. Similarly, the sampling time may be shortened and more clearly defined by utilizing potential step coulometry, allowing a quantitative characterization of reaction dynamics. Both improvements in technique are currently being investigated with utilization of our thin-layer electrochemical cell system in order to achieve an accurate determination of enzyme activity.

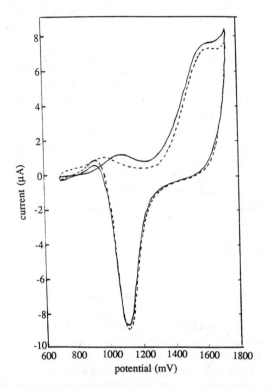

Figure 3.   Cyclic voltammogram of a GOx-PPNVP/PEUU
thin-layer cell in 0.2 M sodium phosphate/5 mM β-D(+)-glucose
(pH 5.2) under air. Reference: RHE. Solid line: first sweep.
Broken line: second sweep.

Our previous work with PPNVP and GOx-PPNVP/PEUU (*6, 12, 25*), which focused on characterization of the surface modified thin films, has demonstrated that a reasonable amount of control can be exerted over the chemistry of plasma modification. The techniques used here to attach active GOx to PEUU may be applied to a wide variety of biomolecules and a wide variety of organic and inorganic substrates. Incorporation of GOx-PPNVP/PEUU into the thin-layer cells extended the potential applicability of GOx-PPNVP/PEUU and similar materials to specific practical applications such as sensing devices. Similarly, the ease of fabrication of the thin-layer cells and the wide variety of electrochemical techniques which are available for use with thin-layer cells warrant further development of this system.

Conclusions

This study on the immobilization of glucose oxidase and the characterization of its activity has demonstrated that a bioactive interface material may be prepared from derivatized plasma polymerized films. UV/Visible spectrophotometric analysis indicated that washed GOx-PPNVP/PEUU (2.4 cm^2) had activity approximately equivalent to that of 13.4 nM GOx in 50 mM sodium acetate with a specific activity of 32.0 U/mg at pH 5.1 and room temperature. A sandwich-type thin-layer electrochemical cell was also used to qualitatively demonstrate the activity of 13.4 nM glucose oxidase under the same conditions. A quantitatively low specific activity value of 4.34 U/mg was obtained for the same enzyme solution by monitoring the hydrogen peroxide oxidation current using cyclic voltammetry. Incorporation of GOx-PPNVP/PEUU into the thin-layer allowed for the detection of immobilized enzyme activity in 0.2 M sodium phosphate (pH 5.2) at room temperature.

Acknowledgement

The authors are grateful to Tracey L. Bonfield, Ph. D. and James M. Anderson, M. D., Ph. D. for their assistance in the RIA binding studies, and to Kandice Kottke-Marchant, M. D., Ph. D. for her assistance in the immunochemical stain binding studies. We acknowledge the financial support of the National Institutes of Health grants HL-40047 and RR02024.

Literature Cited

1. Yasuda, H. K. *Plasma Polymerization*; Academic Press Inc., San Diego, CA **1985**.
2. Griesser, H. J.; Johnson, G.; and Steele, J. G. *ACS Polymer Preprints*, presented at the Boston, MA meeting, **1990**, vol. 31.
3. Pratt, K. J.; Williams, S. K.; and Jarrell, B. E. *J. Biomed. Mat. Res.* **1989**, *23*, 1131.
4. Sipehia, R. *Biomat., Art. Cells, Art. Org.* **1989**, *16*, 955.
5. Sharma, C. P.; and Jayasree, G. *J. Coll. Int. Sci.* **1990**, *137*, 289.
6. Yu, D. and Marchant, R. E. *Macromolecules* **1989**, *22*, 2957.

7. Scheller, F. W.; Pfeiffer, D.; Schubert, F.; Renneberg, R. and Kirstein, D.
    in *Biosensors: Fundamentals and Applications*. Turner, A. P. F.; Karube,
    I.; and Wilson, G. S. Eds.; Oxford University Press, New York, NY
    1987, pg. 315.
8. Tsuge, H.; Natsuaki, O. and Ohashi, K. *J. Biochem.* **1975**, *78*, 835.
9. Jones, M. N.; Manley, P. and Wilkinson, A. *Biochem. J.* **1982**, *203*, 285.
10. Weetall, H. H. and Hersch, L. S. *Biochim. Biophys. Acta.* **1970**, *206*, 54.
11. Danilich, M. J.; Kottke-Marchant, K.; Anderson, J. M and Marchant, R.
    E. *J. Biomat. Sci., Polym. Ed.*, in press.
12. Marchant, R. E.; Yu, D. and Khoo, C. *J. Polym. Sci.,Polym.Chem.*
    1989, *27*, 881.
13. Kohn, J. and Wilchek, M. *FEBS Letters* **1983**, *154*, 209.
14. Ziats, N. P.; Pankowsky, D. A.; Tierney, B. P.; Ratnoff, O. D. and
    Anderson, J. M. *J. Lab. Clin. Med.* **1990**, *116*, 687.
15. Hsu, S. M.; Raine, L. and Fanger, H. *J. Histochem., Cytochem.* **1981**, *29*,
    577.
16. Tsuge, H. and Mitsuda, H. *J. Vitaminol.* **1971**, *17*, 24.
17. Yildiz, A.; Kissinger, P. T.; and Reilley, C. N. *Anal. Chem.* **1968**, *4*,
    1018.
18. Ratner, B. D. and McElroy, B. J., in *Spectroscopy in the Biomedical
    Sciences*, Genbreau, Ed.; CRC press, Boca Raton, FL, **1986**, pg. 107.
19. Marchant, R. E.; Li, X.; Yu, D. and Danilich, M. J. *ACS Polymer
    Preprints*, presented at the Boston, MA meeting, **1990**, vol. 31.
20. Moss, D. W. in *Methods of Enzymatic Analysis, Third ed.*, Bergmeyer,
    H. U. Ed.; Weinheim: Deerfield Beach, CA **1983**, Vol. 1; pp. 7-14.
21. Gloger, M. and Tischer, W. in *Methods of Enzymatic Analysis,Third ed.*,
    Bergmeyer, H. U. Ed.; Weinheim: Deerfield Beach, CA **1983**, Vol. 1;
    p.142.
22. Szucs, A.; Hitchens, G. D. and Bockris, J. O. *J. Electrochem. Soc.* **1989**,
    136, 3748.
23. Danilich, M. J.; Gervasio, D. and Marchant, R. E., submitted.
24. Nikolaeva, N. N.; Khazova, O. A. and Vasiliev, Y. B. *Soviet
    Electrochem.* **1983**, *19*, 934.
25. Marchant, R. E.; Johnson, S. D.; Schneider, B. H.; Agger, M. P. and
    Anderson, J. M. *J. Biomed. Mater. Res.*, **1990**, *24*, 1521.

RECEIVED December 10, 1991

Chapter 8

Role of Polymeric Materials in the Fabrication of Ion-Selective Electrodes and Biosensors

G. J. Moody

School of Chemistry and Applied Chemistry, University of Wales College of Cardiff, P.O. Box 912, Cardiff CF1 3TB, United Kingdom

The scope of ion-selective electrodes (ISEs) has been greatly enhanced by employing a poly(vinyl chloride) matrix to entangle sensor cocktail materials. For ISFET devices an in situ photopolymerisation of monobutyl methacrylate provides a viable poly(butyl methacrylate) calcium sensor film with good gate adhesion properties. One or more enzymes can be chemically immobilized on modified nylon mesh. The resultant matrices are suitable for the amperometric assay of carbohydrates in blood and food products.

Ion-selective electrodes (ISEs) with sensor membranes based on sensor molecules plus suitable plasticising solvent mediators are best fabricated with such components physically entangled in a thin poly(vinyl chloride) (PVC) membrane.

In analogous fashion enzymes with their highly sophisticated natural selectivity have been covalently immobilized (particularly on nylon mesh) to provide long-life amperometric enzyme sensors.

PVC Ion-selective Electrodes

Sensor cocktails comprising either liquid ion-exchangers or neutral carriers and an appropriate plasticising solvent mediator offer the prospect for a bountiful range of ISEs. However, they are preferably constructed in a manner analogous to the classical pH glass electrode. This configuration is simply realised by casting a thin, flexible master membrane (diameter ≈30 mm) from the chosen sensor cocktail and PVC dissolved in tetrahydrofuran by controlled evaporation over 2 days. A small disc (diameter ≈6 mm) is removed, sealed to one end of a hollow PVC tube, and the ISE fabricated with an internal Ag/AgCl reference electrode immersed in the internal filling solution (1).

0097–6156/92/0487–0099$06.00/0

Since the evaluation of the first PVC calcium ISE in 1970, hundreds of other viable PVC models have been employed for diverse analytical purposes. Potentiometric sensing has also gained in popularity by the introduction of flow injection analysis (FIA) techniques and advances in electronics.

Advantages of PVC as an Entanglement Matrix for Cocktails. Important advantages ensue from the use of PVC matrices.

The fluid nature of PVC cocktails allows them to conform to the shape of a surface. Consequently, on evaporation of the tetrahydrofuran the PVC sensor film is left as a particular contour. This has allowed improved designs of ISEs, e.g., coated wires/epoxy, tubular flow-through, micro and all solid-state epoxy models. Thus, a lithium sensor cast on top of a small epoxy base in a flow injection system is suitable for the assay of lithium in the saliva of manic depressive patients (Beswick,C.W., Moody,G.J., Thomas,J.D.R., University of Wales College of Cardiff, unpublished data).

ISEs are easily fabricated.

Many ISEs can be produced from a small quantity of a cocktail.

The polymer network is compatible with most solvent mediators, decan-1-ol being a notable exception.

The entanglement generally reduces the leaching of active components from the matrix and so extends the operational lifetime of the ISE. However, PVC can also be blended with different amounts of vinyl chloride/vinyl acetate/vinyl alcohol copolymer (VAGH) which permits grafting of alkyl phosphate sensors and phosphonate mediators and so further reduces leaching effects.

Master membranes (up to 9 cm in diameter) can be cast with a uniform thickness except at their edges near the glass casting rings.

Membranes provide sufficient mechanical strength to support the column of the internal reference solution in macro models (1) as well as resisting sheer forces in flowing systems.

Disadvantages. The few disadvantages relate to the incompatibility with alkanol type mediator solvents, the limited scope for covalent immobilization of sensor and mediator so as to prevent their leaching and poor adhesion to ISFET gates.

Functionality of Different PVCs. Viable matrices can be realised with a range of PVCs. Thus, the behaviour of eight calcium ISEs comprising the model cocktail based on the calcium salt of bis-[4-(1,1,3,3-tetramethylbutyl)phenyl]phosphoric acid and solvent mediator dioctylphenylphosphonate (DOPP) plus seven standard IUPAC PVCs as well as the widely-used reference PVC, Breon III EP, have been examined (Table I). The master membranes turned out to be clear and flexible and (except for numbers 3, 4, 5 and 7 with $\overline{M}_n$ values <40 000) non-sticky (2). On the basis of the parameters listed there is little to recommend any particular PVC product.

Alternative Polymer Matrices. The prospect for alternative matrix materials (Table II) for calcium ISEs has been examined in conjunction with the model liquid calcium ion-exchanger/DOPP cocktail (2). Evidently functional electrodes can be fabricated with poly(2-methyl propyl methacrylate) (P2-MPMA) but not with the corresponding poly(butyl methacrylate) (PBMA), poly(methyl methacrylate) (PMMA) or

Table I. Some Characteristics of Calcium ISEs Based
on Different PVCs

PVC			Properties of Sensor Membrane		
Membrane Number	T_g /°C	$\overline{M}_n$	Calibration /mV decade^{-1}	Resistance /MΩ	$k_{Ca,Mg}^{pot}$ [a]
1(Breon 111 EP)	98	77 000	28.0	3.0	2.2×10^{-3}
2	102	47 150	32.0	2.0	1.3×10^{-3}
3	85	39 770	29.6	1.6	9.8×10^{-4}
4	86	25 900	30.0	1.9	3.5×10^{-3}
5	86	39 350	30.8	1.1	10^{-3}
6	85	48 520	30.5	2.1	1.3×10^{-3}
7	88	36 310	30.8	2.5	1.3×10^{-3}
8	86	43 700	26.5	1.3	2.2×10^{-3}

SOURCE: Adapted from ref.2.
[a] Separate solution method. $[Mg^{2+}] = 10^{-2}M$.

poly(methyl acrylate) PMA matrices. Thus, although P2-MPMA (like
PVC) is a viable matrix, properties other than just the glass transition temperatures of polymers must be involved in providing the
necessary compatibility with sensor cocktails.

However, such materials are unlikely to usurp the well-tried
poly(vinyl chloride) matrix for most purposes. It is advisable to
employ an established PVC product, e.g., Breon III-EP, Flowell 470 or
Fluka S704, otherwise poor quality ISEs may result even with the best
quality cocktails.

Table II. Some Characteristics of Acrylate-based Membranes

Polymer Type	T_g/°C		Polymer Ca^{2+} ISE	
	Polymer	Polymer/DOPP	Calibration /mV decade^{-1}	$k_{Ca,Mg}^{pot}$ [b]
PVC (Breon 111 EP)	98	−65	26.0	2.0×10^{-3}
P2-MPMA	75	−73	27.4	8.4×10^{-3}
PMMA	108	−85	} Non-functional	
PMA	8	−90		
PBMA[a]	−	−	26.0	8.0×10^{-3}

SOURCE: Adapted from refs.2 and 3.
[a] Fabrication achieved by photolysis of monomer plus cocktail.
[b] Separate solution method. $[Mg^{2+}] = 10^{-3}M$.

Membrane-casting Techniques.   Until recently, PVC membranes have
been exclusively formed by solvent casting techniques but which are
not well-suited to the fabrication of ISFET devices. Membrane components in tetrahydrofuran are difficult to manipulate on a micro scale
and are prone to absorb atmospheric moisture, thus weakening the
adhesion at the sensor-ISFET interface. One innovation which dispen-
ses with the tetrahydrofuran casting stage is based on an in situ
photolysis of the model calcium sensor cocktail admixed with monobutyl methacrylate + benzoyl peroxide + benzoin methyl ether at
340 nm (3). The resultant matrix adhered well to the ISFET gate and
its potentiometric response compared favourably with the analogous
PVC and P2-MPMA ISE (Table II).
    Unfortunately, this novel technique is unsuitable for any UV
absorbing sensor or mediator but could find a significant application
in the realm of ISFET fabrication (3).

Immobilized Enzyme Electrodes

An enzyme electrode may be envisaged as a self-contained analytical
biosensor comprising a thin enzyme layer overlying, for example, a
carbon or platinum anode. The enzymes need to be immobilized and
this is best realised by direct covalent bonding to an insoluble
matrix rather than by physical means. Several viable polymers are
suitable for this purpose.

Nylon Mesh Enzyme Electrodes.   Enzymes can be conveniently immobilized on modified nylon-6,6 mesh **(I)** produced by successive treatments
of nylon-6,6 with dimethylsulphate, lysine spacer and finally benzoquinone or glutaraldehyde coupling agencies:

Thus, glucose oxidase can be randomly immobilised on the modified nylon mesh (**I**). The resultant enzyme membrane (**II**) when held tautly over a platinum anode disc provides a high performance, long life glucose electrode which can be housed in a Stelte cell adapted for flow injection analysis (4).

The influence of different spacer and coupling molecules on the relative performances of seventeen glucose oxidase–nylon electrodes (NGO) fabricated from the same batch of fresh enzyme have been conveniently established in the FIA mode with standard glucose(1 mM):

$$\text{Glucose} + O_2 \xrightarrow{\text{NGO}} \text{Gluconic acid} + H_2O_2$$

The mean currents associated with the oxidation of the hydrogen peroxide at a platinum anode (covered in turn with each type of NGO membrane) and poised at 600 mV vs a Ag/AgCl electrode could then be conveniently compared. (Table III). Evidently lysine is the best spacer irrespective of the coupling agency. However, spacer for spacer the p-benzoquinone based NGOs produced higher currents and wider, linear calibration ranges, e.g. 0.001 to 5 mM glucose compared with 0.001 to 2 mM for the classical glutaraldehyde based NGOs. It is also interesting that without a spacer in the immobilization sequence the activity of the enzyme net is about half that of either of the lysine based NGOs as previously reported by Hornby and Morris (5).

None of the systems lost any enzyme activity during 24 h of continuous pumping of glucose solution (1 mM) at 2.3 mL/min. Moreover the various membranes when stored at $4°C$ in sodium dihydrogen phosphate (pH 7) still responded to substrate (70% of the signal for a new electrode) with intermittent use over a period of about 4 months after the fabrication (4).

TABLE III. Responses of Seventeen Different NGO Electrodes to Glucose (10 mM)

Spacer Molecule	Coupling Agency and Mean Current /nA	
	p-Benzoquinone	Glutaraldehyde
Lysine	5325 (4) [b]	4100 (14) [b]
Asparagine	5031 (6)	3848 (6)
Arginine	5100 (16)	4058 (6)
Ornithine	3896 (5)	3530 (6)
Glutamine	2922 (7)	2661 (5)
No spacer	2370 (5)	2366 (4)
m-Phenylene Diamine	3524 (10)	2888 (5)
p-Phenylene Diamine	4800 (16)	3674 (10)
Blank[a]	–	2010 (10)

SOURCE : Adapted from ref.4.

a No spacer or coupling agency.

b Standard deviation is shown in parenthesis (n = 10).

It is possible that some cross linking reactions could arise between the amino group of the lysine-nylon moiety and one of the aldehyde or one of the keto groups of the respective coupling reagents prior to the final enzyme immobilization stage. Thus an alternative scheme has been devised to prevent such premature cross linking reactions (See scheme below). After the first silanization stage the mono-TPDPS derivative was separated from the di-TPDPS material by column chromatography and oxidised with pyridinium dichloride.

Calibration profiles of the sensor based on the final nylon-enzyme net (III) were disappointing compared with the analogous sensor based on nylon net type II . The lower detection limit was only 0.1 mM glucose and currents produced were about 80% smaller. However, this alternative immobilization scheme serves to illustrate the synthetic versatility of nylon-6,6 in the biosensor field.

It is interesting that the addition of glucose (1 mM) to the glucose oxidase solution prior to the final immobilization step gave an electrode with an improved all-round response (Donlan,A.M.; Moody,G.J.; Thomas,J.D.R., University of Wales College of Cardiff, unpublished data). This is probably due to an improved conformation since the enzymatic reaction continued throughout the immobilization process.

(III)

Multienzyme Nylon Electrodes. Di- and polysaccharides require more than one enzyme to realise the amperometrically detectable hydrogen peroxide and even glucose really needs the back up of mutarotase with glucose oxidase. It is fortunate that all the necessary enzymes can be immobilized simultaneously on just one nylon net. Thus a viable starch electrode has been fabricated (6) from a nylon net immersed in a cocktail of glucose oxidase, mutarotase and amyloglucosidase (Figure 1). Its response to a continuous flow of 0.1% m/v starch remained steady for over a period of 60 h.

Sterilization of Nylon Enzyme Electrodes. Sterilization of in vivo electrodes is essential for clinical use and advisable for applications in the food industry. It was thus of interest to study the behaviour of various carbohydrate sensor membranes before and after irradiation with ^{60}Co-γ radiation (Figure 1). Thus, after each membrane calibration in a batch mode, the membrane was detached, placed in a sealed glass tube with phosphate buffer (pH 7) and irradiated to set doses. The membrane was then reattached to the platinum anode and the electrode recalibrated. Doses of 1.2 Mrad, and even higher in some cases, had little effect as judged by the post-irradiation calibrations of starch, sucrose, lactose, lactic acid and glucose enzyme electrodes respectively (7).

Thus the viability of these carbohydrate nylon net sensors after γ-radiation doses of 1.2 Mrad is of importance because the appropriate sterilization dose can possibly be administered without seriously impairing their subsequent performance.

Advantages of Nylon-6,6. Nylon is now established as a most efficient agency for immobilizing enzymes. Indeed, nylon mesh is to amperometry what PVC is to potentiometry.

Its mechanical strength is sustained in the enzyme immobilized matrix.

It offers considerable prospect for synthetic reactions on the matrix.

Several enzymes can be simultaneously immobilized by simple immersion in the appropriate cocktail. Enzymes could thus also be isolated (by immobilization) from natural media. Over 2000 different enzymes are known!

The nylon mesh alone as well as nylon mesh—enzyme nets are essentially unaffected by γ-radiation doses up to 1.2 Mrad.

Enzyme loadings are high enough to give long life biosensors.

Mediated Enzyme Electrodes. Further improvements in the performance of immobilized enzyme sensors stem from the use of redox mediators which shuttle electrons from the redox centre of the enzyme to the surface of the indicator electrode according to the following reaction sequences depicted for glucose oxidase:

$$\text{Glucose} + \text{GO}_{ox} \longrightarrow \text{Gluconic acid} + \text{GO}_{red}$$

$$\text{GO}_{red} + 2\text{Fecp}_2^+ \longrightarrow \text{GO}_{ox} + 2\text{Fecp}_2 + 2\text{H}^+$$

$$2\text{Fecp}_2 \longrightarrow 2\text{Fecp}_2^+ + 2e$$

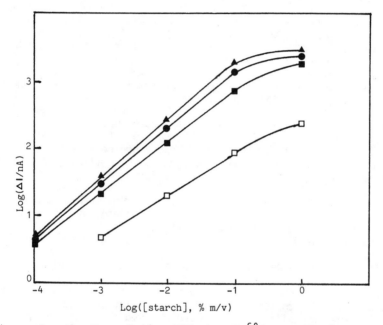

Figure 1.   The Accumulative Effects of $^{60}Co-\gamma$ Radiation on the
Response of a Triple Enzyme Starch Electrode.   ▲ Control; ● 1.2
Mrad;   ■ 2.4 Mrad;   □ 4.8 Mrad.
(Reproduced with permission from ref.6.   Copyright 1990 The
Analyst, London.)

Such a chemically modified electrode system for glucose has been recently fabricated from a ferrocene (Fecp$_2$) mediator–carbon paste–cellulose triacetate mixture packed into the well of an electrode holder and covered with the usual NGO mesh (II) (8). The severe interference from ascorbic acid at 600 mV vs Ag/AgCl was virtually eliminated at 160 mV when the linear glucose calibration ranged from 0.01–70 mM. The inclusion of a viscose acetate exclusion membrane placed between the surface of the ferrocene–paste layer and the outer NGO mesh extended the linear range to between 0.01 and 100 mM. The response times and wash–out times were ≈25 s and ≈45 s, respectively, compared with ≈15 s and ≈30 s for the electrode system without the exclusion membrane in place. These times compare favourably with those of an analogous glucose electrode described by Wang and co-workers (9). Both electrodes could withstand at least 24 h of continuous glucose (1 mM) flow before any loss of enzyme activity or mediator arose.

Alternative Polymers for Immobilizing Enzymes. Vinyl acetate (after hydrolysis with bicarbonate) and polycarbonate have also been chemically modified to accommodate glucose oxidase (Donlan,A.M.; Moody,G.J.; Thomas,J.D.R., University of Wales College of Cardiff, unpublished data).

$$ROC\overset{O}{\underset{O^-}{<}} \xrightarrow[\text{triethoxysilane}]{\text{3-Aminopropyl}} ROSi(OC_2H_5)(OC_2H_5)(CH_2)_3NH_2 + CO_2$$

$$\xrightarrow{\text{Glutaraldehyde}}$$

$$ROSi(OC_2H_5)(OC_2H_5)(CH_2)_3N = CH(CH_2)_3CHO$$

$$\xrightarrow{NH_2-\text{ENZYME}}$$

$$ROSi(OC_2H_5)(OC_2H_5)(CH_2)_3N=CH(CH_2)_3 \quad HC=N-\text{ENZYME}$$

The response times and washout times of electrodes fabricated from these materials are, respectively, <30 s and <50 s but the linear calibrations are restricted, e.g. 0.1-5 mM for the polycarbonate electrode.

Catalase can be easily immobilized on a hydrolysed vinyl acetate-cyanuric chloride membrane:

$$ROH \xrightarrow[\text{chloride}]{\text{Cyanuric}} \xrightarrow{NH_2-\text{CATALASE}}$$

Unlike the enzyme nylon meshes, the viscose acetate-catalase membrane became very brittle after a γ-radiation dose of 1.2 Mrad, a condition which often led to tearing when attempting to reassemble the electrode for recalibration (7).

Applications. Oxidases immobilised on nylon mesh have been successfully employed as amperometric sensors in the FIA mode for a variety of important materials (6,10-13).

A tri-enzyme lactose electrode suitable for determining lactose in milk products can be fabricated by immobilising lactase, glucose oxidase and mutarotase on nylon mesh. Samples were pre-treated to remove protein. Any glucose present was determined with a glucose electrode and subtracted from the total glucose determined with the lactose electrode (12). Good agreement was found between the results obtained by the Boehringer Mannheim lactose test-kit method and the lactose enzyme electrode (Table IV).

Table IV. Analysis of Milk Products with Tri-enzyme Electrode and Boehringer Mannheim (BM) Test-kit Methods

Food Sample	[Lactose]/g per 100 g of Sample	
	Enzyme Electrode[a]	BM Test-kit[b]
Fluid skimmed milk	5.04	5.31
Pasteurised milk	4.87	4.89
Canned evaporated milk	9.87	10.17
Unsweetened condensed milk	8.69	9.13
Dried skimmed milk	27.41	27.64

SOURCE: Adapted from ref.12.
[a] Means of 5 injections (CV <1%).
[b] Means of 2 assays.

The starch content of flours has been similarly analysed with a tri-enzyme electrode comprising amyloglucosidase, mutarotase and glucose oxidase on nylon mesh and a Boehringer Mannheim starch test-kit. Each sample was pre-incubated with soluble α-amylase for 1 h at room temperature before the actual analysis. Again results obtained with the two techniques were in close agreement (6).

Glucose in pre-treated ice-creams, drinks, molasses and flour has also been determined with a glucose oxidase based nylon electrode. The results were similar to those obtained by the Yellow Springs Instrument glucose analyser and a Boehringer Mannheim glucose test-kit (10).

The above FIA systems are based on monitoring the anodic decomposition of hydrogen peroxide at a platinum electrode set at 600-700 mV vs a Ag/AgCl electrode. However, at such high potentials other electroactive species, notably ascorbic acid, uric acid and hypoxanthine, will also be oxidised unless appropriate sample pre-treatment is taken.

Fortunately, in such circumstances hydrogen peroxide can be measured indirectly at much lower potentials using hexacyanoferrate(II) redox mediator admixed in the buffered FIA stream and a peroxidase nylon mesh electrode:

$$H_2O_2 + 2[Fe(CN)_6]^{4-} + 2H^+ \xrightarrow{\text{Peroxidase}} 2H_2O + 2[Fe(CN)_6]^{3-}$$

The hexacyanoferrate(III) so generated can then be reduced at a platinum electrode around -100 mV vs a Ag/AgCl electrode:

$$[Fe(CN)_6]^{3-} + e \longrightarrow [Fe(CN)_6]^{4-}$$

This concept has been employed (11) to measure hypoxanthine extracted from fish with a xanthine oxidase-peroxidase nylon mesh electrode:

$$\text{Hypoxanthine} + O_2 \xrightarrow{\text{Xanthine oxidase}} \text{Uric acid} + H_2O_2$$

The values obtained with the bi-enzyme electrode compared favourably with those for the alternative spectrophotometric method (Table V) and offers the prospect for quality control monitoring in the food industry.

Table V. Analysis of Hypoxanthine in Fish Meats Using Bi-enzyme Electrode and AMC Spectrophotometric Methods

Fish Sample	Fresh Fish μmol g^{-1} (tissue)		Fish Stored at Room Temperature for 20 h μmol g^{-1} (tissue)	
	Electrode [a]	AMC [b]	Electrode [a]	AMC [b]
Rainbow trout	0.32	0.32	0.42	0.39
Herring	0.74	0.88	1.87	1.84
Hake	0.59	0.63	2.19	2.40
Plaice	0.95	0.91	1.89	1.76

SOURCE: Adapted from ref.11.

[a] Mean of 2 assays.
[b] Analytical Methods Committee Method. Single measurement.

Literature Cited.

(1) Moody,G.J.; Oke,R.B.; Thomas,J.D.R.   Analyst (London) 1970 ,95, 910-918.
(2) Saad,B.B.; Moody,G.J.; Thomas,J.D.R.   Analyst (London) 1987 , 112, 1143-1147.
(3) Moody,G.J.; Slater,J.M.; Thomas,J.D.R.   Analyst (London) 1988 , 113, 103-108.
(4) Beh,S.K.; Moody,G.J.; Thomas,J.D.R.   Analyst (London) 1989 , 114, 1421-1425.
(5) Hornby,W.C.; Morris,D.L.   Immobilized Enzymes, Antigens, Antibodies and Peptides: Weetall,H.H., Ed.; Dekker: New York,N.Y., 1975 , 141-169.
(6) Abdul Hamid,J.; Moody,G.J.; Thomas,J.D.R. Analyst (London) 1990 , 115, 1289-1295.
(7) Abdul Hamid,J.; Beh,S.K.; Donlan,A.M.; Moody,G.J.; Thomas,J.D.R. J.Sci.Food Agric. 1991 , 55, 323-326.
(8) Beh,S.K.; Moody,G.J.; Thomas,J.D.R.   Analyst (London) 1991 , 116, 459-462.

(9) Wang,J.; Wu,L.-H.; Lu,Z.; Li,R.; Sanchez,J. Anal.Chim.Acta
     1990 , 228, 257-261.
(10) Moody,G.J.;   Sanghera,G.S.;   Thomas,J.D.R.   Analyst   (London)
     1986 , 111, 605-609.
(11) Moody,G.J.;   Sanghera,G.S.;   Thomas,J.D.R.   Analyst   (London)
     1987 , 112, 65-70.
(12) Abdul Hamid,J.; Moody,G.J.; Thomas,J.D.R. Analyst (London)
     1989 , 114, 1587-1592.
(13) Cosgrove,M.;   Moody,G.J.;   Thomas,J.D.R.   Analyst   (London)
     1988 , 113, 1811-1815.

RECEIVED February 28, 1992

Chapter 9

Electrical Wiring of Flavoenzymes with Flexible Redox Polymers

P. D. Hale[1], L. I. Boguslavsky[1], T. A. Skotheim[1], L. F. Liu[1], H. S. Lee[2], H. I. Karan[3], H. L. Lan[4], and Y. Okamoto[4]

[1]Moltech Corporation, Chemistry Building, State University of New York, Stony Brook, NY 11794–3400
[2]Materials Science Division, Department of Applied Science, Brookhaven National Laboratory, Upton, NY 11973
[3]Division of Natural Science and Mathematics, Chemistry Department, Medgar Evers College, City University of New York, Brooklyn, NY 11225
[4]Department of Chemistry, Polytechnic University, Brooklyn, NY 11201

It is well known that the flavin adenine dinucleotide redox centers of many oxidases are electrically inaccessible due to the insulating effect of the surrounding protein; thus, direct electron transfer from the reduced enzyme to a conventional electrode is negligible. In the present work, a variety of polymeric materials have been developed which can facilitate a flow of electrons from the flavin redox centers of oxidases to an electrode. Highly flexible siloxane and ethylene oxide polymers containing covalently attached redox moieties, such as ferrocene, are shown to be capable of rapidly re-oxidizing the reduced flavoenzyme. The construction and response of amperometric biosensors for glucose, acetylcholine, and glutamate based on these polymeric materials are described, and the dependence of sensor response on the polymer structure is discussed.

Flavoenzyme-Based Amperometric Biosensors

Amperometric biosensors based on flavin-containing enzymes have been studied for nearly 30 years. These sensors typically undergo several chemical or electrochemical steps which produce a measurable current that is related to the substrate concentration. In the initial step, the substrate converts the oxidized flavin adenine dinucleotide (FAD) center of the enzyme into its reduced form ($FADH_2$). Because these redox centers are essentially electrically insulated within the enzyme molecule, direct electron transfer to the surface of a conventional electrode does not occur to a substantial degree. The "classical" methods (1-4) of indirectly measuring the amount of reduced enzyme, and hence the amount of substrate present, rely on the natural enzymatic reaction:

$$\text{substrate} + O_2 \xrightarrow{\text{flavoenzyme}} \text{product} + H_2O_2 \qquad (1)$$

where oxygen is the electron acceptor for the oxidase. The oxygen is reduced by the $FADH_2$ to hydrogen peroxide, which may then be detected electrochemically. Alternatively, one could use the electrode to measure the change in oxygen concentration that occurs during the above reaction. In both of these measuring schemes, this type of sensor has the disadvantage of being extremely sensitive to the ambient oxygen concentration.

Recently, biosensors have been developed which use a non-physiological redox couple to shuttle electrons between the $FADH_2$ and the electrode by the following mechanism:

$$\text{substrate} + E(FAD) \rightarrow \text{product} + E(FADH_2) \tag{2}$$

$$E(FADH_2) + 2M_{ox} \rightarrow E(FAD) + 2M_{red} + 2H^+ \tag{3}$$

$$2M_{red} \rightarrow 2M_{ox} + 2e^- \text{ (at the electrode)} \tag{4}$$

In this scheme, E(FAD) represents the oxidized form of the flavoenzyme and $E(FADH_2)$ refers to the reduced form; the mediating species M_{ox}/M_{red} is assumed to be a one-electron couple. Sensors based on derivatives of the ferrocene/ferricinium redox couple (5-8), on quinone derivatives (9-11), and on electrodes consisting of organic conducting salts such as TTF-TCNQ (tetrathiafulvalene-tetracyanoquinodi-methane) (12-17) have been reported. For some applications, however, sensors based on electron-shuttling redox couples suffer from an inherent drawback: the soluble, or partially soluble, mediating species can diffuse away from the electrode surface into the bulk solution, which would preclude the use of these devices as implantable probes in clinical applications, and restrict their use in other long-term *in situ* measurements (e.g., fermentation monitoring).

With this in mind, several research groups have been investigating systems where the mediating species is chemically bound in a manner which allows close contact between the $FAD/FADH_2$ centers of the enzyme and the mediator, yet prevents the latter from diffusing away from the electrode surface. For instance, an electron transfer relay system has been designed where the mediating species (ruthenium pentaammine or ferrocene derivatives) are chemically attached to the enzyme itself (18-20). The chemical modification of the enzyme can, however, cause a measurable decrease in its activity. More recently, studies have been carried out where the mediating redox moieties are covalently attached to polymers such as poly(pyrrole) (21), poly(vinyl-pyridine) (20,22-24), and in our laboratory, poly(siloxane) (25-32). These systems serve to "electrically wire" the enzyme, facilitating a flow of electrons from the enzyme to the electrode. In the present chapter, we discuss the use of ferrocene-containing siloxane and ethylene oxide polymers as electron relay systems in amperometric biosensors. These polymer systems, shown in Figure 1, can effectively mediate electron transfer from reduced flavoenzymes to a conventional carbon electrode.

Synthesis of the Redox Polymers

Siloxane Polymers. The synthesis of the ferrocene-modified siloxane polymers (**A - E**) has been described previously (25,27,32). Briefly, the methyl(2-ferrocenylethyl)-siloxane polymers were prepared by the hydrosilylation of vinylferrocene with the methylhydrosiloxane homopolymer or the methylhydrosiloxane-dimethylsiloxane copolymers (m:n ratios of 1:1, 1:2, and 1:7.5; see Figure 1) in the presence of chloroplatinic acid as a catalyst. The methyl(9-ferrocenylnonyl)siloxane-dimethylsiloxane (1:2) copolymer was prepared via hydrosilylation of 9-ferrocenyl-1-nonene with the methylhydrosiloxane-dimethylsiloxane (1:2) copolymer. The molecular weight range of these ferrocene-modified siloxane polymers is approximately 5,000-10,000. Purification of the polymers was achieved by reprecipitation from chloroform solution, via dropwise

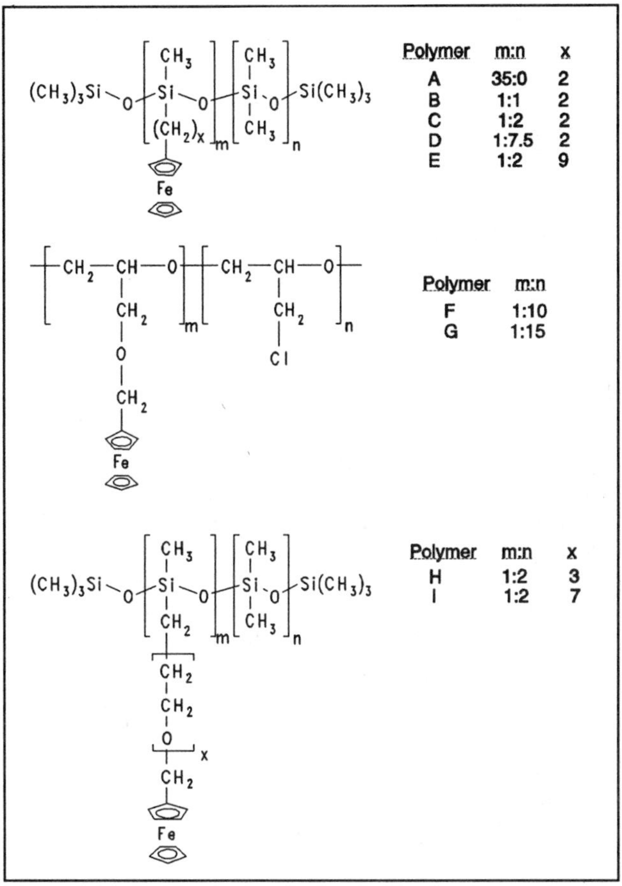

Figure 1. Structures of redox polymers used as electron relay systems in flavoenzyme-based biosensors. Shown are siloxane (top), ethylene oxide (middle), and branched siloxane-ethylene oxide (bottom) polymers.

addition into a large excess amount of acetonitrile at room temperature. This reprecipitation was repeated 2-3 times to ensure that no low molecular weight species (which could act as freely diffusing electron transfer mediators) were present. Thin layer chromatography and high-performance liquid chromatography showed that no oligomeric materials were present in the purified materials.

Ethylene Oxide Polymers. The synthetic procedure for the ferrocene-containing poly(ethylene oxide) materials (**F** and **G**) is outlined in Figure 2. The average spacing between the ferrocene relays can be adjusted by varying the amount of the ferrocene starting material which is reacted with poly(epichloride). After purification, the polymers were characterized using IR and NMR spectroscopies; as above, thin layer chromatography and high-performance liquid chromatography showed that no low molecular weight materials were present in the purified materials.

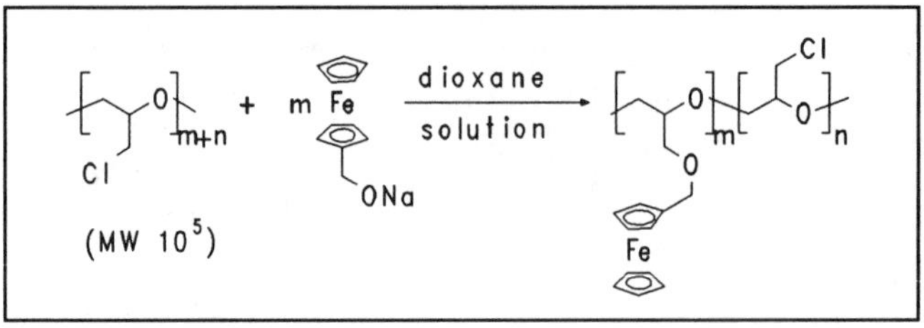

Figure 2. Synthetic procedure for the ferrocene-containing ethylene oxide polymers **F** and **G**.

Siloxane-Ethylene Oxide Branch Copolymers. The ferrocene-ethylene oxide-siloxane polymer (**H** and **I**) was prepared by the hydrosilylation of the terminal vinyl group of the ferrocene-containing ethylene oxide oligomer (Figure 3) with a methylhydrosiloxane-dimethylsiloxane copolymer (m:n ratio of 1:2, m+n≈35) in the presence of chloroplatinic acid as a catalyst. The molecular weight of the resulting ferrocene-modified polymer is approximately 10,000. Purification of the polymer was achieved by reprecipitation from chloroform solution, as described above for the siloxane polymers. As above, the materials were characterized using IR and NMR spectroscopies, thin layer chromatography, and high-performance liquid chromatography, and no low molecular weight oligomeric species were present in the final product.

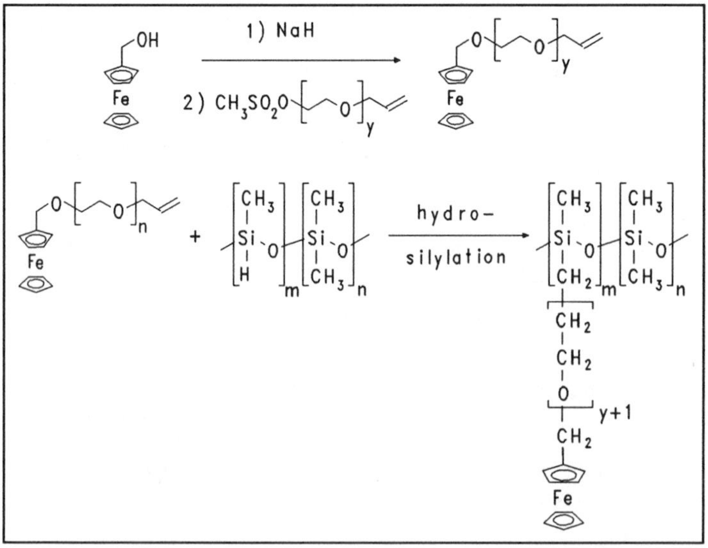

Figure 3. Synthetic procedure for the ferrocene-ethylene oxide-siloxane branch copolymers **H** and **I**.

Sensor Construction

Glucose Sensors. The modified carbon paste for the sensors was made by thoroughly mixing 50 mg of graphite powder (Fluka, product no. 50870) with a measured amount of the ferrocene-containing polymer (the latter was first dissolved in chloroform); in the present work, the molar amount of the redox moiety was the same for all electrodes (1.8 μmole of polymer-bound ferrocene per 50 mg of graphite powder). After evaporation of the solvent, 5 mg of glucose oxidase (Sigma Type VII, from *Aspergillus niger*) and 10 μl of paraffin oil (Fluka, product no. 76235) were added, and the resulting mixture was blended into a paste. It has previously been shown that glucose oxidase retains its activity when incorporated into a carbon paste matrix (25,29-33). The paste was packed into a 1.0 ml plastic syringe (7.0 mm outer diameter; 1.8 mm inner diameter) which had previously been partially filled with unmodified carbon paste, leaving approximately a 2 mm deep well at the base of the syringe. The electrodes were polished by rubbing gently on a piece of weighing paper, which produced a flat shiny surface with an area of approximately 0.025 cm^2. Electrical contact was achieved by inserting a silver wire into the top of the carbon paste. The electrodes were stored under dry conditions at 5°C.

Acetylcholine Sensors. The modified carbon paste for the acetylcholine sensors was made by thoroughly mixing 25 mg of graphite powder with a measured amount of the ferrocene redox polymer (corresponding to 1 μmole of polymer-bound ferrocene), which was first dissolved in chloroform. After evaporation of the solvent, 2.5 mg of choline oxidase (Sigma, from *Alcaligenes* species), 2.8 mg of acetylcholinesterase (Sigma Type VI-S, from electric eel), and 10 μl of paraffin oil were added, and the resulting mixture was blended into a paste. The paste was then packed into an electrode holder and polished, as described above. The electrodes were then coated twice by dipping into a 0.5% aqueous dispersion of Eastman AQ29D poly(ester sulfonic acid); this latter polymer coating (29) improves the enzyme immobilization and provides increased electrode stability. The electrodes were stored under dry conditions at 5°C.

Glutamate Sensors. The glutamate sensors were constructed using graphite rods (Ultracarbon, Bay City, MI), which were polished to a smooth finish and coated with a small amount of redox polymer (0.033 μmole of polymer-bound ferrocene) dissolved in chloroform. After evaporation of the solvent, a 5 μl aliquot of glutamate oxidase (Yamasa Shoyu Co., Japan) solution (100 units/ml in pH 7.0 phosphate buffer) was added to the surface and allowed to dry at room temperature. The electrodes were stored under dry conditions at 5°C.

Electrochemical Methods

Stationary potential measurements were performed using a conventional 3-electrode potentiostat and a strip chart recorder. All experiments were carried out in a conventional electrochemical cell containing pH 7.0 phosphate (0.1 M) buffer with 0.1 M KCl at 23($\pm$2)°C. All experimental solutions were thoroughly deoxygenated by bubbling N_2 through the solution for at least 10 min; a gentle flow of N_2 was also used to facilitate stirring. Substrate samples injected into the cell were also thoroughly deoxygenated. In addition to the modified carbon paste working electrode, a saturated calomel reference electrode (SCE) and a platinum wire auxiliary electrode were employed. In these experiments, the background current was allowed to decay to a constant value before samples of a stock substrate solution were added to the cell.

Results and Discussion

Glucose Sensors. Siloxane polymers are known to be extremely flexible. This flexibility will, of course, be sensitive to the amount of side-chain substitution present along the polymer backbone. For instance, in the homopolymer used in these studies (polymer **A**), the presence of a ferrocenylethyl moiety bound to each silicon subunit should provide an additional degree of steric hindrance, and thus a barrier to rotation about the siloxane backbone, in comparison with the copolymers, which have ferrocene relays attached to only a fraction of the Si atoms. Because these siloxane polymers are insoluble in water, their flexibility is an important factor in their ability to facilitate electron transfer from the reduced enzyme. Relays contained within more rigid redox polymers, such as poly(vinylferrocene), cannot achieve close contact with the enzyme's redox centers and are thus less effective as electron transfer mediators (25,34). The importance of this feature can be seen quite clearly by comparing the mediating ability of the homopolymer **A** with that of copolymers **B-D**, as shown in Figures 4 and 5.

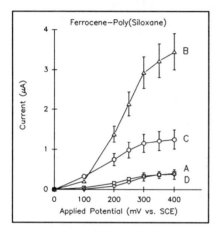 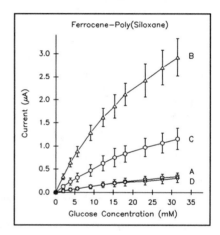

Figure 4. (left) Steady state response to 31.5 mM glucose of the ferrocene-modified poly(siloxane) / glucose oxidase / carbon paste electrodes at several applied potentials. The relay systems are indicated next to each curve, which is the mean result for four electrodes.

Figure 5. (right) Glucose calibration curves for the ferrocene-modified poly(siloxane) / glucose oxidase / carbon paste electrodes at E = +300 mV (vs. SCE). The polymeric relay systems are indicated next to each curve, which is the mean result for four electrodes.

As expected, the dependence of the glucose response on the applied potential (Figure 4) is similar (maximal responses at potential values $\geq$ 300 mV vs. SCE) for sensors containing polymers **A-D** (because each polymer contains the same redox species), but the magnitude of this response is highly dependent on the polymer structure. By, in effect, systematically replacing a fraction of the ferrocenylethyl groups with methyl groups, it is possible to increase the flexibility of the polymer as well as the average spacing between the redox sites. From the results in Figure 4, and from the glucose calibration curves shown in Figure 5, it is apparent that these factors play a key role in the interaction between the polymeric relay system and the FAD redox centers in glucose oxidase. These results show an enhanced sensitivity to glucose for the electrodes containing polymers **B** and **C** as electron relay systems; these polymers have intermediate values for m:n.

On the other hand, when the average spacing between the electron relays becomes very large, as in the case of polymer **D** (m:n = 1:7.5), the increased polymer flexibility does not result in an increase in the measured catalytic current. The total molar amount of the mediating species is kept constant in each electrode (the polymers with larger m:n ratios have fewer redox sites per mole, so more material is necessary), so a change in the glucose response at constant potential can be attributed to either a change in the interaction between the enzyme and the polymer-bound relays or to changes in the interactions between the relays themselves. In the case of polymer **D**, the decrease in sensor response could be due to the lower relay density which results in a less efficient electron transfer between adjacent relays, or to an electrical insulation of the relays by the relatively larger amount of siloxane polymer which prevents close contact between the relays and the $FAD/FADH_2$ centers of the enzyme. From the results shown in Figures 4 and 5, it is clear that a systematic adjustment of the siloxane polymer's m:n ratio can provide an ideal compromise between polymer flexibility and relay density, and an optimal sensor response.

The average spacing between the polymer-bound relays can also be adjusted by changing the length (x in Figure 1) of the alkyl chain onto which they are attached. The effect of increasing the alkyl side chain length is apparent in Figure 6, which shows glucose response curves for sensors containing polymers **C** (x=2) and **E** (x=9) as the electron relay systems. The maximal currents measured with sensors containing polymer **E** are approximately 2 times greater than those measured with sensors containing polymer **C**. Again, the longer alkyl chain may facilitate a more intimate interaction between the ferrocene moieties and the $FAD/FADH_2$ centers of glucose oxidase, or better electrical communication between the relays themselves.

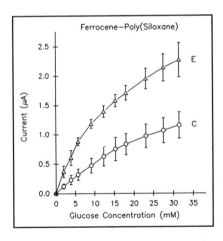

Figure 6. Effect of increased alkyl chain length on glucose sensor response: glucose calibration curves for the ferrocene-modified poly(siloxane) / glucose oxidase / carbon paste electrodes at E = +300 mV (vs. SCE). The polymeric relay systems are indicated next to each curve, which is the mean result for four electrodes.

The glucose response of sensors containing the ferrocene-poly(ethylene oxide) relay systems can be compared directly with the measurements made with sensors containing ferrocene-modified poly(siloxane), as the molar amount of polymer-attached ferrocene and the amount of glucose oxidase used were the same in both studies. It is clear from this comparison (Figures 7 and 8) that polymer F can mediate electron transfer from reduced glucose oxidase more efficiently than the relay systems based on poly(siloxane). The glucose response of sensors containing polymer F (at +300 mV vs. SCE) is over twice as large as that of sensors containing ferrocene-modified siloxane polymers, while the response of sensors containing polymer G is similar to that measured with the best poly(siloxane)-based electrode. The improved response of the poly(ethylene oxide)-based sensors could be due to the hydrophilic nature of this polymer (siloxane polymers, on the other hand, are quite hydrophobic), which allows the ferrocene moieties to achieve a close contact with the enzyme molecules. The lower response of sensors containing polymer G (relative to those containing polymer F) could again be due to the lower density of ferrocene moieties in that polymer, resulting in a decreased ability to mediate the electron transfer from the reduced enzyme to the electrode; we intend to synthesize several new poly(ethylene oxide) systems which have higher relay densities than polymers F and G in order to investigate this point further.

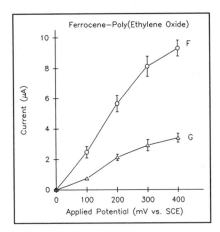 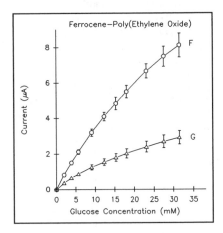

Figure 7. (left) Steady state response to 31.5 mM glucose of the ferrocene-modified poly(ethylene oxide)/glucose oxidase/carbon paste electrodes at several applied potentials. The relay systems are indicated next to each curve. Each curve is the mean result for four electrodes.
Figure 8. (right) Glucose calibration curves for the ferrocene-modified poly(ethylene oxide)/glucose oxidase/carbon paste electrodes at E = +300 mV (vs. SCE). The relay systems are indicated next to each curve. Each curve is the mean result for four electrodes.

A further improvement in sensor response is obtained when the ferrocene-siloxane-ethylene oxide polymers (H and I) are used as the electron relay system, as shown in Figures 9 and 10. These materials are based on the hydrophobic siloxane backbone, yet the hydrophilic ethylene oxide side chains, onto which the ferrocene moieties are attached, allow the electron relays to achieve a close interaction with the enzyme molecules.

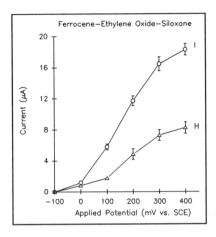

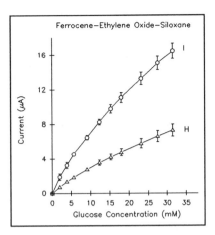

Figure 9. (left) Steady state response to 31.5 mM glucose of the ferrocene-siloxane-ethylene oxide polymer/glucose oxidase/carbon paste electrodes at several applied potentials. The polymeric relay systems are indicated next to each curve. Each curve is the mean result for four electrodes.

Figure 10. (right) Glucose calibration curves for the ferrocene-siloxane-ethylene oxide polymer/glucose oxidase/carbon paste electrodes at E = +300 mV (vs. SCE). The polymeric relay systems are indicated next to each curve. Each curve is the mean result for four electrodes.

The results shown in Figures 7 and 9 also indicate that the sensors based on the poly(ethylene oxide) and siloxane-ethylene oxide branch polymer systems can operate efficiently at relatively low applied potentials. In fact, the sensors containing these polymers show steady-state glucose responses at a potential of +100 mV (vs. SCE) which are similar to the response of the best poly(siloxane)-based sensor at +300 mV. This is an important consideration because lower operating potentials are often advantageous in real measurements, where easily oxidizable interfering species are usually present.

The linear response range of the glucose sensors can be estimated from a Michaelis-Menten analysis of the glucose calibration curves. The apparent Michaelis-Menten constant K_M^{app} can be determined from the electrochemical Eadie-Hofstee form of the Michaelis-Menten equation, $i = i_{max} - K_M^{app}(i/C)$, where i is the steady-state current, i_{max} is the maximum current, and C is the glucose concentration. A plot of i versus i/C (an electrochemical Eadie-Hofstee plot) produces a straight line, and provides both K_M^{app} (-slope) and i_{max} (y-intercept). The apparent Michaelis-Menten constant characterizes the enzyme electrode, not the enzyme itself. It provides a measure of the substrate concentration range over which the electrode response is approximately linear. A summary of the K_M^{app} values obtained from this analysis is shown in Table I.

It is clear from these K_M^{app} values, and from the glucose calibration curves, that the response of the sensors begins to deviate from linearity even at glucose concentrations below 10 mM (the response to glucose is expected to be strictly linear for concentrations approximately less than or equal to $0.1K_M^{app}$). We have previously found (29) that the linear range of similar glucose sensors can also be increased substantially (K_M^{app} values greater than 200 mM) by using an additional polymer coating on the surface of the sensor. As one would expect, such a polymer coating provides an additional resistance to the diffusion of glucose to the active enzyme layer, and the overall kinetics would be controlled by this diffusion process.

Table I. Apparent Michaelis-Menten Constants and Maximum Current Densities for Redox Polymer Based Glucose Biosensors[a]

Polymer	K_M^{app} (mM) [b]	i_{max} (μA) [b]
A	71	1.12
B	32	5.80
C	42	2.70
D	43	0.73
E	16	3.28
F	49	20.2
G	27	5.55
H	51	18.6
I	56	45.9

[a] Under N_2-saturated conditions at an applied potential of +300 mV vs. SCE.
[b] Each value is the mean result for four electrodes.

Acetylcholine Sensors. The general scheme for determination of the neurotransmitter acetylcholine is outlined in Figure 11. In this scheme, acetylcholine is first converted catalytically to choline by the enzyme acetylcholinesterase. The choline produced reduces the FAD redox centers of choline oxidase, and electron transfer from these centers to the electrode is facilitated by the polymeric relay system.

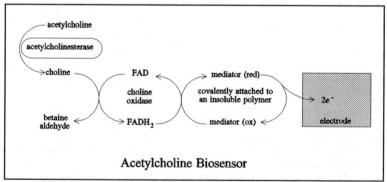

Acetylcholine Biosensor

Figure 11. Schematic representation of an amperometric biosensor for acetylcholine based on polymeric electron relay systems.

Figures 12-14 show the steady-state current dependence of the acetylcholine sensors on substrate concentration; these sensors contained polymers C, F, and I, respectively, as the electron relay systems. For an applied potential of +250 mV vs. SCE, the time required to reach 95% of the steady-state current was typically 10-15 sec after addition of the acetylcholine sample. At lower potentials, the response time was slightly slower. For these systems, a detection limit (as defined by a signal-to-noise ratio of approximately 2) of approximately 0.5 to 1.0 μM was achieved under N_2-saturated conditions. The response of the sensors to choline was nearly identical to the acetylcholine response, which demonstrates the efficient conversion of acetylcholine to choline by acetylcholinesterase.

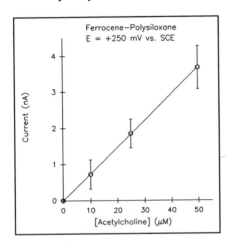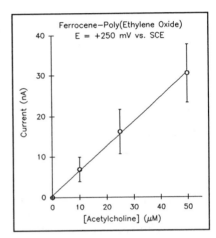

Figure 12. (left) Steady-state current response of acetylcholine sensors based on polymer C, in pH 7.0 phosphate buffer under N_2-saturated conditions. Each point is the mean result for five electrodes.

Figure 13. (right) Steady-state current response of acetylcholine sensors based on polymer F, in pH 7.0 phosphate buffer under N_2-saturated conditions. Each point is the mean result for five electrodes.

It is clear from these results that the ability of the redox polymers to mediate electron transfer from reduced choline oxidase is dependent upon the structure of the polymer backbone. The trend in mediating efficiency is qualitatively the same as that found for the glucose sensors: siloxane-ethylene oxide branch polymer > poly(ethylene oxide) > poly(siloxane).

Glutamate Sensors. Glutamate occurs in unusually high concentrations in the brain and has been shown to stimulate neuronal activity (35). The role of this species as a neurotransmitter is not completely understood, however, and there has been a great deal of research aimed at studying its excitatory function in the brain. A selective biosensor for rapid determination of glutamate would be of great importance to neurochemical researchers.

Figure 15 shows the steady-state current dependence of the glutamate electrode (based on polymer I) on glutamate concentration at an applied potential of +350 mV (vs. SCE). The time required to reach 95% of the steady-state current was typically less

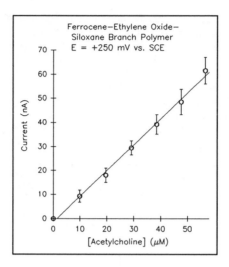

Figure 14. Steady-state current response of acetylcholine sensors based on polymer **I**, in pH 7.0 phosphate buffer under N_2-saturated conditions. Each point is the mean result for five electrodes.

than 10 sec after addition of the glutamate sample; a steady-state response was achieved in less than 1 min. The rapid response under stationary potential conditions indicates that the ferrocene-containing polymer can efficiently mediate electron transfer from the FAD centers of the enzyme to the electrode. The lower detection limit was approximately 0.01 mM, which is similar to that found using the manual Glukometer device (36). The use of substrate recycling techniques (37,38) may lower this detection limit considerably.

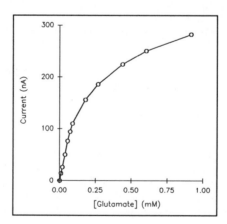

Figure 15. Typical steady-state current response for a ferrocene-ethylene oxide-siloxane polymer (**I**) / glutamate oxidase / graphite rod electrode at an applied potential of +350 mV vs. SCE.

Conclusion

The covalent attachment of electron transfer mediators to siloxane or ethylene oxide polymers produces highly efficient relay systems for use in amperometric sensors based on flavin-containing oxidases. It is clear from the response curves that the biosensors can be optimized through systematic changes in the polymeric backbone. The results discussed above, as well as those described previously (25-32), show that the mediating ability of these flexible polymers is quite general and that it is possible to systematically tailor these systems in order to enhance this mediating ability.

Acknowledgments

We are grateful to the National Science Foundation (Grant No. ISI-8960426), the National Institute of Diabetes and Digestive and Kidney Diseases (Grant No. 1-R43-DK42379-01), and the National Institute of Mental Health (Grant No. 1-R43-MH46764-01) for support of this research.

Literature Cited

(1) Clark, L. C. In *Biosensors: Fundamentals and Applications*, Turner, A. P. F.; Karube, I.; Wilson, G. S., Eds.; Oxford University Press: New York, 1987; Chapter 1.

(2) Clark, L. C.; Lyons, C. *Ann. N. Y. Acad. Sci.* **1962**, *102*, 29-45.

(3) Jönsson, G.; Gorton, L. *Anal. Lett.* **1987**, *20*, 839-855.

(4) Heider, G. H.; Sasso, S. V.; Huang, K. -m.; Yacynych, A. M.; Wieck, H. J. *Anal. Chem.* **1990**, *62*, 1106-1110.

(5) Cass, A. E. G.; Davis, G.; Francis, G. D.; Hill, H. A. O.; Aston, W. J.; Higgins, I. J.; Plotkin, E. V.; Scott L. D. L.; Turner, A. P. F. *Anal. Chem.* **1984**, *56*, 667-671.

(6) Lange, M. A.; Chambers, J. Q. *Anal. Chim. Acta* **1985**, *175*, 89-97.

(7) Iwakura, C.; Kajiya, Y.; Yoneyama, H. *J. Chem. Soc., Chem. Commun.* **1988**, 1019-1020.

(8) Jönsson, G.; Gorton, L.; Petterson, L. *Electroanalysis* **1989**, *1*, 49-55.

(9) Ikeda, T.; Hamada, H.; Senda, M. *Agric. Biol. Chem.* **1986**, *50*, 883-890.

(10) Ikeda, T.; Shibata, T.; Senda. S. *J. Electroanal. Chem.* **1989**, *261*, 351-362.

(11) Kulys, J. J.; Čénas, N. K. *Biochim. Biophys. Acta* **1983**, *744*, 57-63.

(12) Kulys, J. J.; Samalius, A. S.; Švirmickas, G.-J. S. *FEBS Lett.* **1980**, 7-10.

(13) Kulys, J. J. *Biosensors* **1986**, *2*, 3-13.

(14) Albery, W. J.; Bartlett, P. N.; Craston, D. H. *J. Electroanal. Chem.* **1985**, *194*, 223-235.

(15) McKenna, K.; Brajter-Toth, A. *Anal. Chem.* **1987**, *59*, 954-958.

(16) Hale, P. D.; Wightman, R. M. *Mol. Cryst. Liq. Cryst.* **1988**, *160*, 269-279.

(17) Hale, P. D.; Skotheim, T. A. *Synth. Met.*, **1989**, *28*, 853-858.

(18) Degani, Y.; Heller, A. *J. Phys. Chem.* **1987**, *91*, 1285-1289.

(19) Degani, Y.; Heller, A. *J. Am. Chem. Soc.* **1988**, *110*, 2615-2620.

(20) Heller, A. *Acc. Chem. Res.* **1990**, *23*, 128-134.

(21) Foulds, N. C.; Lowe, C. R. *Anal. Chem.* **1988**, *60*, 2473-2478.

(22) Degani Y.; Heller, A. *J. Am. Chem. Soc.* **1989**, *111*, 2357-2358.

(23) Gregg, B. A.; Heller, A. *Anal. Chem.* **1990**, *62*, 258-263.

(24) Pishko, M. V.; Katakis, I.; Lindquist, S.- E.; Ye, L.; Gregg, B. A.; Heller, A. *Angew. Chem. Int. Ed. Engl.* **1990**, *29*, 82-84.

(25) Hale, P. D.; Inagaki, T.; Karan, H. I.; Okamoto, Y.; Skotheim, T. *J. Am. Chem. Soc.* **1989**, *111*, 3482-3484.

(26) Inagaki, T.; Lee, H. S.; Hale, P. D.; Skotheim, T. A.; Okamoto, Y. *Macromolecules* **1989**, *22*, 4641-4643.

(27) Inagaki, T.; Lee, H. S.; Skotheim, T. A.; Okamoto, Y. *J. Chem. Soc., Chem. Commun.* **1989**, 1181-1183.

(28) Hale, P. D.; Inagaki, T.; Lee, H. S.; Karan, H. I.; Okamoto, Y.; Skotheim, T. A. *Anal. Chim. Acta* **1990**, *228*, 31-37.

(29) Gorton, L.; Karan, H. I.; Hale, P. D.; Inagaki, T.; Okamoto, Y.; Skotheim, T. A. *Anal. Chim. Acta* **1990**, *228*, 23-30.

(30) Hale, P. D.; Inagaki, T.; Lee, H. S.; Skotheim, T. A.; Karan, H. I.; Okamoto, Y. In *Biosensor Technology: Fundamentals and Applications*, Buck, R. P.; Hatfield, W. E.; Umaña, M.; Bowden, E.F., Eds.; Marcel Dekker: New York, 1990; Chapter 14.

(31) Hale, P. D.; Boguslavsky, L. I.; Inagaki, T.; Lee, H. S.; Skotheim, T. A.; Karan, H. I.; Okamoto, Y. *Mol. Cryst. Liq. Cryst.* **1990**, *190*, 251-258.

(32) Hale, P. D.; Boguslavsky, L. I.; Inagaki, T.; Karan, H. I.; Lee, H. S.; Skotheim, T. A.; Okamoto, Y. *Anal. Chem.* **1991**, *63*, 677-682.

(33) Wang, J.; Wu, L.-H.; Lu, Z.; Li, R.; Sanchez, J. *Anal. Chim. Acta* **1990**, *228*, 251-257.

(34) Chambers, J. A.; Walton, N. J. *J. Electroanal. Chem.* **1988**, *250*, 417-425.

(35) Cooper, J. R.; Bloom, F. E.; Roth, R. H., *The Biochemical Basis of Neuropharmacology*; Oxford University Press: New York, 1986, pp. 161-170.

(36) Wollenberger, U.; Scheller, F. W.; Böhmer, A.; Passarge, M.; Müller, H.-G. *Biosensors* **1989**, *4*, 381-391.

(37) Schubert, F.; Kirstein, D.; Scheller, F.; Appelqvist, R.; Gorton, L.; Johansson, G. *Anal. Lett.* **1986**, *19*, 1273-1288.

(38) Yao, T.; Yamamoto, H.; Wasa, T. *Anal. Chim. Acta* **1990**, *236*, 437-440.

RECEIVED November 4, 1991

Chapter 10

Permselective Coatings for Amperometric Biosensing

Joseph Wang

Department of Chemistry, New Mexico State University, Las Cruces, NM 88003

Access to the surface of amperometric biosensors can be controlled by coverage with an appropriate permselective film. Such coatings effectively exclude coexisting interferences, and thus greatly improve the selectivity and stability. Discriminative films based on different transport properties are described. Structural factors and fundamental interactions that govern the transport through such films are discussed. Future prospects are examined.

Amperometric biosensors satisfy many of the requirements for clinical assays, environmental monitoring or process control. Such sensors offer excellent sensitivity, fast response, selectivity toward electroactive species, miniaturization and low cost. However, there are still problems of stability and selectivity associated with the utility of amperometric devices. Despite the inherent specificity accrued from the biological recognition process, co-existing electroactive species may result in overlapping current signals. In addition, adsorption of surface-active macromolecules can cause a gradual fouling of the sensing surface.

One promising avenue to impart higher selectivity and stability to amperometric sensors is to cover the sensing surface with an appropriate permselective coating. The selective permeation and protection offered by such films can thus greatly promote routine applications of biosensing devices. The following sections examine the requirements, utility, and possibilities of using permselective films for electrochemical biosensing. The physical and chemical features that determine the permeability of these coatings are also discussed. While the concept is presented in the context of amperometric measurements, it could be easily extended to other (nonelectrochemical) sensing schemes.

Discriminative Films

Permselective films greatly enhance the selectivity and stability of amperometric probes by rejecting from the surface undesired (interfering) constituents, while allowing transport of the target analyte (Figure 1). An effective separation step is thus performed in situ on the sensing surface. Different avenues to control the access to the surface, based on various discriminative coatings, have been explored (Table I). Often, the coverage of the surface with the permselective film is used for a

0097–6156/92/0487–0125$06.00/0

Table I. Discriminative coatings for amperometric biosensors

Transport mechanism	Permselective Film	Ref.
Size exclusion	Cellulose acetate	1
	Base-hydrolyzed cellulose acetate	2
	Phase-inversion cellulose acetate	3
	Polyaniline, Polypyrrole	4
	Polyphenol	5
	Gamma radiated poly(acrylonitrile)	6
Charge exclusion	Nafion	7, 8
	Poly(vinylpyridine)	9
	Poly(ester-sulfonic acid)	10, 11
Polarity	Phospholipid	12, 13
Mixed control	Cellulose acetate - Nafion	14
	Cellulose acetate-poly(vinylpyridine)	15

simultaneous entrapment of the biocomponent (and its cofactor), as well as of another "active" moiety (e.g. electrocatalyst). The polymerization/casting conditions (e.g. pH, solvent) should be compatible with the requirements for the enzyme activity/stability. Attention should be given to possible changes in the film permeability in the presence of these immobilized reagents. In designing effective membrane barriers, one should attempt to maintain a fast and sensitive response for the analyte through a facile transport of this target species. As such, the optimal film thickness represents a compromise between effective rejection of interferences, and high sensitivity and speed. The transport of analytes and interferents through a given coating is determined by their diffusion coefficients in the film and their distribution ratios (between the film and solution). The improved selectivity is thus being achieved by taking advantage of analyte properties such as charge, size, polarity or shape.

Size exclusion films are particularly attractive in connection with oxidase-based enzyme electrodes. Such films greatly facilitate the anodic detection of the small hydrogen peroxide product. Cellulose-acetate (CA) modification has been used to build size-exclusion selectivity into amperometric sensors. Sittampalam and Wilson (1) illustrated the utility of CA coatings for minimizing surface fouling by proteins adsorption during the detection of hydrogen peroxide. Wang and Hutchins (2) demonstrated that base hydrolysis of CA coatings opens up their pores (through hydrolytic removal of acetate functionalities). Hence, different permeabilities were obtained by changing the hydrolysis time (Figure 2). Efforts to further extend the molecular-weight cut-off of porous CA films (up to 1500 daltons), based on a phase inversion method, were described by Kuhn et al (3). Selectivity based on molecular size can be achieved also through gamma radiation cross-linking of certain films (6). The permeability may be affected by variations in the radiation dose; response times are relatively long. The gamma irradiation scheme was used also for immobilizing enzymes (e.g. lactate oxidase) in polymeric layers (16).

Of particular interest is the ability to control and manipulate the permeability of size-exclusion films to meet specific biosensing needs. Electropolymerization processes are particularly suitable for creating controllable size-exclusion films (4).

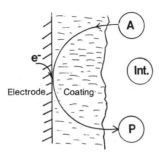

Figure 1. A permselective coating for amperometric sensing (A, analyte; P, product; Int., interferent).

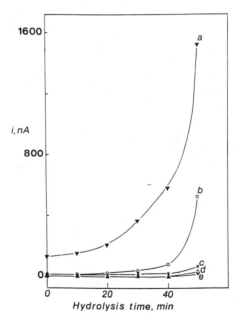

Figure 2. Dependence of the current response on the hydrolysis time of the CA film for phenol (a), acetaminophen (b), estriol (c), NADH (d), and potassium ferrocyanide (e). Flow injection amperometric operation. (Reproduced from ref. 2. Copyright 1985 American Chemical Society.)

This can be accomplished by controlling the amount of charge consumed during the anodization process (through the use of different polymerization times or monomer concentrations). The electropolymerization process has been very useful for a simultaneous physical entrapment of enzymes (17). In particular, glucose sensors based on electropolymerized 1,2-diaminobenzene couple an effective permselective response (Figure 3), with a very fast (1s!) response and thermal stabilization of the immobilized enzyme (18, 19). Note also (from Figure 3) the protection of the surface from foulants present in the serum solution. Extension of the dynamic linear range (through control of the film thickness) is another attractive advantage. Polypyrrole/ glucose-oxidase systems have received considerable attention (17,20), with additional advantages attained through the coimmobilization of redox mediators (e.g. ferrocene, quinone) (21) or electrocatalytic centers (e.g. platinum microparticles) (22). Other enzymes, e.g. cholesterol oxidase, have also been successfully incorporated within polypyrrole films (23). Other electropolymerized films, useful for one-step enzyme entrapments, include poly-N-methylpyrrole, polyacetylene, polyaniline or polyphenol.

Discriminative properties based on solute charge can offer additional selectivity improvements. As a result of electrostatic interactions, charged coatings offer a facile transport of oppositely charged ionic species while excluding co-ionic interferences. In particular, the common anionic interferences, ascorbic and uric acids, have been excluded by negatively-charged perfluorinated (Nafion) and polyester (Eastman Kodak AQ) ionomeric films (7, 8, 10, 11, 24). Effective prevention of surface fouling and immobilization of glucose oxidase is also offered by these sulfonated coatings (25,26). The dispersion of the AQ polymers in aqueous media is advantageous for the enzyme casting task. Because of their hydrophobic character, Nafion and AQ ionomers possess large ion-exchange affinity for organic cations relative to simple inorganic ones. Positively charged coatings, such as the cationic polyvinylpyridine (PVP), can be used to repel cationic interferences (9). The charge (and hence the permeability) of these polyelectrolytes are strongly dependent upon the solution pH. Analogous improvements in the selectivity can be obtained for ionomer-film coated potentiometric enzyme electrodes (27). In particular, the rejection of endogeneous ammonium and potassium ions greatly facilitated the use of enzyme electrodes based on detection of liberated ammonium ions.

Permselective films based on other transport properties should be suitable for amperometric biosensing. In particular, cast layers of lipids can facilitate the detection of hydrophobic compounds (at the underlying electrode) through the rejection of hydrophilic interferences (Figure 4) (12, 13). Depending on the charge of the lipid (presence of anionic phosphate groups) electrostatic interactions can also contribute to the overall permeation. Various approaches to increase the mechanical stability of lipid coatings have been explored (13). The selectivity improvements were illustrated for lipid modified glucose electrodes, based on optimizing the amount of the asolectin lipid in the enzyme layer (28). In addition to ascorbic and uric acids, reduced interferences was reported for acetaminophen, tyrosine and glutathione. Loading of the lipid film with redox active lipophilic substances or the use of alkylthiol monolayer assemblies may offer further advantages.

Additional improvements can be achieved through the use of multilayers (based on different overlaid films). Such combination of the properties of different films has been documented with bilayers of Nation/CA (14) and Nafion/collagen (29). The former allows selective measurements of the neurotransmitter dopamine in the presence of the slightly larger epinephrine and the anionic ascorbic acid (Figure 5). In addition to bilayers, mixed (composite) films, such as PVP/CA (15) or polypyrrole/Eastman Kodak AQ (30) layers can offer additional permselectivity advantages, such composites exhibit properties superior to those of their individual components. Also promising are sensor arrays, based on electrodes coated with

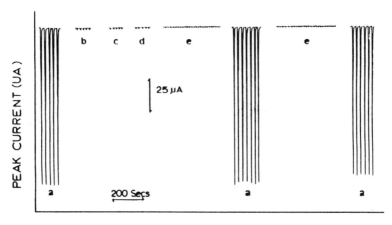

Figure 3. Current response of a poly(1,2-diaminobenzene)-coated electrode to hydrogen peroxide (a), ascorbic acid (b), uric acid (c), cysteine (d), and control human serum (e). Flow injection amperometric operation. (Reproduced from ref. 18. Copyright 1990 American Chemical Society.)

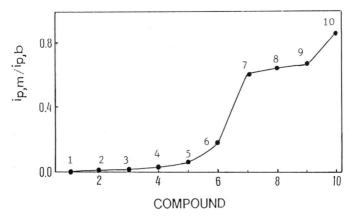

Figure 4. Permeability of a phospholipid/cholesterol film for different solutes: ascorbic acid (1), tyrosine (2), uric acid (3), acetaminophen (4), cysteine (5), desipramine (6), perphenazine (7), trimipramine (8), promethazine (9), and chlorpromazine (10). (Adapted from ref. 13.)

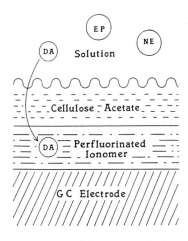

Figure 5. A Nafion/cellulose acetate bilayer coated electrode. (Reproduced from ref. 14. Copyright 1986 American Chemical Society.)

different permselective films (i.e. tuned toward different analytes), and operated in connection with statistical pattern recognition methods (*31*).

Future Prospects

The above discussion illustrates the biosensing opportunities provided by permselective coatings. Besides electrochemical transducers, such films can greatly benefit mass, optical or thermal sensing devices. A redox polymer may be coupled to the permselective film (in a mixed or multilayer configurations) to establish electrical communication between the underlying electrode and the enzyme. Recent work by Karube's group (*32*) has illustrated that an efficient direct electron transfer can be obtained between the active site of glucose oxidase, entrapped with an electropolymerized layer of N-methylpolypyrrole film on gold, and the electrode. Immobilization of other bioreagents (e.g. antibodies) within permselective films is also envisioned. Further improvements will be achieved by understanding the exact mechanism by which solutes permeate through coatings (*33*), and by engineering novel surface microstructures with unique environments for molecular interactions and consequently molecular recognition. One promising avenue is to produce defects within monolayer coatings that recognize target molecules based on their characteristic shape. In addition to a template approach, it is possible to functionalize the polymeric network for effective molecular recognition and permselective response. Highly stable inorganic (filtering) layers may offer additional advantages. The unique structural characteristics of zeolites and clays hold a great promise in this direction. Further insights into the structural factors and fundamental interactions that govern the transport through permselective films can be gained through high-resolution surface techniques (e.g. scanning tunneling microscopy)(*34*). These developments should be coupled to improved reproducibility and long-term stability of the coatings. As our ability to precisely manipulate the surface microstructures (and hence to transport properties) continues to grow, one can expect new and powerful biosensing opportunities.

Acknowledgments

The author gratefully acknowledges the financial support of the Petroleum Research Fund, administrated by the American Chemical Society.

Literature Cited

1. Sittampalam, G.; Wilson, G.S. *Anal. Chem.* **1983**, *55*, 1608.
2. Wang, J.; Hutchins, L.D. *Anal. Chem.* **1985**, *57*, 1536.
3. Kuhn, L.S.; Weber, S.G.; Ismail, K.Z. *Anal. Chem.* **1989**, *61*, 303.
4. Wang, J.; Chen, S.P.; Lin. M.S. *J. Electroanal. Chem.* **1989**, *273*, 231.
5. Ohsaka, T.; Hirokawa, T.; Miyamoto, H.; Oyama, N. *Anal. Chem.* **1987**, *59*, 1758.
6. De Castro E.S.; Huber, E.W.; Villarroel, D.; Galiatsatos, C.; Mark. J.E.; Heineman, W.R.; Murray, P.T. *Anal. Chem.* **1987**, *59*, 134.
7. Szentirmay, M.N.; Martin, C.R. *Anal. Chem.* **1984**, *56*, 1898.
8. Wang, J.; Tuzhi, P.; Golden, T. *Anal. Chim. Acta* **1984**, *194*, 129.
9. Wang, J.; Golden, T.; Tuzhi, P. *Anal. Chem.* **1987**, *59*, 740.
10. Wang, J.; Golden, T. *Anal. Chem.* **1989**, *61*, 1397.
11. Fortier, G.; Beliveau, R.; Leblond, E.; Belanger, D. *Anal. Lett.* **1990**, *23*, 1607.
12. Garcia, O.J.; Quintela, P.A.; Kaifer, A.E. *Anal. Chem.* **1989**, *61*, 979.

13. Wang. J.; Lu, Z. *Anal. Chem.* **1990**, *62*, 826.
14. Wang, J.; Tuzhi, P. *Anal. Chem.* **1986**, *58*, 3257.
15. Wang, J.; Tuzhi, P. *J. Electrochem. Soc.* **1987**, *134*, 586.
16. Hajizadeh, K.; Halsall, H.B.; Heineman, W.R. *Anal. Chim. Acta* **1991**, *243*, 23.
17. Umana, M.; Waller, J. *Anal. Chem.* **1986**, *58*, 2979.
18. Sasso, S.V.; Pierce, R.J.; Walla, R.; Yacynych, A. *Anal. Chem.* **1990**, *62*, 1111.
19. Malitesta, C.; Palmisano, F.; Torsi, L.; Zambonin, P.G. *Anal. Chem.* **1990**, *62*, 2735.
20. Foulds, N.C.; Lowe, C.R. *J. Chem. Soc.* Faraday Trans. **1986**, *82*, 1259.
21. Kajiya, Y.; Sugai, H.; Iwakura, C.; Yoneyama, H. *Anal. Chem.* **1991**, *63*, 49.
22. Belanger, D.; Brassard, E.; Fortier, G. *Anal. Chim. Acta* **1990**, *228*, 311.
23. Kajiya, Y.; Tsuda, R.; Yoneyama, H. *J. Electroanal. Chem.* **1991**, *301*, 155.
24. Gorton, L.; Karan, H.I.; Hale, P.D.; Inagaski, Y.; Okamoto, Y.; Skotheim, T.A. *Anal. Chim. Acta* **1990**, *223*, 23.
25. Harrison, D.J.; Turner, R.F.B.; Baltes, H.P. *Anal. Chem.* **1988**, *60*, 2002.
26. Wang, J.; Leech, D.; Ozsoz, M.; Martinez, S.; Smyth, M.R. *Anal. Chim. Acta* **1991**, *245*, 139.
27. Rosario, S.A.; Cha, G.S.; Meyerhoff, M.E.; Trojanowicz, M. *Anal. Chem.* **1990**, *62*, 2418.
28. Amine, A.; Kauffmann, J.M.; Patriarche, G.J.; Guilbault, G.G. *Anal. Lett.* **1989**, *22*, 2403.
29. Bindra, D.S.; Wilson, G.S. *Anal. Chem.* **1989**, *61*, 2566.
30. Wang, J.; Sun, Z. .; Lu, Z. *J. Electroanal. Chem.*, in press.
31. Wang. J.; Rayson, G.; Lu, Z.; Wu, H. *Anal. Chem.* **1990**, *62*, 1924.
32. de Taxis du Poet, P.; Miyamoto, S.; Mukarami, T.; Kimura, J.; Karube, I. *Anal. Chim. Acta* **1990**, *235*, 255.
33. Saveant, J.M. *J. Electroanal. Chem.* **1991**, *302*, 91.
34. Yaniv, D.R.; McCormick, L.; Wang, J.; Naser, N. *J. Electroanal. Chem.*, in press.

RECEIVED November 5, 1991

ELECTROPOLYMERIZED THIN FILMS

Chapter 11

Polypyrrole Film Electrode Incorporating Glucose Oxidase

Electrochemical Behavior, Catalytic Response to Glucose, and Selectivity to Pharmaceutical Drugs

Zhisheng Sun and Hiroyasu Tachikawa

**Department of Chemistry, Jackson State University,
Jackson, MS 39217–0510**

Polypyrrole thin film doped with glucose oxidase (PPy-GOD) has been prepared on a glassy carbon electrode by the electrochemical polymerization of the pyrrole monomer in the solution of glucose oxidase enzyme in the absence of other supporting electrolytes. The cyclic voltammetry of the PPy-GOD film electrode shows electrochemical activity which is mainly due to the redox reaction of the PPy in the film. Both in situ Raman and in situ UV-visible spectroscopic results also show the formation of the PPy film, which can be oxidized and reduced by the application of the redox potential. A good catalytic response to the glucose and an electrochemical selectivity to some hydrophilic pharmaceutical drugs are seen at the PPy-GOD film electrode.

In recent years the electrochemistry of the enzyme membrane has been a subject of great interest due to its significance in both theories and practical applications to biosensors (1-5). Since the enzyme electrode was first proposed and prepared by Clark et al. (6) and Updike et al. (7), enzyme-based biosensors have become a widely interested research field. Research efforts have been directed toward improved designs of the electrode and the necessary membrane materials required for the proper operation of sensors. Different methods have been developed for immobilizing the enzyme on the electrode surface, such as covalent and adsorptive couplings (8-12) of the enzymes to the electrode surface, entrapment of the enzymes in the carbon paste mixture (13), etc. The entrapment of the enzyme into a conducting polymer has become an attractive method (14-22) because of the conducting nature of the polymer matrix and of the easy preparation procedure of the enzyme electrode. The entrapment of enzymes in the polypyrrole film provides a simple way of enzyme immobilization for the construction of a biosensor. It is known that the PPy-

0097–6156/92/0487–0134$06.00/0
© 1992 American Chemical Society

GOD film electrode exhibits a catalytic response to glucose. However, the PPy enzyme film prepared by the electrochemical method is normally doped with other anionic species from the supporting electrolytes (*15, 19-20*). The electrochemical behavior of this type of PPy enzyme film is dominated by the PPy film doped with small anions. A PPy-GOD film electrode has also been prepared on a printed platinum electrode without using other supporting electrolytes, and its catalytic response to glucose has been reported (*14*). However, the nature of this type of PPy-GOD film has not been fully understood. It is well known that holoenzyme glucose oxidase consists of an apoenzyme protein and coenzyme such as flavin adenine dinucleotide (FAD) (*1-3, 23, 24*). Proteins are macromolecules which are built by successive condensations of amino acids. In a basic medium (pH is higher than isoelectric point), the proteins are negatively charged. The coenzyme FAD itself contains phosphate groups and can dissociate into anions. This structural feature makes the glucose oxidase to be a polyelectrolyte (*1*).

In this paper we report the electrochemical polymerization of the PPy-GOD film on the glassy carbon (GC) electrode in enzyme solution without other supporting electrolytes and the electrochemical behavior of the synthesized PPy-GOD film electrode. Because the GOD enzyme molecules were doped into the polymer, the film electrode showed a different cyclic voltammetric behavior from that of a polypyrrole film doped with small anions. The film electrode has a good catalytic behavior to glucose, which is dependent on the film thickness and pH. The interesting result observed is that the thin PPy-GOD film electrode shows selectivity to some hydrophilic pharmaceutical drugs which may result in a new analytical application of the enzyme electrode.

Experimental Section

Chemicals. Glucose oxidase (GOD, EC 1.1.3.4, Aspergillus niger, Type II, 25 U/mg), β-D(+) glucose, uric acid, ascorbic acid, and other pharmaceutical drugs were used as received from Sigma Chemical Co. Pyrrole (99% purity, Aldrich Chemical Co. Inc.) was purified by refluxing with zinc metal and distilling under vacuum. Other chemicals were analytical reagents and were used as received. Doubly distilled water was used as a solvent in the experiment.

Electrochemical Measurements. Electrochemical polymerization and measurements were carried out by an EG&G PARC Model 273 potentiostat/galvanostat. The PPy film was formed by either controlling the potential or constant current method, in which the film was prepared by controlling oxidizing current of 50 μA in the solution of 0.05 g/ml GOD and 0.35 M pyrrole without other supporting electrolytes. The film thickness was controlled by the charge passed and was estimated to be 1000 Å by passing 48 mC/cm^2 (*27*). An electrochemical cell with three electrodes was used for both the polymerization and electrochemical measurements of PPy films. For the reference electrodes, an Ag/AgCl electrode was used for polymerization, and a SCE was used for the electrochemical measurements. A platinum wire with a large area was used as a counter electrode, and a glassy carbon (GC) electrode (BAS MF 2012, 3.0 mm dia) was used as a working electrode. Prior to the electrochemical polymerization, the GC electrode was polished with 0.05 micron alumina powder (Buehler Ltd.) for 5 min, sonicated for 5 min and then washed thoroughly by distilled water. Phosphate buffer (PB) solutions were used for studying the pH effects of the film electrodes.

Spectroscopic Measurements. A Beckman Model 5230 spectrophotometer was used to record in situ UV-visible spectra of the PPy films, which were electrochemically deposited on the indium-tin oxide (ITO) coated glass (Delta Technologies). For Raman measurements a Spex Model 1403 double spectrometer, a DM1B Datamate, and a Houston Instrument DMP-40 digital plotter were employed. Details of the experimental setup for in situ Raman spectroscopy are described elsewhere (26).

Results and Discussion

Polypyrrole Film Formation in Glucose Oxidase Enzyme Solution. Cyclic voltammograms recorded in the GOD and pyrrole solution showed an anodic peak current (E_{pa} = 1.08 V), which suggested the polymerization of pyrrole in the above solution. However, the polymerization potential moved toward the more positive direction compared to the polymerization potential of PPy doped with Cl⁻ (E_{pa} < 1.0 V). This is due to the fact that the polymerization is more difficult to take place in enzyme solution than in Cl⁻ solution because the enzyme solution is a much weaker electrolyte than NaCl; it may also be due to the less conductive nature of the PPy-GOD film as compared to that of the PPy-Cl film. The polymerization current level was much lower in the enzyme solution than in the Cl⁻ solution because of the poor charge-transport property of the enzyme protein molecules. It was found that the constant current method was more suitable than the controlled potential method for making the PPy-GOD film on the GC electrode.

Cyclic Voltammetric Behavior of the PPy-GOD Film. Figure 1 shows the cyclic voltammetric curves of a PPy-GOD film (4000 Å) in phosphate buffer solution with pH 7.4 at different scan rates. Both anodic and cathodic peaks should correspond to the redox reactions of PPy chains. The peak potentials, which were recorded at the scan rate of 200 mV/s, were -380 mV and -200 mV for cathodic and anodic peaks, respectively. This is similar to the potential shifts of the PPy film doped with large anions (27) such as poly(p-styrenesulfonate). Enzyme protein molecules are composed of amino acid and have large molecular size, which can not move out freely from the PPy-GOD film by the application of the reduction potential. In order to balance the charge of the PPy-GOD film, cations must move into the film, and redox potentials move toward a more negative potential. This behavior is different from the one observed for the PPy-GOD film, which was prepared in the solution of GOD with KCl as supporting electrolytes (*20, 22*). In this case the redox potential of the PPy film was near 0 V because the PPy was mainly doped with Cl⁻ and the GOD was entrapped in the PPy-Cl film. Both cathodic and anodic peak currents for the redox reaction of PPy in the PPy-GOD film increased with increased film thickness, suggesting that the PPy-GOD film was certainly formed on the GC electrode surface. The results also suggest that the PPy-GOD film has reasonably good conductivity.

Effect of Cations. Figure 2 shows the cyclic voltammetric curves of the PPy-GOD film in the solutions with different cations. It can be seen that the redox current of PPy-GOD is dependent on the cations in the solution. When the PPy chain is reduced, huge enzyme molecules cannot move out and cations must move into the film in order to balance the charge. Since the large cations such as TBA^+ have difficulty in moving into the film, the redox current of the PPy decreased in the supporting electrolytes with large cations. The order of the redox current level of the PPy-GOD film in supporting electrolytes of different cations is as follows.

$$TBA^+ < TMA^+ < K^+, Na^+, Li^+$$

For cations such as K^+, Na^+, and Li^+, no significant change of current was observed, since all of these cations can move freely in the film.

Effect of Anions. Figure 3 shows the cyclic voltammetric curves of the PPy-GOD film in the solution of different anions. In the solutions with ClO_4^-, PO_4^{3-}, NO_3^-, and Cl⁻ anions, the redox potentials of the PPy-GOD film were approximately at the same potential. A small peak current also appeared near +0.1 V in the solutions which contain Cl⁻ and Br⁻. This current may be caused by the partial doping of small anions into the film forming PPy-Cl (or PPy-Br) during the redox reaction of the PPy-GOD film. In the BF_4^- solution, the redox potential of the PPy-GOD film shifted toward a more positive potential. The observed redox current in the positive potential range may be due to the PPy-BF_4. It is likely that the PPy interacts more strongly with BF_4 than with GOD.

Effect of pH. It is expected that the pH of the solution will affect the electrochemical behavior of the PPy-GOD because both the activity and the structure of enzyme GOD are pH dependent. The cyclic voltammetric measurements of the PPy-GOD film electrode were carried out in phosphate buffer solutions of different pHs. The cathodic peak current of the PPy-GOD film was at approximately -380 mV (scan rate: 100 mV/s) when pHs of the solution were between 6 and 11. When the pH decreased from 6 to 2, the peak currents of the PPy-GOD film shifted toward a more positive potential. Figure 4 shows the relationship between the cathodic peak potential of the PPy-GOD film and the pH of the solution (curve 1). Curve 2 in Figure 4 shows the relationship between the cathodic peak potential of the PPy-Cl film and the pH. Contrary to the behavior of the PPy-GOD film, the redox potential of the PPy-Cl film was constant in the pH ranging between 2 and 8.

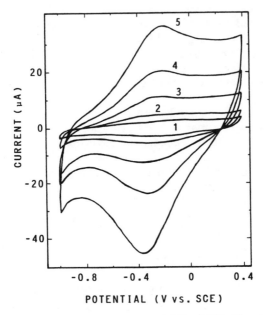

Figure 1. Cyclic voltammetric curves of PPy-GOD film (4000 Å) with different scan rates in 0.05 M phosphate buffer (pH 7.4) solution. Scan rates (mV/s): (1) 10, (2) 20, (3) 50, (4) 100, and (5) 200.

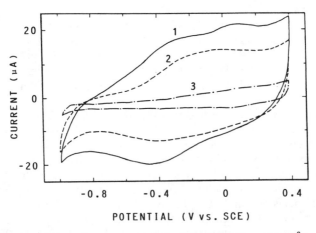

Figure 2. Cyclic voltammograms of the PPy-GOD film (4000 Å) electrode in solutions with different cations: (1) LiCl, (2) (TMA)Cl, and (3) (TBA)Cl. Concentration of supporting electrolytes: 0.05 M. Scan rate: 100 mV/s.

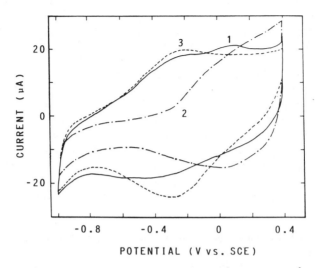

Figure 3. Cyclic voltammograms of the PPy-GOD film (4000 Å) electrode in several solutions with different anions: (1) NaCl, (2) NaBF$_4$, and (3) PO$_4^{3-}$ (pH 7.4). Concentration of the supporting electrolytes: 0.05 M. Scan rate: 100 mV/s.

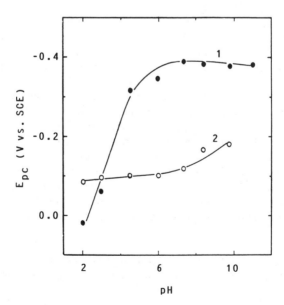

Figure 4. Relationship between cathodic peak potential and pH: (1) PPy-GOD (4000 Å) and (2) PPy-Cl (1000 Å).

The observed pH dependency of the redox potential of the PPy-GOD film reflects the difference between the PPy doped with inorganic anions and the PPy doped with biological protein molecules. It is known that the protein is structurally an amphiphilic macromolecule. When the pH is more than the isoelectric point, a protein dissociates to form negatively charged species, which can interact with PPy^+ forming PPy-GOD. When the pH is lower than the isoelectric point, the protein becomes a protonated form with a positive charge, and it may be separated from the PPy^+. In the acidic PB solution, the PPy^+ should interact with $H_2PO_4^-$ forming $PPy-H_2PO_4$ when the oxidation potential is applied. For the reason described above, the CV of the PPy-GOD film in low pH solutions is similar to that of the PPy doped with small anions (the redox potential of the PPy is more positive than that of the PPy-GOD). As can be seen in curve 1 in Figure 4, E_{pc} value started to increase around pH 4-5, which is close to the isoelectric point of protein, which has been reported to be 4.2 (*28, 29*).

UV-Visible and Raman Spectroscopies. In situ UV-visible absorption spectra of a 5000 Å PPy-GOD film, which was formed on an ITO coated glass, were recorded in the PB solution (pH 7.4). The spectra recorded at both the oxidation (0.4 V) and the reduction (-1.0 V) potentials showed an absorption peak near 380 nm, which is due to the PPy. When the PPy was reduced at -1.0 V, the absorbance in the wavelength range of 500-800 nm decreased, and the absorbance at 380 nm increased. The observed spectral changes of the PPy-GOD film during the redox reaction were similar to those of the PPy film doped with ClO_4^- ($PPy-ClO_4$) (*27*).

In situ Raman spectra of 2000 Å PPy-GOD film, which was deposited on a GC electrode, were recorded in the 0.05 M PB (pH 7.4) solution. Figure 5 shows Raman spectra of the as-polymerized film and of those recorded at -1.0 V and + 0.4 V. In the spectrum for the as-polymerized PPy-GOD film, there are several bands which may be due to the PPy (*27, 30, 31*); 922, 988, 1044, 1345, and 1552 cm^{-1}. The 1345 cm^{-1} band could also be due to the GOD (*2*). However, there has been no clear evidence for the observation of resonance Raman spectrum of GOD itself (*2*), and the observed Raman spectrum for the as-polymerized PPy-GOD film may be totally due to the PPy. Virdee and Hester (*30*) have previously observed similar spectrum changes for a $PPy-SO_4$ film by applying the redox potentials. At any rate, the above spectroscopic data confirmed that the PPy-GOD film was formed on the GC electrode. Both UV-visible and Raman spectroscopic data were also consistent with the CV data, which indicated the presence of the redox reaction of the PPy in the PPy-GOD film.

Catalytic Response of the PPy-GOD Film Electrode to Glucose. For the catalytic activity of the PPy-GOD film, the following reactions should take place.

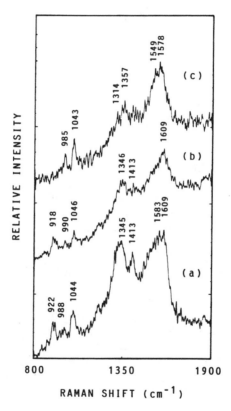

Figure 5.   In situ Raman spectra of PPy-GOD film (2000 Å on GC) in
0.05 M PB (pH 7.4) solution: (a) as-polymerized, (b) at 0.4 V, and (c) at
-1.0 V. λ_{ex} = 514.5 nm.

$$\text{Glucose} + O_2 \xrightarrow{\text{GOD}} \text{Gluconic acid} + H_2O_2$$

$$H_2O_2 \rightarrow 2H^+ + O_2 + 2e^-$$

H_2O_2 produced in the enzyme catalytic reaction can be detected by the oxidation of H_2O_2 at the electrode. As shown in Figure 6, the catalytic current was observed at the PPy-GOD film when the glucose was added to the PB solution. The catalytic response of the PPy-GOD film electrode to the glucose showed a linear relationship at a low glucose concentration, but the catalytic response deviated from the linear relationship at a high concentration of glucose as shown in Figure 6b. The apparent Michaelis-Menten constant K_M' was obtained to be ~15 mM from the electrochemical Eadie-Hofstee form of the Michaelis-Menten equation (32, 33, 34). This value is lower than those obtained for the PPy-GOD film prepared with small anions (14). Other factors which affected the catalytic response of the PPy-GOD film to glucose were film thickness, pH etc. The details of these effects are described in the following sections.

Effect of Film Thickness. As shown in Figure 7, the electrochemical response of the film electrode was dependent on the film thickness. The catalytic response current increased when the film thickness increased from 0 to 100 Å, then it decreased when the film thickness further increased. The maximum current was observed when the film thickness was near 100 Å. When the film was thick (>1000 Å), the PPy-GOD film still showed a catalytic response to glucose, but the current was low and remained at a constant level. The above results may be explained by the diffusion of H_2O_2 in the PPy-GOD film toward the GC surface where the H_2O_2 is oxidized. Although a thick PPy-GOD film should have more enzyme sites which are supposed to generate a high catalytic current, the diffusion of H_2O_2 may be more difficult in the thick film than in the thin film. Since the conductivity of the PPy-GOD film is not very high, particularly under the experimental condition used (see section Effect of Overoxidation), contribution of the catalytic current which was due to the oxidation of H_2O_2 in the PPy-GOD film may be less significant than that of H_2O_2 oxidized at the GC surface.

Effect of pH. The relationship between the catalytic current of a PPy-GOD film (2000 Å) and the pH of the solution was recorded in the pH range of 3-11. When the pH was less than 3, no catalytic current was observed, but the current was increased by increasing the pH from 3 to 7. When pH was more than 7.5, the current decreased. The PPy-GOD film electrode showed a good catalytic response to glucose in the solutions of pH ranging between 6 and 8.

Effect of Overoxidation. It has been reported that the PPy film doped with small anions can be overoxidized by applying 1.0 V or higher and the

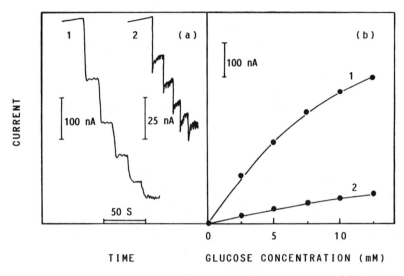

Figure 6. Catalytic response of PPy-GOD film to glucose. (a) Current response at PPy-GOD film electrodes with successive additions of 2.5 mM glucose solution: (1) 80 Å film and (2) 2000 Å film, in PB buffer (pH 7.4) at 1.0 V. (b) Calibration curves. 1: 80 Å film, 2: 2000 Å film.

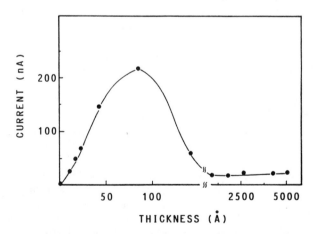

Figure 7. Relationship between catalytic current and thickness of PPy-GOD film with 2.5 mM glucose in PB buffer (pH 7.4) solution.

overoxidation of the PPy film has resulted in the loss of its conductivity (*35, 36*). Since the amperometric response current of the PPy-GOD film with glucose was measured by applying 1.0 V to the electrode, the possible overoxidation of the PPy-GOD film was investigated. Cyclic voltammograms were recorded from the PPy-GOD film after applying 1.0 V for a certain period of time. Cyclic voltammetric data showed that the overoxidation of the PPy-GOD film certainly took place by applying 1.0 V for several minutes. However, the degree of overoxidation of the PPy-GOD film was much smaller than that of the PPy film doped with small anions such as PPy-Cl (33% of the initial PPy redox current remained on the PPy-GOD film after 15 min of oxidation at 1.0 V, compared with 5% of initial PPy redox current that remained on the PPy-Cl film after 7 min of oxidation at 1.0 V). The observed different degrees of overoxidation for the above two PPy films may be due to the high resistivity of the PPy-GOD film.

Amperometric response currents from the PPy-GOD film electrode with glucose were also recorded after overoxidizing the electrode at 1.0 V for a certain period of time. The results showed that the overoxidation of the PPy-GOD film did not decrease the amperometric response current of the PPy-GOD film electrode with glucose. The results suggested that the GOD, which was immobilized on the GC electrode surface by the electropolymerization of the pyrrole with GOD, remained on the electrode surface after the overoxidation of the PPy-GOD film.

Electrochemical Behavior of PPy-GOD Film Electrode with Pharmaceutical Drugs. Although the chemically modified electrode has been developed for more than a decade, and many kinds of materials have been used for the modification of the electrode surface, the enzyme modified electrode has rarely been used for the study of electroactive species. This is probably due to the fact that the enzyme is not electronically conductive and also it is difficult to immobilize an enzyme on the electrode surface. So far, biological lipids have been used to modify the electrode, and the modified electrode shows a selectivity for hydrophobic molecules because the lipid molecule is hydrophobic (*37-39*).

An interesting result that was observed during the experiment was that the thin PPy-GOD film electrode showed more sensitivity to hydrophilic pharmaceutical drugs such as ascorbic acid, uric acid, dopamine, and tyrosine than to hydrophobic drugs such as desipramine and chlorpromazine. Figure 8 shows the cyclic voltammetric curves at a PPy-GOD thin film (70 Å) in ascorbic acid (Figure 8a) and in desipramine (Figure 8b). Oxidation peak currents were observed at the GC electrodes in both desipramine and ascorbic acid solutions with 0.05 M PB. At a 70 Å PPy-GOD film electrode, the oxidation peak of desipramine was not observed but the oxidation current of ascorbic acid was observed. This selectivity of the PPy-GOD film electrode for different chemicals may be explained by the hydrophilic affinity of biological enzyme molecules. The enzyme consists of the amino acid which is hydrophilic and helps hydrophilic drugs to pass through the film. Table I lists response currents

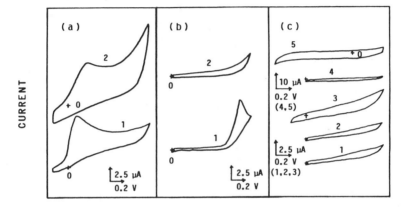

POTENTIAL (V vs. SCE)

Figure 8. Electrochemical behaviors of pharmaceutical drugs at PPy-GOD film electrodes and at GC electrode: (a) At 20 Å PPy-GOD film electrode in 0.5 mM ascorbic acid with 0.05 M PB (pH 7.4) solution. (b) At 20 Å PPy-GOD film electrode in 0.5 mM desipramine with 0.05 M PB (pH 7.4) solution. 1: at GC electrode, 2: at PPy-GOD film electrode, (c) At 1000 Å PPy-GOD film in 0.05 M PB solution (pH 7.4) with (1) 0.5 mM uric acid, (2) 0.5 mM tyrosine, (3) 1 mM ascorbic acid, (4) 0.5 mM promethazine, and (5) 1 mM $K_4Fe(CN)_6$. Scan rate: 50 mV/s.

Table I. Response Current of Pharmaceutical Drugs at PPy-GOD Film Electrode

	Drugs	i_{GC} $(\mu A)^a$	$i_{PPy\text{-}GOD}$ $(\mu A)^b$	$i_{PPy\text{-}GOD}/i_{GC}$ $(\%)$
1.	Trimipramine	10.0	0.5	5
2.	Desipramine	2.8	0.25	9
3.	Chlorpramazine	8.5	3.8	45
4.	Promethazine	10.5	6.0	57
5.	Perphenazine	8.8	5.0	57
6.	Tyrosine	5.1	3.3	65
7.	L-Cysteine	3.8	2.5	66
8.	Catechol	6.6	5.0	76
9.	Acetaminophen	4.3	4.0	93
10.	Dopamine	6.0	6.0	100
11.	Uric acid	4.3	4.4	102
12.	L-Dopa	4.5	5.5	122
13.	Ascorbic acid	3.8	6.0	158

[a]Response currents were recorded by applying 1.0 V under stirring condition. The concentration of the drugs was 0.1 mM in the PB solution (pH 7.4).
[b]Thickness of the PPy-GOD film was 30 Å. Other conditions were the same as those in footnote a.

of pharmaceutical drugs at both the GC and a 30 Å PPy-GOD film electrode. The results showed that the anodic currents observed for the hydrophilic drugs were much higher than those of the hydrophobic drugs. As shown in Figure 8c, no electrochemical activities were observed at thick PPy-GOD films (more than 300 Å) in the presence of pharmaceutical drugs and other redox pairs such as $Fe(CN)_6^{4-}/Fe(CN)_6^{3-}$. It is interesting to compare the above results with the CV of a 4000 Å PPy-GOD film (see Figure 1), which indicated that the PPy-GOD film had a reasonably good conductivity. The absence of the charge-transfer reactions of the electroactive species in the solution at the 300 Å PPy-GOD film may indicate that the surface of the PPy-GOD film is mostly covered by the GOD molecules, which are non-conductive. The PPy-GOD film showed the ex situ conductivity value *(40)* of $\sim 10^{-6}$ S cm^{-1}, which was considered to be too small to be able to have the kind of CVs which were shown in Figure 1. The observed low ex situ conductivity value also suggested that the charge transfer between the surface of the PPy-GOD film and the mercury contact *(40)* may be restricted due to the presence of the non-conductive GOD molecules at the film surface.

Conclusions

The PPy-GOD film can be prepared on the GC electrode by the electrochemical polymerization of pyrrole in the glucose oxidase enzyme solution without other supporting electrolytes. The negatively charged enzyme molecule can serve as a dopant in the PPy film. The CV data showed that when the PPy-GOD film was reduced, enzyme molecules remained in the film, and small cations in the solution could move into the film to balance the charge. Both in situ UV-visible and in situ Raman spectroscopies showed that the spectral changes of the PPy-GOD film by the application of the redox potentials reflected the oxidation and the reduction of PPy in the PPy-GOD film. The CV behavior of the PPy-GOD film was affected by supporting electrolytes of different cations and anions in the solution. The redox potential of the PPy-GOD film was pH dependent, but that of the PPy-Cl film was not. The cathodic and anodic peak potentials of the PPy-GOD film increased positively when the pH decreased from 6 to 2. The intersection point of the peak potential was near pH 4.5, which corresponds to the isoelectrical point of the protein molecule. The film also showed a good catalytic response to glucose. The catalytic currents were affected by both the pH in the solution and the film thickness. In the pH range 6-8, the PPy-GOD film showed the highest response current to glucose, and when pH was less than 3, the film lost its activity. The relationship between the response current and film thickness revealed that the response current was controlled by the diffusion of H_2O_2 in the film. The PPy-GOD thin films (30-70 Å) showed an electrochemical activity for hydrophilic pharmaceutical drugs, but the thick films (300-1000 Å) did not show any electrochemical activity for the electroactive species including the hydrophilic drugs due to the non-conductive nature of the GOD molecules at the surface of the PPy-GOD film electrode.

Acknowledgment

This work was supported in part by the Office of Naval Research and the National Institutes of Health (Grant SO6GM08047).

Literature Cited

1.   Degani, Y.; Heller, A. *J. Am. Chem. Soc.* **1989**, *111*, 2357.
2.   Holt, R. E.; Cotton, T. M. *J. Am. Chem. Soc.* **1987**, *109*, 1841.
3.   Degani, Y.; Heller, A. *J. Phys. Chem.* **1987**, *91*, 1285.
4.   Degani, Y.; Heller, A. *J. Am. Chem. Soc.* **1988**, *110*, 2615.
5.   Hale, P. D.; Inagaki, T.; Karan, H. I.; Okamoto, Y.; Skotheim, T. A. *J. Am. Chem. Soc.* **1989**, *111*, 3482.
6.   Clark, Jr., L. C.; Lyons, C. *Ann. N.Y. Acad. Sci.* **1962**, *102*, 29.
7.   Updike, S. J.; Hicks, G. P. *Nature*, **1967**, *214*, 986.
8.   Nanjo, M.; Guilbault, G. G. *Anal. Chim. Acta* **1974**, *73*, 367.
9.   Guilbault, G. G.; Montalvo, J. G. *J. Am. Chem. Soc.* **1969**, *91*, 2164.
10. *Methods in Enzymology*; Mosbach, K., Ed.; Academic Press: New York, NY, 1976; Vol. 44.
11. Guilbault, G. G.; *Analytical Uses of Immobilized Enzymes*; Marcel Dekker: New York, NY, 1984; pp.78-92.
12. Bourdillon, C.; Bourgeois, J. -P.; Thomas, D. *Biotechnol. Bioeng.* **1979**, *21*, 1877.
13. Wang, J.; Wu, L.; Lu, Z.; Li, R.; Sanchez, J. *Anal. Chim. Acta* **1990**, *228*, 251.
14. Foulds, N. C.; Lowe, C. R. *J. Chem. Soc., Faraday Trans. 1* **1986**, *82*, 1259.
15. Pandey, P. C.; *J. Chem. Soc., Faraday Trans. 1* **1988**, *84*, 2259.
16. Foulds, N. C.; Lowe, C. R. *Anal. Chem.* **1988**, *60*, 2473.
17. Tomiya, E.; Karube, I. *Sensors and Actuators* **1989**, *18*, 297.
18. Iwakura, C.; Kajiya, Y.; Yoneyama, H. *J. Chem. Soc. Chem. Commun.* **1988**, 1019.
19. Trojanowics, M.; Matuszewski, W.; Podsiadla, M. *Biosensors & Bioelectronics* **1990**, *5*, 149.
20. Bélanger, D.; Nadreau, J.; Fortier, G. *J. Electroanal. Chem.* **1989**, 274.
21. Yabuki, S.; Shinohara, H.; Aizawa, M. *J. Chem. Soc., Chem. Commun.* **1989**, 945.
22. Umaña, M.; Waller, J. *Anal. Chem.* **1986**, *58*, 2979.
23. Helmprecht, H. L.; Friedman, L. T. *Basic Chemistry for the Life Sciences*; McGraw-Hill: New York, NY, 1977; pp 399.
24. Frey, P. A. In *Structure and Function of Coenzymes*; Zubay, G., Ed.; Biochemistry; Addison-Wesley: Reading, MA, 1983, pp 197-241.
25. Diaz, A. F.; Castillo, J. I.; Logan, J. A.; Lee, W.-Y. *J. Electroanal. Chem.* **1981**, *129*, 115.
26. Kahl, J. L.; Faulkner, L. R.; Dwarakanath, K.; Tachikawa, H. *J. Am. Chem. Soc.* **1986**, *108*, 5434.

27. Choi, C. S.; Tachikawa, H. *J. Am. Chem. Soc.* **1990**, *112*, 1757.
28. Bentley, R. In *The Enzymes*; Boyer, P. D.; Lardy, H. A.; Myrback, H. Eds.; Academic Press: New York, NY, 1973, Vol. 7.
29. Kotowski, J.; Janas, T.; Tien, H. *Bioelectrochem. Bioenerg.* **1988**, *19*, 277.
30. Virdee, H. R.; Hester, R. E. *Croat. Chem. Acta* **1988**, *61*, 357.
31. Furukawa, Y.; Tazawa, S.; Fujii, Y.; Harada, I. *Synth. Met.* **1988**, *24*, 329.
32. Hale, P. D.; Boguslavsky, L. I.; Inagaki, T.; Karan, H. I.; Lee, H. S.; Skotheim, T. A. *Anal Chem.* **1991**, *63*, 677.
33. Gregg, B. A.; Heller, A. *Anal. Chem.* **1990**, *62*, 258.
34. Kamin, R. A.; Wilson, G. S. *Anal. Chem.* **1980**, *52*, 1198.
35. Beck, F.; Braun, P.; Oberst, M. *Ber. Bunsenges. Phys. Chem.* **1987**, *91*, 967.
36. Witkowski, A.; Freund, M. S.; Brajter-Toth, A. *Anal. Chem.* **1991**, *63*, 622.
37. Chastel, O.; Kauffmann, J. M.; Patriarche, G. J.; Christian, G. D. *Anal. Chem.* **1989**, *61*, 170.
38. Garcia, O. J.; Quintela, P. A.; Kaifer, A. E. *Anal. Chem.* **1989**, *61*, 979.
39. Wang, J.; Lu, Z. *Anal. Chem.* **1990**, *62*, 826.
40. Mao, H.; Pickup, P. G. *J. Electroanal. Chem.* **1989**, *265*, 127.

RECEIVED October 22, 1991

Chapter 12

Analytical Applications of the Electrochemical Quartz Crystal Microbalance

A. Robert Hillman[1], David C. Loveday[1], Marcus J. Swann[1], Stanley Bruckenstein[2], and C. Paul Wilde[3]

[1]School of Chemistry, University of Bristol, Bristol BS8 1TS, England
[2]Department of Chemistry, State University of New York, Buffalo, NY 14214
[3]Department of Chemistry, University of Ottawa, Ottawa, Ontario K1N 6N5, Canada

The electrochemical quartz crystal microbalance (EQCM) has considerable potential as a sensor. The strategy is to use exchange of mobile species between a surface immobilised polymer film and a solution as a probe of solution composition. Film composition changes are monitored gravimetrically. The quartz crystal microbalance offers generality of detection by mass and high sensitivity. Selectivity is sought via choice of polymer coating and electrochemical control variables. In this paper we discuss the thermodynamics and kinetics of the exchange process and conditions under which the EQCM technique is applicable.

Background Modification of electrode surfaces with polymer films has been the subject of considerable recent research (1). Potential areas of application include electrical/optical devices, surface protection, energy conversion/storage, electrocatalysis and electroanalysis. In this paper we discuss issues relevant to the use of polymer modified electrodes in sensors. The strategy upon which we focus here involves the exchange of mobile species (the target species and, possibly, interferants) between a polymer film (the sensor) and its bathing solution (the analyte). Extraction based on ion-exchange into a charged polymer film has been exploited for determination of metal species (2),(3) and neurotransmitters (4). Complexation by an appropriately chosen ligand is also possible (5), and the specificity of antibody interactions has been exploited (6). Following the immobilisation of the target species on the electrode, voltammetric determination of electroactive ions is straightforward. Although electroinactive ions may be determined indirectly, via their effect on the competitive binding of electroactive ions (7), a more direct approach is clearly preferable.

Uptake of solution species by the surface-modifying polymer film offers

0097–6156/92/0487–0150$06.00/0

the opportunity for pre-concentration. The extent to which this may enhance sensitivity may be determined by both thermodynamic and kinetic factors. When thermodynamic effects predominate (if the system is allowed to come to equilibrium), activity effects are significant (8). Equality of activity of mobile species will, in general, result in the transfer of neutral species, notably solvent, between the solution and polymer phases. Furthermore, activity effects may contribute to the observed (2) non-linearity of concentration-based isotherms for mobile species partition into the polymer film.

Kinetic effects may be superimposed on the thermodynamic requirements. At first sight, slow solution / polymer exchange is expected to be deleterious, since it lowers sensitivity (2),(3),(9). However, we demonstrate here the principles of a method by which differences in mobile species transfer rates may be turned to advantage.

Selectivity is sought by appropriate choice of the polymer, based on established chemical principles. For ion-exchange, there is a crude selectivity based on charge-type (9), provided the film is permselective (see below). Additionally, some selectivity between ions of similar charge type may be based on charge number. For complexation, the use of selective ligands is common in analytical chemistry. In practice, more than one type of interaction may contribute to the selectivity pattern (2)(4).

The Electrochemical Quartz Crystal Microbalance (EQCM) The resonant frequency of a quartz crystal oscillator is perturbed from its base value (f_0) by attached overlayers. For thin, rigid films the measured change in resonant frequency (Δf) with attached mass (ΔM) is described by the Sauerbrey equation (10):

$$(\Delta f/Hz) = -(2/\rho v)f_0^2(\Delta M/g\ cm^{-2}) \qquad [1]$$

where ρ is the density of the quartz and v is the wave velocity. The quartz crystal microbalance (QCM) technique has been used for many years (11) to monitor the deposition of materials (e.g. metals) from the vapour phase onto solid substrates. More recently, it was shown that the crystal continues to oscillate when one face is exposed to a liquid (12),(13). In the EQCM, the electrode exposed to the solution is used as the working electrode in an electrochemical cell. The capabilities of the EQCM have been reviewed recently (14),(15),(16). Of primary concern here is the ability to monitor the exchange of mobile species between a polymer film and its bathing electrolyte (17), (18), (19), (20), (21), (22), (23), (24), (25), (26), (27), (28), (29).

The EQCM can detect overlayer formation, dissolution or composition changes. When these can be related to the composition of the ambient medium, this provides the basis of a sensor. Key advantages are the generality of detection by mass and the high sensitivity of the EQCM: *in situ* mass changes of 1 ng cm^{-2} can routinely be detected. Several applications of the QCM to sensing molecules of biological importance have been described. The adsorption of cholesterol in thin films on a QCM has been used as a model for biological chemoreception in lipid bilayers (30), and the question of partition coefficient considered. Chemical amplification of the QCM sensitivity has been neatly demonstrated using antibody reactions (6). Another interesting example involved glucose detection via binding to hexokinase in a polyacrylamide film on a QCM (31). Here the sensitivity was markedly greater than predicted on the basis of equation [1], a phenomenon ascribed to changes in polymer rheological properties consequent upon uptake of the target species.

In this paper we discuss three issues related to our ability to exploit the undoubted attractions of the EQCM technique: (a) the extent of mobile species uptake as a function of solution concentration; (b) the use of transient measurements to obtain (additional) selectivity and (c) the need to establish that the criteria are satisfied for the Sauerbrey equation (equation [1]) to be used to convert measured frequency changes to mass changes. Of these, (a) and (c) have been demonstrated (see previous paragraph) to be directly relevant to QCM-based biosensors. The concept of using transient measurements in this context has not yet been explored, but is a natural development.

Experimental

The instrumentation (24)-(26) and the general technique (12) have been described previously. Polythionine (PTh) films were deposited by the electrochemical polymerisation (at 1.1V) of thionine from aqueous solutions of 0.05 mol dm^{-3} HClO$_4$ (24). Polyvinylferrocene (PVF) films were electrodeposited (at 0.7V) from solutions of PVF in CH$_2$Cl$_2$ / 0.1 mol dm^{-3} tetrabutylammonium perchlorate (TBAP) (30). Polybithiophene (PBT) films were deposited by the electrochemical polymerisation (at 1.225V) of 2,2'-bithiophene in CH$_3$CN / 0.1 mol dm^{-3} tetraethylammonium tetrafluoroborate (TEAT) (26). After deposition, all modified electrodes were transferred to monomer/polymer-free solutions (see figure legends for composition) for characterisation. The electrodes (area 0.23 cm^2) on the quartz crystals were Au for PTh and PVF, and Pt for PBT. Polymer coverages (reported in terms of moles of electroactive sites, Γ/mol cm^{-2}) were determined by integration of slow scan rate (< 5 mV s^{-1}) voltammetric current responses. For different films, these were typically in the ranges 3-10 nmol cm^{-2} (PTh), 5-12 nmol cm^{-2} (PVF), and 16-38 nmol cm^{-2} (PBT).

Experimental values of mass changes, ΔM / ng cm^{-2}, are normalised by multiplying by the Faraday and dividing by the charge passed, Q / μC cm^{-2}. The resulting "normalised mass change", $\Delta MF/Q$, then corresponds to the mass change associated with redox switching of one mole of redox sites. This facilitates comparison of data for different films (different polymer coverages).

Results and Discussion

Thermodynamics of mobile species uptake. Operationally, one seeks a linear, or at least single valued, relationship between the film mass change and analyte composition. In this section we stress the importance of characterising the mass change / composition relationship, and illustrate circumstances under which the desired behaviour will *not* prevail.

Activity effects. The exchange of trace ions in solution with others in the polymer film might, simplistically, be expected to lead to a linear uptake/solution concentration relationship. Unfortunately, this is seldom the case. The thermodynamic restraint is that of electrochemical potential. Thus electroneutrality is not the sole constraint on the ion exchange process. A second thermodynamic requirement is that the activity of mobile species in the polymer and solution phases be equal. (Temporal satisfaction of these two constraints is discussed below, with reference to Figure 4.) The rather unusual, high concentration environment in the polymer film can lead to significant - and unanticipated - activity effects (8).

This has been demonstrated in EQCM studies of PTh film redox switching in $HClO_4$ solutions of different concentration. Under conditions where thionine reduction is a $2e/3H^+$ process, electroneutrality alone predicts uptake of one anion (and three protons) per Th site: a film mass increase of 102.5 g molTh^{-1}, *independent of solution composition*. Experimentally, the mass change is less than 20 g mol^{-1} in 1 mol dm^{-3} $HClO_4$, decreases as the electrolyte is diluted, and even becomes negative at pH > 2! The variation of mass change with concentration is attributable to activity effects. Hydronium perchlorate is included within the film, to an extent dependent on polymer redox state and solution concentration.

Co-ordination effects. Analogous experiments in weak acid solutions (this time at constant total acid concentration, varying pH by additions of base) yield quite different results, as illustrated in Figure 1. The "simple" prediction of electroneutrality is a mass gain of 62 g mol^{-1} ($CH_3CO_2^- + 3H^+$) independent of solution composition. The experimental result is a mass *loss*: 16 g mol^{-1}, independent of anion (acetate) concentration, for pH < 5, decreasing somewhat in magnitude at higher pH.

Gas phase experiments, involving exposure of PTh films to dry N_2, water saturated N_2 and CH_3CO_2H saturated N_2 show reversible uptake / loss of water, but irreversible uptake of acetic acid. The strength of the interaction between the undissociated acid and the polymer is the key to the behaviour in Figure 1. At low pH, one (or more) co-ordination sites on each oxidised thionine redox site are occupied by an acetic acid molecule. Upon polymer redox site switching, one HA per redox site dissociates, totally satisfying the demand for counter ion, and partially satisfying the demand for protons. In short, there is no counter ion ingress, because it was present in the film all the time. The remaining two protons required by the reaction are supplied from solution. When their entry is taken into account, the net mass loss of 16 g mol^{-1} indicates loss of a species of mass 18. Water is the obvious candidate. We therefore propose that the half-reaction in low pH acetic acid solutions is:

$$[(TH^+A^-) \cdot (H_2O) \cdot X \cdot (HA)]_p + 2e + 2H_3O^+_s =$$

$$[(TH_4^{2+}(A^-)_2) \cdot X]_p + 3H_2O_s \qquad [2]$$

where X may be either acetic acid or water, and the number of X's is determined by the maximum coordination number.

When the solution pH approaches the pK_a for acetic acid, the supply of HA within the solution, and thus partitioned into the film, decreases. Consequently, the counter ion must increasingly be supplied from solution, and the mass change moves in a positive direction.

The key result in terms of a sensor is that specific interactions, as sought for biochemical sensors (6), may be sufficiently strong that a coordination-type model applies. Note that this does not contradict the activity arguments of the previous section, but is a special case within the general thermodynamic framework. Under these special circumstances, the polymer will be "saturated" with the target species, and film composition will *not* depend on solution concentration, except at a very low level.

Permselectivity: real and apparent. Permselectivity is a key issue when the extraction process is based on ion-exchange (31). A straightforward EQCM illustration is provided by the behaviour of PVF films upon redox switching in aqueous $NaClO_4$ solutions (22). At electrolyte concentrations below 1 mol dm^{-3}, the normalised mass change is independent of composition, and indicates ion and solvent entry upon oxidation of neutral PVF. At higher electrolyte concentrations, permselectivity fails and salt also enters the film upon oxidation.

The interpretation of the behaviour of PBT is more subtle. Overall mass changes upon total PBT oxidation / reduction are similar to the counter ion ("dopant") molar mass, for example $F\Delta M/Q = 93$ g mol^{-1} in 0.01-0.1 mol dm^{-3} $Et_4N^+BF_4^-/CH_3CN$ compared to $m_{BF_4^-} = 87$ g mol^{-1}. These results *apparently* imply permselectivity with little or no solvent transfer at low electrolyte concentration, and permselectivity failure at high electrolyte concentration. As we show in the next section, this apparent permselectivity is entirely fortuitous, and results from a compensating combination of mobile species transfers. The message here is that a combination of thermodynamic and kinetic data is required to unequivocally attribute the mass change to the relevant species' transfers.

Kinetics of mobile species uptake. In the previous section, we considered the overall mass changes associated with complete oxidation / reduction of a film, which was then allowed to come to equilibrium with the solution phase. In this section we consider the time course of the transformation, i.e. the transfer of species *during* the oxidation / reduction. As is common in kinetic studies, it is often more convenient to consider fluxes, rather than populations. For this purpose, we will consider current, i, and the time differential of the mass, $\overset{\bullet}{M}$, as well as charge, Q, and mass change, ΔM.

In analyzing kinetic EQCM data, we have found it convenient (32)(33) to define a function $\overset{\bullet}{\Phi}$:

$$\overset{\bullet}{\Phi}_j = \overset{\bullet}{M} + i\,(m_j/z_jF) \qquad\qquad [3]$$

$\overset{\bullet}{\Phi}_j$ is defined for a particular ion j, of molar mass m_j and charge z_j (including sign). It is a weighted sum of the mass and charge fluxes, with units of g cm^{-2} s^{-1}. The definition of $\overset{\bullet}{\Phi}_j$ is such that the contribution of the j-th ion's mass is zero. We are free to choose j as any of the ions present, enabling us to systematically eliminate the contribution of each ion to the mass flux. This mathematical device is related to selectivity, which requires determination of changes in *individual* species populations in the film.

A more general discussion of the significance of $\overset{\bullet}{\Phi}_j$ and the analogous integrated function Φ_j (/g cm^{-2}), defined in terms of the mass change and charge, is presented elsewhere (32)(33). Here we make two observations on the nature of $\overset{\bullet}{\Phi}_j$ before exploiting it to interpret kinetic data.

Firstly, we recognise that mobile species can be divided into charged and net neutral species. The net neutrals contribute to the mass, but not charge, response. The Φ function provides a means of separating ion and neutral species transfers. In the case of a film immersed in a single electrolyte, Φ_j ($\overset{\bullet}{\Phi}_j$) represents the population change (flux) of neutral species (salt and/or solvent).

Secondly, if a given j-th ion is the only mobile species (permselectivity without solvent transfer), Φ_j is zero throughout the redox transformation. Furthermore, $\overset{\bullet}{\Phi}_j$ must also be zero throughout the transformation. The latter is a much more stringent test, since it can detect compensatory transfers of species with different mobility.

Apparent permselectivity and compensatory motion.   Although normalised mass change data for the "doping/undoping" of PBT films were very similar to those predicted by permselectivity in the absence of solvent transfer (cf. electroneutrality), the differences were real.   Furthermore, systematic variation of the anion or cation or deuteration of the solvent produced consistent trends in the departure from "simple" behaviour. Insight into the overall processes involved (the thermodynamics) is gained by considering the kinetics of mobile species transfer.

One way in which this can be done is by considering the shape of the ΔM vs. E and ΔM vs. Q curves, rather than the overall changes in ΔM and Q. For a permselective film, electroneutrality requires 1:1 correlation between the electron and counter ion (here, anion) fluxes. If no solvent is transferred, this demands that the ΔM vs. Q plot be linear and free of hysteresis, *regardless of the shape of the ΔM vs. E plot*. The slope of this plot will be F/m_{A^-}, where m_{A^-} is the counter ion molar mass.

Here we illustrate the utility of the $\overset{\bullet}{\Phi}_j$ function defined in the previous section. By "correcting off" the counter ion contribution to the mass flux (see the form of equation [3]), neutral species fluxes are highlighted: they are the departure of $\overset{\bullet}{\Phi}_j$ from zero.

Experimental mass changes during PBT doping/undoping do <u>not</u> conform to the $\overset{\bullet}{\Phi}_j = 0$ requirement. Figure 2 contains representative data for tetraethylammonium hexafluorophosphate (TEAPF) as the electrolyte. In accord with the general activity constraint (see above), solvent and salt do transfer. For PBT, these net neutral species transfers are in opposite directions.

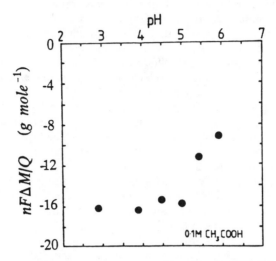

Figure 1: Normalised mass change for a PTh film in CH_3CO_2H solutions as a function of pH at fixed total acetate concentration (0.1 mol dm^{-3}). (Reproduced from ref. 25. Copyright 1990 American Chemical Society.)

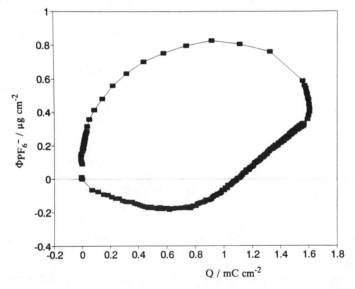

Figure 2: Chronoamperometric and chronogravimetric data for a PBT film immersed in 0.1 mol dm^{-3} TEAPF. The potential was stepped from 0V to 1.05V held at 1.05V for 15s, then stepped back to 0V. Values of $\Phi_{PF_6^-}$ were calculated according to equation [3]. Charge data are referred to the initial condition at 0V.

The magnitudes of these opposing component mass changes can be rather similar. This fortuitous balance gives the illusion of permselectivity if one considers only the overall mass change. Note that the data of Figure 2 suggest salt exit is faster than solvent entry (during oxidation) and that salt entry is faster than solvent exit (during reduction). In fact, this may indicate (partial) maintenance of electroneutrality by co-ion on short time scales following a potential step.

The key results of this section are twofold. First, kinetic data can reveal otherwise hidden thermodynamic information. Second, time can be used as a variable to probe the transfers of individual species, even if several are transferred. This latter aspect is explored in the following section.

Kinetic permselectivity. When permselectivity is not achieved in a thermodynamic sense (for example at high electrolyte concentration), we propose that it may be possible to achieve it kinetically. We aim to exploit the differing rates of mobile species transfer. In a transient experiment the response on a short time scale will be dominated by the fastest moving species. The converse will be true on a long time scale.

We suggest that this approach might be exploited at two levels. Firstly, field assistance of ion transfers (migration) will lead to their being more rapid than neutral species transfers. Secondly, size effects (for ions or neutral species) will lead to a diversity of transport rates. These effects are likely to be more pronounced in the confined geometry of polymer films than for the same species in solution. The extent to which transfer of a given species dominates the net transfer process (on a given time scale) will depend on its availability, i.e. solution concentration.

We first illustrated this type of effect during rapid scan voltammetry of PVF films in concentrated $NaClO_4$ solutions, where the overall redox switching process involves ingress of counter ion, salt and solvent upon oxidation (34). Quantitative treatment of such effects is better explored using a potential step, i.e. chronoamperometry. EQCM data from such an experiment are shown in Figure 3. For comparison purposes, data for analogous experiments in 0.1 and 3 mol dm^{-3} electrolyte, where the polymer is / is not permselective, are superimposed.

The data in the first frame are the (normalised) current, representing the electron flux. In the second frame, the overall mass flux (transfer of charge-compensating ions as well as neutral species) is shown. The third frame corresponds to the difference between the first two. The quantity $\Phi_{ClO_4^-}$, defined by equation [3] with $m_j = 99.5$ g mol^{-1} and $z_j = -1$, represents the flux of all species other than perchlorate. In the case of the dilute (concentrated) electrolyte, this corresponds to the mass flux of water (water+salt).

The presence of kinetic permselectivity is demonstrated by comparison of the mass and charge fluxes, in Figure 3. In each case (either electrolyte, either direction of change), the initial slope of the mass flux / current plot corresponds closely to that anticipated for transfer of one counter ion (no salt or solvent) per electron transferred (dashed lines have slope F/99.5). Transfers of the net neutral species, salt and solvent, are purely diffusive and only contribute significantly to the EQCM response at longer times.

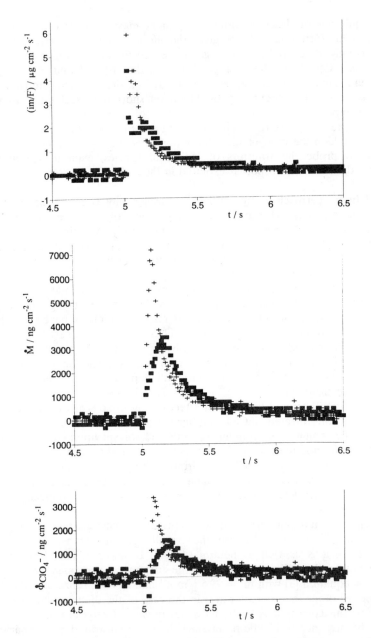

Figure 3: Chronoamperometric data (first frame) and chronogravimetric data (second frame) for a PVF film immersed in 0.1 (■) and 3.0 (+) mol dm^{-3} NaClO$_4$. The potential was stepped from 0V to 0.7V at t=5s. Values of $\Phi_{ClO_4^-}$ (third frame) were calculated according to equation [3]. (Adapted from reference (33), with permission.)

An extreme case of differing transfer rates is anticipated for proton transfer in aqueous media. This is neatly illustrated by EQCM data for the oxidation of PTh in strong acid media at pH2 (see above). Here the overall mass change is zero, as a result of compensatory motion of ions, acid and solvent. However, during a transient experiment, protons will dominate electroneutrality maintenance on short time scales, while anions (ClO_4^-) will have the opportunity to participate on longer time scales. The difference in tranport rates of the mobile species is graphically demonstrated (35) by the transient mass excursion: a rapid mass increase, followed by a slower decrease as the slower-moving species transfer to satisfy the ultimate requirements of thermodynamics.

Non-rigidity. All the foregoing EQCM data analyses employed equation [1] to convert measured frequency changes (Δf) to mass changes (ΔM). This procedure is appropriate for rigid films. If the polymer is extensively solvent-swollen, this requirement may not be satisfied (36), in which case the simple analytical utility of the EQCM technique is invalidated. The importance of establishing rigidity *in the medium and under the conditions employed* cannot therefore be overemphasised. We now present data for ferrocene-based polymer films, showing how rigidity may be a function of both polymer and solution compositions.

Exposure to a liquid of one of the electrodes (bare Au or Pt) of the EQCM results in the establishment of a liquid modulation layer. The thickness of this layer is (12)

$$x_L = (\nu_L/f_0)^{\frac{1}{2}} \qquad [4]$$

where ν_L is the kinematic viscosity of the liquid. For 10 MHz AT-cut crystals exposed to aqueous solutions, x_L is of the order of 300nm. Effectively, the EQCM "weighs" this layer of liquid, and the resonant frequency of the crystal decreases, to an extent dependent on electrolyte composition. (Typical frequency changes in aqueous media are 6-7kHz, although additional contributions due to entrapped liquid in surface roughness may raise this number by several kilohertz.) Frequency changes due to the deposition of a rigid film are then simply additive. In the event that the film is non-rigid, this is no longer the case.

The key question is "How does one establish film rigidity?" One approach involves analysis of the shape, as well as position, of the resonance. Essentially, one determines the Q-factor for the crystal: broadening signals non-rigidity. It is important to appreciate that immersion of the crystal in the liquid significantly broadens the resonance. Small changes in Q-factor associated with incomplete polymer film rigidity may therefore be difficult to detect, and certainly complex to quantify.

An alternative approach is to compare the frequency data with electrochemical (coulometric) data. This is exemplified for PVF and a co-polymer with vinylpyrrolidone, poly(vinylferrocene-co-vinylpyrrolidone) (20:80 PVF-co-PVP). Table I shows *in* and *ex situ* frequency data for an electrode before and after coating with 20:80 PVF-co-PVP.

Table I. EQCM data for 20:80 PVF-co-PVP in various environments

Experiment	Δf /Hz [a]	$(\Delta f_0 - \Delta f)$ /Hz [b]	ΔM_{app} /μg [c]
Uncoated, dry crystal in air	-1050		
Uncoated crystal in deposition solution	-8025	-6975	7.673
Coated crystal, oxidised film in deposition solution	-11,234	-10,184	11.202
Coated crystal, oxidised film, in air	-14,367	-13,317	14.649

a   Frequency difference vs. reference crystal.
b Δf_0 refers to uncoated, dry crystal.
c   Apparent mass change, assuming rigidity (equation [1]). Entries 2 - 4 refer to liquid modulation layer, liquid modulation layer+polymer and polymer, respectively.

The data of column 4 show the apparently paradoxical result that the mass of polymer alone (measured in air) is greater than that of the polymer plus liquid modulation layer. This is clear evidence that the oxidized PVF-co-PVP film is non-rigid in CH_2Cl_2. For this particular film, the dry mass of oxidised copolymer (including counter ion required by electroneutrality) corresponds to deposition of 29 nmol of ferrocene sites. The deposition process involved passage of 3.32 mC of charge, i.e. 34 nmol of ferrocene sites were oxidized in total. This implies an (average) deposition efficiency for this experiment of 85%.

If one were (erroneously) to assume a rigid film and calculate the extent of deposition by simple subtraction of the liquid contribution from the liquid+polymer data, the result would be ca. 9 nmol of ferrocene sites, i.e. factor of 3-4 in error.

For comparison, analogous experiments with the PVF homopolymer showed that Sauerbrey equation-based calculations of coverage from the limiting frequency change during deposition underestimated the coverage by a factor of about two (30). Comparison of coverage estimates based on dry polymer coated electrode frequencies and frequency changes associated with redox cycling in water (ion and solvent exchange) are in excellent ($\pm 10\%$) agreement. This indicates rigidity in aqueous media and air, but not in CH_2Cl_2.

Conclusions

Exchange of species between a solution and a polymer film is an established means of probing solution composition. The quartz crystal microbalance can monitor such exchange processes with high sensitivity. When combined with selectivity via electrochemical control and appropriate choice of polymer, the EQCM becomes an attractive sensor. In order that the potential advantages of the EQCM can be realised, certain criteria must be met.

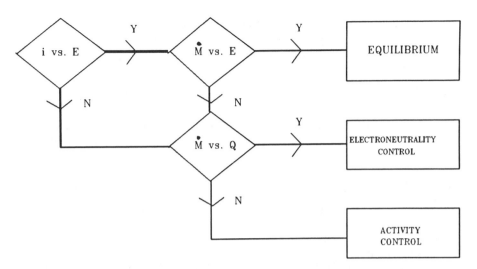

Figure 4: Flow chart for using EQCM data to distinguish electroneutrality and activity control under transient conditions from each other and from equilibrium. Absolute values of time-dependent quantities (i and $\dot{M}$) are employed.

Firstly, when relating (gravimetrically determined) equilibrium film composition to solution composition, activity effects must be taken into account. These effects may cause solvent and other neutral molecules, as well as the target species, to enter/leave the film. The importance of medium effects cannot be overemphasised here. A special case is a co-ordination model, where favourable interaction between polymer and target species results in saturation of the film except at very low concentration. The film mass is then independent of solution composition. This situation is likely for systems where a strong, specific polymer/analyte interaction has been synthetically designed into the polymer.

Secondly, selectivity is not always achievable. For example, permselectivity of ion-exchanging polymer films fails at high electrolyte concentration. We have shown that even if permselectivity is not *thermodynamically* found, measurements on appropriate time scales in transient experiments can lead to *kinetic* permselectivity. To rationalise this behaviour we recall that the thermodynamic restraint, electrochemical potential, can be split into two components: the electrical and chemical terms. These conditions may be satisfied on different time scales. Dependent on the relative transfer rates of ions and net neutral species, transient responses may be under electroneutrality or activity control.

Figure 4 shows how these cases may be distinguished on the basis of EQCM data. (The equivalent plot for integrated data is obtained by replacing i with Q and $\dot{M}$ with ΔM in the ordinates of the plots.) In each decision box, one asks whether the designated plot (for a complete redox cycle) is hysteresis-free. We commonly find that both i vs. E and $\dot{M}$ vs. Q plots show hysteresis, indicating activity control. We suggest that neutral species transfers are slower than ion transfers, since the latter can be enhanced by migration.

Finally, we rely on the Sauerbrey equation for the conversion of frequency data to mass data (and thus solution composition). This procedure is only valid for rigid polymer films. We therefore regard establishment of rigidity as vital.

Acknowledgments

We thank the SERC (GR/E/32946), NATO (86/0830) and the Air Force Office of Scientific Research (87-0037) for financial support. MJS thanks the SERC for a research studentship.

References

1.   Murray, R.W. In *Electroanalytical Chemistry*; Bard, A.J., Ed.; Marcel Dekker: New York, NY 1984, Vol. 13; p.192.
2.   Szentirmay, M.R.; Martin, C.R. *Anal. Chem.* **1984**, *56*, 1898.
3.   Guadalupe, A.R.; Abruna, H.D. *Anal. Chem.*, **1985**, *57*, 142.
4.   Nagy, G; Gerhardt, G.A.; Oke, A.F.; Rice, M.E.; Adams, R.N.; Moore, R.B.; Szentirmay, M.N.; Martin, C.R. *J. Electroanal. Chem.* **1985**, *188*, 85.
5.   Wier, L.M.; Guadalupe, A.R.; Abruna, H.D. *Anal. Chem.* **1985**, *57*, 2009.
6.   Ebersole, R.C.; Ward, M.D. *J. Am. Chem. Soc.*, **1988**, *110*, 8623.
7.   Bruce, J.A.; Wrighton, M.S. *J. Am. Chem. Soc.* **1982**, *104*, 74.
8.   Bruckenstein, S.; Hillman, A.R. *J. Phys. Chem.*, **1988**, *92*, 4837.
9.   Cox, J.A.; Kulesza, P.J. *Anal. Chim. Acta*, **1983**, *154*, 71.
10.   Sauerbrey, G.Z. *Z. Phys.*, **1959**, *155*, 206.
11.   *Applications of Piezoelectric Crystal Microbalances*; Lu, C.; Czanderna, A.W., Eds.; *Methods and Phenomena*; Elsevier: New York, NY, 1984; Vol.7.
12.   Bruckenstein, S; Shay, M. *Electrochim. Acta*, **1985**, *30*, 1295.
13.   Kanazawa, K.K.; Gordon, J.G. *Anal. Chim. Acta*, **1985**, *175*, 99.
14.   Schumacher, R. *Angew. Chemie (Int. Ed. English)* **1990**, *29*, 329.
15.   Deakin, M.R.; Buttry, D.A. *Anal. Chem.* **1989**, *61*, 1147A.
16.   Buttry, D.A. In *"Applications of the Quartz Crystal Microbalance to Electrochemistry"*; Bard, A.J., Ed.; *"Electroanalytical Chemistry"*; Marcel Dekker Inc.: New York, NY, 1991, Vol. 17; pp.1-85.
17.   Kaufman, J.H.; Kanazawa, K.K.; Street, G.B. *Phys. Rev. Lett.* **1984**, *53*, 2461.
18.   Orata, D.; Buttry, D.A. *J. Am. Chem. Soc.* **1987**, *109*, 3574.
19.   Baker, C.K.; Reynolds, J.R. *J.Electroanal.Chem.* **1988**, *251*, 307.
20.   Reynolds, J.R.; Sundaresan, N.S.; Pomerantz, M.; Basak, S.; Baker, C.K. *J.Electroanal.Chem.* **1988**, *250*, 355.
21.   Baker, C.K.; Reynolds, J.R. *Synth. Met.* **1989**, *28*, C21.
22.   Hillman, A.R.; Loveday, D.C.; Bruckenstein, S. *J. Electroanal. Chem.*, **1989**, *274*, 157.
23.   Hillman, A.R.; Loveday, D.C.; Swann, M.J.; Eales, R.M.; Hamnett, A.; Higgins, S.J.; Bruckenstein, S.; Wilde, C.P. *Faraday Disc. Chem. Soc.* **1989**, *88*, 151.

24. Bruckenstein, S.; Wilde, C.P.; Shay, M.; Hillman, A.R. *J. Phys. Chem.*, **1990**, *94*, 787.
25. Bruckenstein, S.; Wilde, C.P.; Hillman, A.R. *J. Phys. Chem.* **1990**, *94*, 6458.
26. Hillman, A.R.; Swann, M.J.; Bruckenstein, S. *J. Electroanal. Chem.*, **1990**, *291*, 147.
27. Kelly, A.J.; Ohsaka, T.; Oyama, N.; Forster, R.J.; Vos, J.G. *J. Electroanal. Chem.* **1990**, *287*, 185.
28. Inzelt, G. *J. Electroanal. Chem.* **1990**, *287*, 171.
29. Naoi, K.; Lien, M.; Smyrl, W.H. *J. Electrochem. Soc.* **1991**, *138*, 440.
30. Hillman, A.R.; Loveday, D.C.; Bruckenstein, S. *Langmuir*, **1991**, *7*, 191.
31. Buck, R.P. *Electrochemistry of Ion-Selective Electrodes* In *Comprehensive Treatise of Electrochemistry*; White, R.E.; Bockris, J.O.; Conway, B.E.; Yeager, E., Eds.; Plenum Pub. Co.: New York, NY, 1984; Vol.8; pp 137-248.
32. Hillman, A.R.; Swann, M.J.; Bruckenstein, S. *J. Phys. Chem.* **1991**, *95*, 3271.
33. Hillman, A.R.; Loveday, D.C.; Bruckenstein, S. *J. Electroanal. Chem.* **1991**, *300*, 67.
34. Hillman, A.R.; Loveday, D.C.; Bruckenstein, S.; Wilde, C.P. *J.C.S. Faraday Transactions*, **1990**, *86*, 437.
35. Bruckenstein, S.; Hillman, A.R.; Swann, M.J. *J. Electrochem. Soc.*, **1990**, *137*, 1323.
36. Borjas, R.; Buttry, D.A. *J. Electroanal. Chem.*, **1990**, *280*, 73.

RECEIVED December 10, 1991

Chapter 13

Electrochemically Prepared Polyelectrolyte Complex of Polypyrrole and a Flavin-Containing Polyanion

Use as a Biosensor

H. F. M. Schoo and G. Challa

Laboratory of Polymer Chemistry, University of Groningen, Nijenborgh 16, 9747 AG Groningen, Netherlands

Pyrrole was polymerized electrochemically in an aqueous medium on a platinum electrode in the presence of a polyanion, containing covalently bound flavin units. Depending on the applied potential the polypyrrole-films thus formed were powdery and non-adherent (V < 700 mV), smooth and adherent (700 mV < V < 1000 mV) or brittle and uneven (V > 1000 mV). Besides the applied potential, medium pH and added low molar mass salt influenced the morphology and composition of the PPy-layer. The flavin-containing polyanion was incorporated in the film as a dopant. Depending on the medium pH and amount of low molar mass salt added, the film formed by this method contained various amounts of polymer-bound flavin. Cyclic voltammograms confirmed the presence of electrochemically active flavin in the layer. The oxidation of 1-benzyl-1,4-dihydronicotinamide (BNAH) was used as a model reaction to test the catalytic activity of the immobilized flavin. The modified electrode showed fast current response on addition of BNAH, not only at an applied potential of 0.9 V (oxidation of H_2O_2) but also in the absence of oxygen at 0.3 V. This might indicate that a direct transfer of electrons takes place between the flavin units and the (PPy-)electrode. Cycling of the PPy-film through its reduced and oxidized state does not lead to any loss of the flavin containing polyanion. In the case of low molar mass flavin-containing dopants most of the flavin is released from the film upon reduction of the PPy.

One of the main motivations for the development of chemically modified electrodes is the introduction of (electro-) catalytic species onto the electrode surface. These modified electrodes can be used to improve specificity and product yields in electrochemical synthesis or as the basis for a biosensor. Catalysts can be attached through (irreversible) adsorption onto a suitable substrate (1). These systems mostly consist of an electrode covered with a monolayer of the electroactive species. In many cases this method does not lead to effective systems, due to instability of the monolayer or low loading with the catalyst. Direct modification of the electrode surface by covalent

0097–6156/92/0487–0164$06.00/0

binding of active moieties is also possible, but it often has drawbacks such as laborious synthesis and poor stability of the modified electrode.

A more favourable approach is the incorporation of the active species in an electrically conducting polymer layer which then acts as an (electrical) intermediate between the electrode surface and the catalyst. Polypyrrole is considered to be especially suitable because it is acceptably stable under ambient conditions (2), has a high conductivity and can be easily prepared electrochemically from a great variety of solvent systems, including aqueous solutions (3-5). The catalytic species that have been applied in such polypyrrole-based systems comprise metal particles (6-9), metal chelates (10-13) (with anionic side groups) and enzymes (14-18).

When redox enzymes are employed, direct electrical communication between the redox centres and the electrode is often inhibited by the insulating protein shell, surrounding the active centre of the enzyme. In this case low molar mass redox couples may be employed (19,20) as mediators, but unless they are covalently bound (21) (to the enzyme) practical applications will be strongly limited due to release of the mediator into the surrounding medium, with subsequent loss of responsivity of the electrode system. In many cases the redox centre of the applied enzymes is flavin. Some efforts have been made to obtain a modified electrode, containing immobilized flavine units (22-25), but in all cases the catalyst was bound covalently to the electrode surface. Another interesting point is the possibility to incorporate polyanions as dopants in polypyrrole films (26-28), which are very effectively immobilized in the conducting polymer film. Here we describe the use of a catalytically active polyanionic dopant, containing covalently bound flavin moieties, in the formation of a flavin-containing electrode and its application as a biosensor.

EXPERIMENTAL

Apparatus and materials. All electrochemical polymerizations, amperometric measurements and cyclic voltammetry were carried out with an EG&G Princeton Applied Research Potentiostat model 273. Pyrrole was purified by vacuum distillation and by passage over neutral alumina prior to electropolymerization. The polyanions (1) used as dopants are shown in Figure 1, and were purified twice by precipitation from methanol in 0.1 M HCl. Detailed information about their synthesis and catalytic

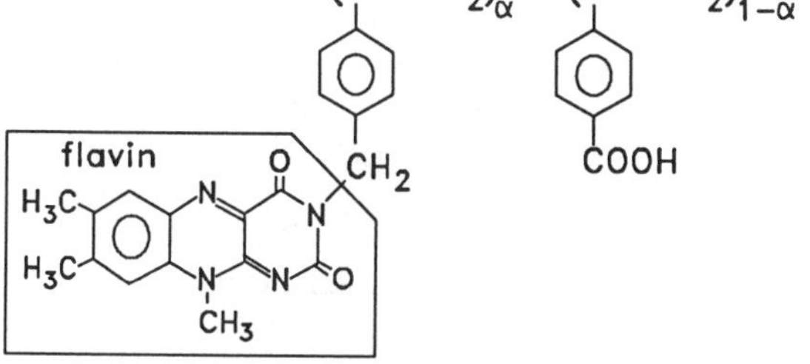

Figure 1. Flavin-containing polyanions 1 ($\alpha=0.10$)

properties will be published elsewhere (*29*). Sodium p-toluenesulphonate (NaOTs, Aldrich) was used without purification. 1-Benzyl-1,4-dihydronicotinamide (BNAH) was synthesized as described elsewhere (*30*), and purified twice by crystallization from EtOH/H$_2$O.

Procedures. Electropolymerizations were performed potentiostatically at +0.8 V (vs. Ag/AgCl) in an undivided cell with polished platinum electrodes and a saturated calomel reference electrode. For UV/VIS measurements on PPy-films Indium-Tinoxide (ITO) electrodes were used. For polymerization reactions aqueous (doubly distilled) solutions were thermostatted at 25.0°C and contained 0.1 M of pyrrole, 0.05 M COO(H)-groups (polymer **1**) and 0-0.1 M sodium-p-toluenesulphonate. Solution pH was adjusted prior to the reaction, and kept constant during the reaction by addition of appropriate amounts of concentrated solutions of HCl or NaOH. Prior to and during the polymerization nitrogen was bubbled through the stirred solution. PPy/(**1**)-films were rinsed with water/MeCN (5:1) mixtures and stored in doubly distilled water. Determination of the amount of polyanion **1** immobilized in the PPy-film, was performed indirectly by determining the amount of the polyanion left in the solution after the electropolymerization. The polyanion was recovered from the solution by complexation to an anion-exchanging macroporous polymer disk, containing quaternary ammonium groups (a detailed description of this procedure can be found elsewhere (*31*)). After washing with water/MeCN (5:1) (removal of pyrrole and soluble reaction products), the polyanion was decomplexed by flushing with acidic water/MeOH (1:2). The amount of polyanion could then be determined by means of UV/VIS-spectroscopy. Cyclic voltammetry and amperometric response experiments were performed in 0.05 M Tris buffer, Ph=8.0.

RESULTS AND DISCUSSION

Polymerization of pyrrole in aqueous medium. The polymerization of pyrrole was performed in aqueous medium. This imposes some restrictions upon the reaction conditions. The electropolymerization of pyrrole in the presence of polymer **1** in aqueous medium appeared to be very sensitive to the applied potential, especially when considering the effect upon the morphology of the PPy-film formed. Figure 2 shows typical current transient curves for the polymerization, starting with a clean, polished electrode. When potentials below 700 mV are applied, the resulting current is rather small and very brittle non-adherent films are formed, and in some cases only part of the electrode was covered with PPy. At potentials above 1000 mV transient currents initially are high, but decrease rapidly. The films formed are uneven and tend to lift off the electrode. If the potential is kept between these two values, smooth adhesive films were formed. In this case macromolecular assemblies are formed that might be described as polyelectrolyte complexes (*32*) (shown schematically in Figure 3).

Influence of polymerization conditions upon incorporation of flavin-containing polyanion. The amount of polyanion (**1**) incorporated as dopant in a PPy film during electropolymerization can easily be controlled by changing the ionic strength (low molar mass salt) and/or the pH of the monomer solution. The correlation between the amount of polymer-bound flavin incorporated in the film and the concentration of the added low molar mass salt, sodium-p-toluenesulphonate (NaOTs), at pH=7 is shown

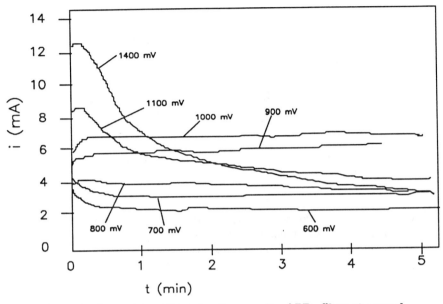

Figure 2. Current transients for the growth of PPy films at several (constant) potentials at pH=8.0

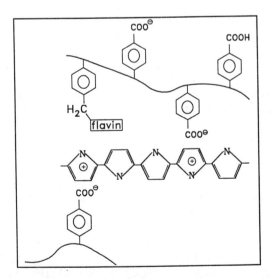

Figure 3. Schematic representation of the macromolecular complexes of PPy and polymer (1)

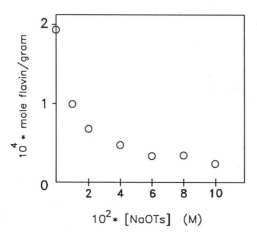

**Figure 4. Amount of incorporated flavin vs. the concen-
tration of low molar mass salt (sodium-p-toluenesulphonate) in the solution (pH=7)**

in Figure 4. It can be seen that the addition of a low molar mass dopant anion has a pronounced influence upon the amount of polymer-bound flavin immobilized in the PPy film. Assuming that the degree of doping of the PPy remains about the same, this decrease in uptake of polymer **1** must be compensated by NaOTs. One might expect that the polymeric dopant would be preferred over low molar mass anions due to cooperative effects, which are known to be of great importance in polyelectrolyte interactions (*33*). It seems, however, that the low molar mass dopant is incorporated much easier than its polymeric antagonist, since low concentrations of NaOTs (about 0.01 M; about one fifth of the concentration of polyanionic carboxyl groups) lead to a large decrease (factor 2) of polyanion incorporation. This might be due to kinetic restrictions during the polymerization and/or doping process, which are obviously more important for the polymeric dopant.

Another important variable which determines the amount of incorporated polymeric dopant is the medium pH. As can be seen in Figure 5 the amount of polymer-bound flavin incorporated in the PPy-film decreases with higher medium pH during electropolymerization. This effect can be explained in terms of the degree of ionization of the polyanion (**1**). Since (**1**) is a weak polyacid, at low pH only few of the carboxyl groups are dissociated, whereas at high pH nearly all groups are ionized. This means that more of the polymer has to be incorporated at lower pH in order to provide a sufficient amount of dopant (-COO⁻ groups). It is therefore not surprising that the addition of small amounts of NaOTs (0.01 M) at pH<7 practically inhibited the incorporation of polymeric dopant. Changing the medium pH also has a pronounced effect on the morphology of the film formed. Especially at higher pH the films became smoother and adhered stronger to the electrode. If the pH was increased further than the values shown in Figure 5, a reversed effect was found, however, probably due to side reactions of the monomer.

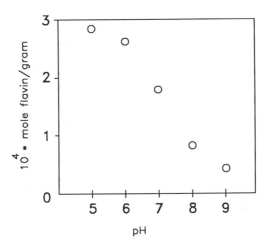

Figure 5. Amount of incorporated polymer-bound flavin vs. medium pH (No low molar mass dopant added)

The PPy-films doped with (**1**) were characterized by cyclic voltammetry (Figure 6) and UV/VIS-spectroscopy (Figure 7). The immobilized polymer-bound flavin moieties showed very good electrochemical activity with $E_{1/2}$= -0.496 V (vs SCE), which is a little more negative than values (-0.45 V) found in the literature for free flavins (*34*). If we compare the UV/VIS-spectra (on ITO-electrodes, Figure 7) of

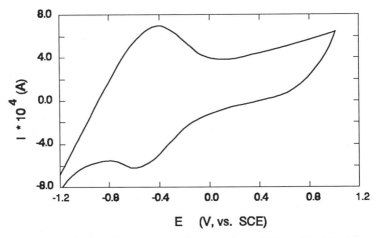

Figure 6. Cyclic Voltammogram of PPy/I-film, 50mV/s

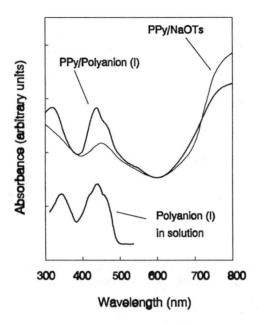

Figure 7. UV/VIS-spectra of PPy films doped with polyanion (1) or NaOTs (inset: aqueous solution of (1) at pH=8)

PPy-films doped with either NaOTs or polymer (**1**), they appear to be quite similar. Both show the typical absorbance profile of polypyrrole. The latter also shows two characteristic bands of the flavin units bound to polymer (**1**) (λ_{max}=438 and 335 nm).

The catalytic activity of the immobilized flavin was determined using the oxidation of an NADH-analog, namely 1-benzyl-1,4-dihydronicotinamide (BNAH), as a model reaction (Figure 8). If a potential of +0.9 V is applied to the system, hydrogen peroxide, which is formed in the aerobic oxidation of BNAH by flavin, can be oxidized

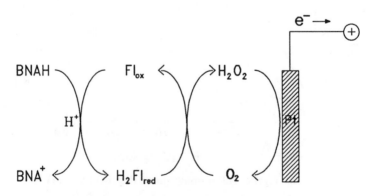

Figure 8. Model reaction cycle: oxidation BNAH

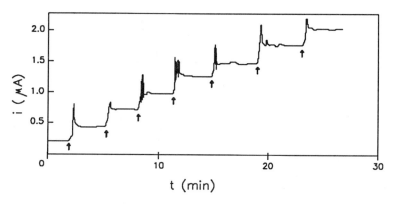

Figure 9. The response of a PPy/(1)-electrode to additions of BNAH to the bulk solution. E=0.9 V

at the electrode. The response of the electrode (after stabilization) to the addition of aliquots of BNAH is shown in Figure 9. We believe that both response time and sensitivity are satisfactory.

Quite remarkable is the fact that the electrode also showed a response to the addition of BNAH if no oxygen was present in the system. Response currents are lowered by a factor 2, but still were very well measurable. Even when the applied potential was decreased to +0.3 V (high enough to oxidize H_2Fl_{red} but too low to oxidize H_2O_2 on a bare electrode), a response was measured. This means that the flavin units are oxidized directly by the PPy-electrode, a process that is very unlikely to take place if enzymes are applied with the redox-active sites buried deep inside a protein shell. So, incorporation of catalytically active flavin-containing polyanions as dopants in polypyrrole can be utilized to synthesize a biosensor, in which no intermediate substances are required for connection of the flavin moieties to the electrode. One might envisage a system in which an NADH-producing enzyme would also be incorporated in the electrode system. In this case the activity of the enzyme could be monitored using the (electrochemical) oxidation of NADH via the polyanion-bound flavin moieties at relatively low potentials.

The thickness of the PPy-film has an effect on the response of the electrode towards the substrate, and can be controlled by the amount of charge passed through the system during synthesis of the conducting polymer. As can be seen in Table I, the response of the electrode to BNAH (slope of the calibration curve) increases with the thickness of the film up to about three Coulombs of charge passed. If thicker layers are deposited the response is only slightly lowered. This suggests that the transport of electrons from the (reduced) flavin to the electrode does not depend upon the diffusion of a reactive species (H_2O_2) to the platinum surface, which would limit the current as the film thickness is increased.

The incorporated flavin-containing polyanions are immobilized very efficiently in the polypyrrole film. Figure 10 shows the current response curve of PPy-films doped with polymer (1) or with a low molar mass species, 3-(p-methylbenzoic acid)-7,8,10-trimethylisoalloxazine (2) respectively, for a freshly prepared film and for a PPy-film

Table I. Charge passed during synthesis vs. response currents

Charge (C)	Slope of calibration curve ($\mu A\ dm^3\ mmol^{-1}$)
0	2
0.5	6
0.9	16
1.9	25
2.3	32
2.7	37
3.5	35
3.9	34
4.6	30
6.0	29
7.0	29

that has been cycled through its reduced and oxidized state three times in an aqueous (KCl) solution. In the case of the low molar mass flavin anion nearly complete loss of activity towards the substrate is found, showing that the anion is almost entirely lost from the PPy-film upon dedoping. This was confirmed by UV/VIS-measurements of

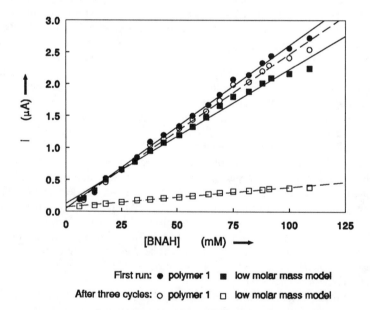

First run: ● polymer 1   ■ low molar mass model

After three cycles: ○ polymer 1   □ low molar mass model

Figure 10. Current as a function of the concentration of substrate (BNAH) for polymer dopant (1) and low molar mass dopant (2) for the first run and after three reduction-oxidation cycles

the surrounding solution. If a flavin-containing polyanion is used, hardly any loss of response is found. UV/VIS-spectra showed no flavin in the solution. Apparently the polymeric dopant cannot be ejected from the film, thus retaining the catalytically active moieties on the electrode. This concept might also be applied in other sensor systems in which presently a low molar mass mediator or cofactor is used. In these cases the loss of the (soluble) species to the bulk of the solution usually results in inactivation of the sensor. Attaching of these molecules to a polymeric carrier might lead to 'self-supporting' systems that are independent of the surrounding medium. This would considerably improve the practical applicability of such sensor systems.

CONCLUSIONS

We have shown that it is possible to incorporate polyanions with covalently bound flavin as dopant in a polypyrrole film during the electrochemical polymerization reaction of pyrrole. The amount of catalytically active polyanion present in the film depends on the composition of the reaction medium during polymerization. The effects of changing the medium pH and/or amount of added low molar mass dopant are described. The modified electrodes contain electrochemically active flavin, which is able to react with the model substrate (BNAH). Amperometric measurements showed linear response to BNAH up to 0.1 M. The electrode kept showing a measurable response if anaerobic conditions and/or lowered potentials were applied. This might be explained by assuming direct electron transfer from the reduced flavin moieties to the (polypyrrole) electrode. The polyanion-bound flavin moieties are immobilized very efficiently in the PPy-film. Even upon reduction of the polypyrrole no release of polymer-bound flavin could be detected. This concept of retaining an electrochemically active molecule on the surface of a modified electrode by binding it to a polymeric dopant might also prove to be very useful in other sensor systems in which a mediator is required.

Acknowledgements. We wish to express our gratitude to E.W. Meijer, E.E. Havinga and I. Rotte at the Philips Research Laboratory and to J.B Roedelof of the Inorganic Chemistry Department of the University of Groningen for their kind assistance during the setup of the electrochemical experiments.

REFERENCES

1. O. Miyawaki and L.B. Wingard, *Ann. N.Y. Acad. Sci.* **1984**, *434*, 520
2. A.F. Diaz and B. Hall, *IBM Jl Res. Dev.* **1983**, *27(4)*, 342
3. B. Zinger, *Synth. Met.* **1989**, *28(1-2)*, C37-C42
4. P. Huelser and F. Beck, *J. Appl. Electrochem.* **1990**, *20*, 596
5. S. Dong, J. Ding and R. Zhan, *J. Chem. Soc., Faraday Trans. 1* **1989**, *85*, 1599
6. S. Holdcroft and B. L. Funt, *J. Electroanal. Chem. Interfacial Electrochem.* **1988**, *240*, 89
7. F. Vork and E. Barendrecht, *Synth. Met.* **1989**, *28(1-2)*, C121-C126
8. F. T. A. Vork and E. Barendrecht, *Electrochim. Acta* **1990**, *35*, 135
9. L. Coche, B. Ehui, D. Limosin and J. C. Moutet, *J. Org. Chem.* **1990**, *55*, 5905

10.   F. Bedioui, P. Moisy, J. Devynck, L. Salmon and C. Bied-Charreton, *J. Mol. Catal.* **1989**, *56*, 267

11.   E. Barendrecht, A. Elzing, L. J. J. Janssen, A. Van der Putten, W. Visscher and F. Vork, *Makromol. Chem., Macromol. Symp.* **1987**, *8*, 211

12.   P. Moisy, F. Bedioui, J. Devynck, L. Salmon and C. Bied-Charreton, *New J. Chem.* **1989**, *13*, 511

13.   F. Bedioui, P. Bernard, P. Moisy, C. Bied-Charreton and J.Devynck, *Mater. Sci. Forum* **1989**, *42*, 221

14.   L. D. Couves and S. J. Porter, *Synth. Met.* **1989**, *28(1-2)*, C761-C768

15.   Y. Degani and A. Heller, *J. Am. Chem. Soc.* **1989**, *111*, 2357

16.   R. E. Holt and T. M. Cotton, *J. Am. Chem. Soc.* **1989**, *111*, 2815

17.   J. R. Li, M. Cai, T. F. Chen and L. Jiang, *Thin Solid Films* **1989**, *180*, 205

18.   W. Schuhmann, R. Lammert, B. Uhe and H. L. Schmidt, *Sens. Actuators* **1990**, *B1(1-6)*, 537-41

19.   A.E.G. Cass, G. Davis, G.D. Francis, H.A.O. Hill, W.J. Aston, I.J. Higgins, E.V. Plotkin, L.D.L. Scott and A.P.F. Turner, *Anal.Chem.* **1984**, *56*, 667

20.   A.L. Crumbliss, H.A.O. Hill and D.J. Page, *J.Electroanal.Chem.* **1986**, *206*, 327

21.   Y. Degani and A. Heller, *J. Am. Chem. Soc.* **1988**, *110*, 2615

22.   L.B. Wingard and J.L. Gurecka, *J.Mol.Catal* **1980**, *9*, 209

23.   O. Miyawaki and L. B. J. Wingard, *Ann. N. Y. Acad. Sci.* **1984**, *434*, 520

24.   L. B. J. Wingard, *Gov. Rep. Announce. Index (U.S.) 86(14)* **1986**, Abstr. No. 631,142

25.   C. N. Durfor, M. L. Bowers and B. A. Yenser, *U.S. Patent 4,797,181* (10 Jan. **1989**) to GTE Lab.Inc.

26.   T. Iyoda, A. Ohtani, T. Shimidzu and K. Honda, *Synth. Met.* **1987**, *18*, 747

27.   J. R. Reynolds, C. K. Baker and M. Gieselman, *Polym. Prepr.* **1989**, *30*, 151

28.   W. Janssen and F. Beck, *Polymer* **1989**, *30*, 353

29.   H.F.M. Schoo and G. Challa, submitted to *Macromolecules*

30.   D. Mauzerall and F.H. Westheimer, *J. Am. Chem. Soc.* **1955**, *77*, 2261

31.   H.F.M. Schoo, G. Challa, D.C. Sherrington and B. Rowatt, accepted for publication by Reactive Polymers

32.   A.S. Michaels, *Ind.Eng.Chem.* **1965**, *57*, 35

33.   E. Tsuchida and K. Abe in: "*Developments in ionic polymers 2*", A. Wilson and E. Prosser (Eds.), **1986**, chapter 5

34.   V.I. Birss, H. Elzanowska and R.A. Turner, *Can.J.Chem.* **1988**, *66*, 86

RECEIVED October 22, 1991

Chapter 14

Development of Polymer Membrane Anion-Selective Electrodes Based on Molecular Recognition Principles

Antonio Florido[1], Sylvia Daunert, and Leonidas G. Bachas[2]

Department of Chemistry and Center of Membrane Sciences, University of Kentucky, Lexington, KY 40506–0055

The incorporation of vitamin B_{12} derivatives into plasticized poly(vinyl chloride) membranes has resulted in the development of several ion-selective electrodes (ISEs). The response of the electrodes has been related to principles of molecular recognition chemistry. In addition, ISEs have been prepared by electropolymerization of a cobalt porphyrin. These electrodes have selectivity properties that are controlled by both the intrinsic selectivity of the metalloporphyrin and the characteristics of the polymer film (e.g., pore size).

Several of the polymer membrane anion-selective electrodes described in the literature use quaternary ammonium salts as ion carriers (ionophores) (1). These electrodes respond according to the Hofmeister series ($ClO_4^- > SCN^- > I^- > NO_3^- > Br^- \sim N_3^- > NO_2^- > Cl^- > HCO_3^- \sim$ acetate) (2, 3), which is the order of relative lipophilicity of the anions. Therefore, in strict terms, electrodes that respond according to this series could be considered "nonselective".

In order to develop selective electrodes, it is necessary to introduce specific interactions between the ionophore and the anion of interest. This can be achieved by designing an ion carrier whose structure is complementary to the anion. This type of design can be based on molecular recognition principles, such as the ones that involve complementarity of shape and charge distribution between the ion and the ionophore.

Molecular recognition can be controlled, among other factors, by the various types of interactions that take place between the ion and the ion carrier. These include charge, hydrogen bonding, and π-π electron interactions (4, 5). Another significant factor is the spatial arrangement of the recognition elements, and how that relates to the size and shape of the anion (4-6). Further, in ionophores that are metal complexes (e.g., metalloporphyrins, corrins, etc.) the ligand field stabilization energy plays also an important role. Finally, it should be noted that because the lipophilicity of the ion controls its partition from the aqueous sample phase into the

---

[1]Current address: Departament d'Enginyeria Química, Universitat Politècnica de Catalunya, 08028 Barcelona, Spain
[2]Corresponding author

0097–6156/92/0487–0175$06.00/0

polymer membrane, the overall selectivity displayed by an ISE is determined by both the selective interactions of ions with the ionophore and the intrinsic lipophilicity of the ions.

In this paper, we report the development of ISEs that have been designed by using molecular recognition principles. Specific examples include the development of polymer membrane anion-selective electrodes based on hydrophobic vitamin B_{12} derivatives and a cobalt porphyrin. The selectivity patterns observed with these electrodes can be related to differences in the structure of the various ionophores, and to properties of the polymer film.

Anion-Selective Electrodes Based on Hydrophobic Vitamin B_{12} Derivatives

Vitamin B_{12} is a corrin containing a cobalt(III) center that has a coordination number of six. The four equatorial coordination sites are occupied by the nitrogens of the corrin ring, and the two axial sites by a cyano group and the dimethylbenzimidazole ribonucleotide part (proximal base) of the corrin ring of the vitamin (Figure 1). The ion-exchange properties of this vitamin are governed by the coordination of the metal with the axial ligands (7-9).

ISEs Based on Cobyrinates. Vitamin B_{12} is hydrophilic and, therefore, it is necessary to modify it chemically in order to use it as ionophore in polymer membrane-based ISEs. Different hydrophobic derivatives that lack the nucleotide part of the vitamin (cobyrinates) have been prepared for this purpose. These compounds, although structurally similar to vitamin B_{12}, have a quite different coordination behavior from that of the vitamin (7).

A hydrophobic cobyrinate (Figure 2, structure 2) was used to prepare solvent polymeric membranes (10). The typical membrane composition was 1% (w/w) ionophore, 66% (w/w) plasticizer and 33% (w/w) polymer. Electrodes prepared with this ionophore, dioctyl sebacate (DOS) and poly(vinyl chloride) (PVC) presented, at pH 6.6, the selectivity pattern shown in Figure 3. The response of the electrodes was near-Nernstian for salicylate, thiocyanate, and nitrite. Their selectivity behavior clearly deviates from that of the Hofmeister series, with nitrite being the anion that presents the larger deviation.

In dicyanocobalt(III) a,b,c,d,e,g-hexamethyl-f-stearylamide cobyrinate (derivative 3) the six peripheral amide groups of vitamin B_{12} have been replaced with methyl ester groups, and the proximal base of the vitamin at the f-position with a stearylamide group (11). Electrodes prepared with this ionophore and DOS as the plasticizer were also selective for thiocyanate and nitrite over the rest of the anions tested. The main anionic interferent was salicylate. In all cases, the response of the electrodes to the preferred anions was sub-Nernstian. Overall, the selectivity pattern obtained with ionophore **3** is similar to that of **2** and to that of the hydrophobic cobyrinate-based electrodes reported previously (3, 12, 13). This observation suggests that in all cobyrinate ionophores the anions interact with the cobalt(III) center, and that the side chains of the corrin ring have a small effect on the selectivity of this interaction.

ISE Based on a Hydrophobic Cobalamin. Cobalamin **4** (Figure 2) is a hydrophobic vitamin B_{12} derivative that retains a proximal base (an imidazole ring), which provides the molecule with a resemblance to the original vitamin (Figure 1). The selectivity pattern of electrodes prepared with ionophore **4** at pH 5.5 is shown in Figure 4. These electrodes exhibit an anion selectivity pattern that differs from that of the Hofmeister series and from that of the previously reported cobyric acid derivatives (see section above). Indeed, electrodes based on **4** are selective for

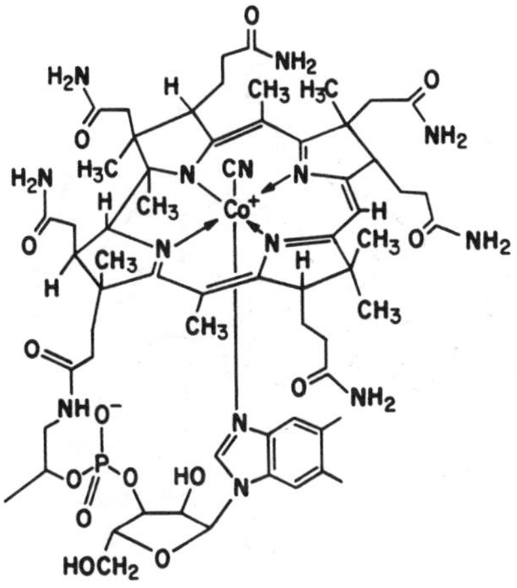

Figure 1. Structure of vitamin B$_{12}$.

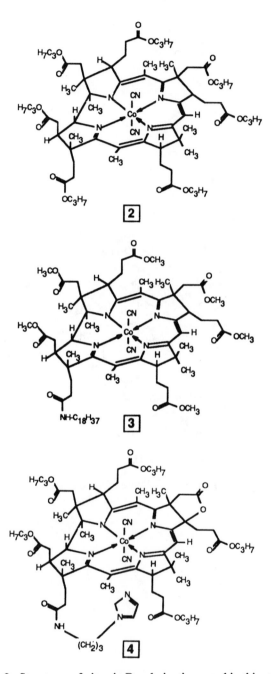

Figure 2. Structures of vitamin B$_{12}$ derivatives used in this study.

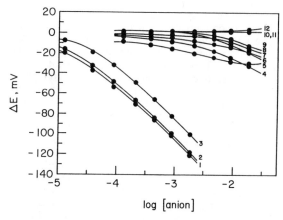

Figure 3. Selectivity pattern of an ISE based on ionophore **2**. The electrode was exposed to the following anions: salicylate (1), thiocyanate (2), nitrite (3), perchlorate (4), iodide (5), benzoate (6), bromide (7), bicarbonate (8), hydrogen phosphate (9), nitrate (10), chloride (11), sulfate (12). (Reproduced with permission from ref. 10. Copyright 1989 Alan R. Liss.)

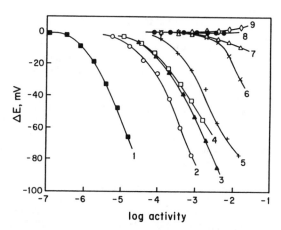

Figure 4. Selectivity pattern of an ISE based on ionophore **4**. The electrode was exposed to the following anions: iodide (1), thiocyanate (2), salicylate (3), perchlorate (4), nitrite (5), phosphate (6), nitrate (7), sulfate (8), and chloride (9). (Adapted from ref. 14.)

iodide over a variety of anions. The response toward iodide was Nernstian. Small interferences were presented by thiocyanate, salicylate, perchlorate, and nitrite (*14*).

It is interesting to compare the selectivity pattern of electrodes based on hydrophobic derivatives of cobyric acid (e.g., ionophore **2**) with that of ISEs based on ionophore **4**. This different behavior might be explained by differences in the molecular structure of the corresponding ion carriers. As depicted in Figure 2, the proximal base and the lactone ring of ionophore **4** are the main differences between this and the cobyric acid derivatives. Even though both factors may influence the properties of the ionophore, the proximal base plays an important role in controlling the selectivity of the ISEs. This is illustrated by the scheme in Figure 5. In both cases, the aquocyano derivative is the active form of the ionophore and acts as a positively charged carrier (*3, 12, 14*). In addition, there is evidence that, at solutions of low pH, one of the nitrogens of the imidazole ring of **4** is protonated (*14*), thereby, resulting in an ionophore that has two positively charged centers. Displacement of the water molecule (or of a coordinated anion) by anions at the sample-membrane interface causes a change in the electrode potential. Therefore, the observed selectivity for iodide may be explained by considering that the size of this anion may enable it to interact simultaneously with both positive sites of the ionophore (i.e., the Co and the N of the imidazole ring). Consequently, coordination chemistry, as well as the spatial arrangement of the recognition elements are important in determining the selectivity of electrodes based on this vitamin B_{12} derivative. On the contrary, the recognition of anions by ionophores **2** and **3** appears to be only a result of the coordination chemistry of the metal center.

Nitrogen Oxide Gas Sensor

Although the ISEs based on cobyrinates have good selectivity for nitrite over several anions, they also respond to salicylate and thiocyanate. To eliminate this interference, the nitrite-selective electrode based on ionophore **2** was placed behind a microporous gas-permeable membrane (GPM) in a nitrogen oxide gas-sensor mode (*15*). NO_x was generated from nitrite in the sample at pH 1.7 and, after crossing the GPM, was trapped as nitrite by an internal solution that was buffered at pH 5.5 (0.100 M MES-NaOH, pH 5.5, containing 0.100 M NaCl). The internal solution was "sandwiched" between the nitrite-selective electrode and the GPM.

Figure 6 shows the selectivity behavior of this NO_x gas sensor. The sensor had a sub-Nernstian response toward nitrite, with slopes in the range of -45 to -50 mV/decade. Further, the response observed with salicylate and thiocyanate was diminished substantially, as compared to that obtained with the original nitrite-selective electrode (Figure 3). In addition, the gas sensor described here does not suffer interferences from nitrate, bicarbonate, acetate, benzoate, or chloride. These excellent selectivity properties of the sensor are a combination of the selectivity characteristics of the nitrite-selective electrode and the additional discrimination provided by the GPM.

This NO_x sensor has several advantages over the commercially available Severinghaus-type gas sensor, which uses as sensing element a pH electrode placed behind a GPM (*16-18*). In the Severinghaus-type sensor, gases in the sample diffuse through the GPM and change the pH of the internal solution. This solution is unbuffered and contains a relatively high concentration of sodium nitrite (*19*). The main limitation of the commercial nitrogen oxide sensor is that other acidic species can cross the membrane and alter the pH of the internal solution (*20*). In contrast, the described gas sensor monitors changes in the nitrite concentration of the internal solution rather than variations in pH. Therefore, because the nitrite-selective electrode demonstrates very little response to other anions such as bicarbonate, benzoate, etc. (see Figure 3), an overall better selectivity was obtained with this gas sensor (Figure 6). In addition, the gas sensor based on the nitrite-selective electrode

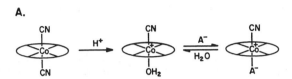

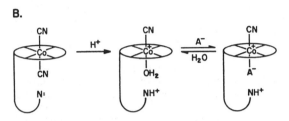

Figure 5. Schematic representation of the effect of pH on the coordination properties of ionophores **2** and **3** (A), and ionophore **4** (B).

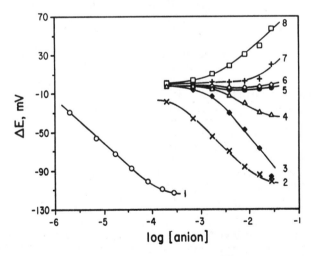

Figure 6. Selectivity pattern of the NO$_x$ gas sensor. The sensor was exposed to 0.010 M H$_2$SO$_4$ containing the following anions: nitrite (1), salicylate (2), thiocyanate (3), benzoate (4), nitrate (5), chloride (6), bicarbonate (7), acetate (8). (Adapted from ref. 15.)

had a better detection limit (4×10^{-7} M) than the commercial sensor (2×10^{-6} M). This can be attributed to an increased amount of NO_x being trapped as nitrite by the buffered solution (15, 21).

Anion-Selective Electrodes Based on Electropolymerized Porphyrin Films

The ISEs discussed thus far use ionophores dissolved in plasticized polymer membranes. Electrodes have also been prepared by covalently attaching the ionophore to suitable polymer matrices (22, 23). Recently, ISEs have been prepared by direct electrodeposition of appropriate ionophore monomers on electrode surfaces, such as glassy carbon, platinum, etc. (24). The monomers used contain groups that can be electrochemically oxidized or reduced to form a polymer. In ISEs based on this type of electropolymerized films the ionophore is firmly attached to the electrode surface. This should provide for two distinct advantages. First, electrodes with extended lifetimes should be obtained because there should be no leaching of the ionophore from the polymer matrix. Second, the ability to control the polymerization conditions should provide an additional parameter that can be used to adjust the selectivity pattern displayed by these electrodes. This is analogous to the "imprinting" principle used in the field of molecular recognition; i.e., a cavity is created by polymerization of monomers that surround a template molecule (25). In order to test the above two hypotheses and in view of the structural similarities between cobalt porphyrins and cobyrinates, a cobalt(II) tetrakis(*o*-aminophenyl)porphyrin, [Co(*o*-NH$_2$)TPP], (Figure 7) was chosen as the ionophore monomer.

[Co(*o*-NH$_2$)TPP] was polymerized onto glassy-carbon electrodes from an electrolytic solution containing the monomer and tetraethylammonium perchlorate as the electrolyte by cycling the electrode potential in the oxidative way (26). The polymerization of this monomer appears to proceed through a radical cation of the porphyrin with a mechanism similar to that of the oxidative electropolymerization of aniline (27).

The potentiometric behavior of electrodes based on these films was studied (Figure 8). These ISEs presented sub-Nernstian slopes for thiocyanate (from -40 to -53 mV/decade, depending on the buffer used), and had detection limits of 5×10^{-7} M. The response time of the electrodes was typically less than 25 s. The selectivity pattern observed was: thiocyanate > perchlorate > iodide > nitrite ~ salicylate ~ bromide > chloride > bicarbonate > phosphate. This anion-selectivity behavior does not follow the Hofmeister series, with thiocyanate and nitrite being the ions that deviate the most from it. This indicates that there is a selective interaction of the immobilized porphyrin with the two anions.

With respect to lifetimes, after 2 months the electrodes still had good slopes and detection limits. During this time the electrodes were stored at room temperature in a 0.01 M thiocyanate solution. The lifetime of these electropolymerized electrodes (at least 2 months) is a significant improvement over that of PVC-based electrodes that use cobalt porphyrins (28, 29). The latter electrodes were selective toward thiocyanate, but the slopes of the calibration curves deteriorated substantially in less than one month. The improved lifetimes of the poly[Co(*o*-NH$_2$)TPP] electrodes may be attributed to the covalent fixation of the ionophore to the polymeric matrix; i.e., there is a reduced leaching of the ionophore out of the membrane.

It should be noted that, although both the electropolymerized films and the PVC electrodes that use cobalt porphyrins as ionophores respond primarily to thiocyanate, their selectivity properties are quite different. Specifically, the electropolymerized [Co(*o*-NH$_2$)TPP] films demonstrate a much larger discrimination for thiocyanate

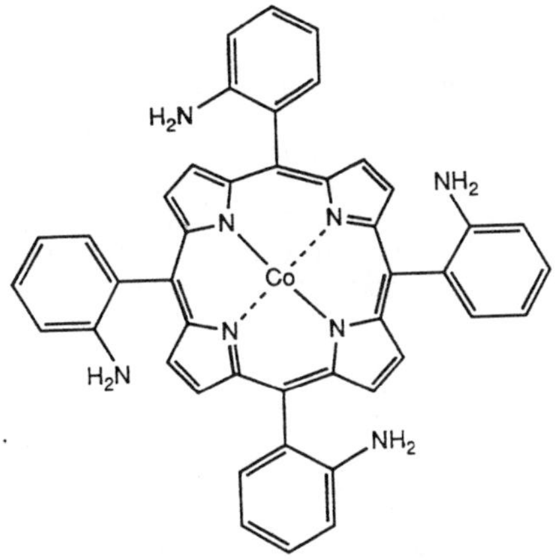

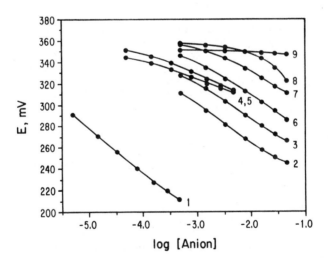

Figure 7. Structure of cobalt(II) tetrakis(*o*-aminophenyl)porphyrin.

Figure 8. Selectivity pattern of an ISE based on an electropolymerized cobalt porphyrin film. The electrode was exposed to the following anions: thiocyanate (1), perchlorate (2), iodide (3), nitrite (4), salicylate (5), bromide (6), chloride (7), bicarbónate (8), phosphate (9). The baseline potential was 364 mV. (Adapted from ref. 26.)

over iodide (k_{SCN^-,I^-}^{pot} = 5.1x10^{-4}; determined by the matched-potential method) than the PVC-based electrodes (selectivity coefficients of 0.78 and 0.13, from references *30* and *28*, respectively). This may be due to differences in the ion-recognition properties of the porphyrin imposed by its immobilization in the matrix. Indeed, the porphyrin monomers are held in the membrane in a three-dimensional cross-linked structure. Such an arrangement should provide an additional control on selectivity by functioning as a "sieve" for ions. This molecular sieving property of electropolymerized porphyrin films has been confirmed by Murray and co-workers (*31*). It should be noted that it may be feasible to control this sieving effect by appropriate selection of the electrolyte during the electropolymerization reaction (*32*). In that respect the size and shape of the counter ion in the electrolyte may act as a template and result in electropolymerized films with different size pores (i.e., "imprinting"). Consequently, the ion-recognition properties of the electrodes could be altered by changing the electropolymerization conditions.

Conclusions

In summary, it has been demonstrated that ISEs can be designed by employing molecular recognition principles. In particular, the feasibility of using hydrophobic vitamin B$_{12}$ derivatives and electropolymerized porphyrin films in the development of polymer membrane anion-selective electrodes has been demonstrated. The studies indicated that the changes in the selectivity of these ISEs can be explained by the difference in structure of the ionophores. In addition, it was shown that by electropolymerization of a cobalt porphyrin, anion-selective electrodes can be prepared that have extended lifetimes compared with PVC-based ISEs, which use a similar compound as the ionophore.

Acknowledgments. This work was supported in part by grants from the National Science Foundation (R11-86-10671, DMR-8900979 and DMR-9000782), the NATO Scientific Affairs Division (CRG 890610), and the Society for Analytical Chemists of Pittsburgh. Acknowledgment is also made to the donors of the Petroleum Research Fund, administered by the American Chemical Society, and to CICyT, Madrid (project MAT90-0886). A.F. thanks CIRIT, Generalitat de Catalunya, for a research grant to support his stay at the University of Kentucky.

Literature Cited

1. Yu, R. *Ion-Sel. Rev.* **1986**, *8*, 153-171.
2. Hofmeister, F. *Arch. Exp. Pathol. Pharmakol.* **1888**, *24*, 247.
3. Schulthess, P.; Ammann, D.; Kräutler, B.; Caderas, C.; Stepánek, R.; Simon, W. *Anal. Chem.* **1985**, *57*, 1397-1401.
4. Cram, D. J. *Science* **1988**, *240*, 760-767.
5. Hamilton, A. D. *J. Chem. Educ.* **1990**, *67*, 821-828.
6. Sutherland, I. O. *J. Chem. Soc., Faraday Trans. 1* **1986**, *82*, 1145-1159.
7. Pratt, J. M. *Inorganic Chemistry of Vitamin B$_{12}$*; Academic Press: New York, 1972.
8. Thusius, D. J. *J. Am. Chem. Soc.* **1971**, *93*, 2629-2635.
9. Hasinoff, B. B. *Can. J. Chem.* **1974**, *52*, 910-914.
10. Daunert, S; Witkowski, A.; Bachas, L. G. *Prog. Clin. Biol. Res.* **1989**, *292*, 215-225.
11. Florido, A.; Daunert, S.; Bachas, L. G. *Electroanalysis* **1991**, *3*, 177-182.
12. Schulthess, P.; Ammann, D.; Simon, W.; Caderas, C.; Stepánek, R.; Kräutler, B. *Helv. Chim. Acta* **1984**, *67*, 1026-1032.

13. Stepánek, R.; Kräutler, B.; Schulthess, P.; Lindemann, B.; Ammann, D.; Simon, W. *Anal. Chim. Acta* **1986**, *182*, 83-90.
14. Daunert, S.; Bachas, L. G. *Anal. Chem.* **1989**, *61*, 499-503.
15. O'Reilly, S. A.; Daunert, S.; Bachas, L. G. *Anal. Chem.* **1991**, *63*, 1278-1281.
16. Severinghaus, J. W.; Bradley, A. F. *J. Appl. Physiol.* **1958**, *13*, 515-520.
17. Riley, M. In *Ion-Selective Electrode Methodology*; Covington, A. K., Ed.; CRC Press: Boca Raton, FL; 1980, pp 1-21.
18. Ross, J. W.; Riseman, J. H.; Krueger, J. A. *Pure Appl. Chem.* **1973**, *36*, 473-487.
19. Instruction Manual, Nitrogen Oxide Electrode, Model 95-46. Orion Research, Inc., Boston, MA, 1987.
20. Hassan, S. S. M.; Tadros, F. S. *Anal. Chem.* **1985**, *57*, 162-166.
21. Meyerhoff, M. E.; Fraticelli, Y. M.; Opdycke, W. N.; Bachas, L. G.; Gordus, A. D. *Anal. Chim. Acta* **1983**, *154*, 17-31.
22. Thomas, J. D. R. *Anal. Chim. Acta* **1986**, *180*, 289-297.
23. Daunert, S.; Bachas, L. G. *Anal. Chem.* **1990**, *62*, 1428-1431.
24. Dong, S.; Wang, Y. *Electroanalysis* **1989**, *1*, 99-106.
25. Sellergren, B.; Lepistö, M.; Mosbach, K. *J. Am. Chem. Soc.* **1988**, *110*, 5853-5860.
26. Daunert, S.; Wallace, S.; Florido, A.; Bachas, L. G. *Anal. Chem.* **1991**, *63*, 1676-1679.
27. White, B. A.; Raybuck, S. A.; Bettelheim, A.; Pressprich, K.; Murray, R. W. In *Inorganic and Organometallic Polymers*; Zeldin, M., Wynne, K. J., Allcock, H. R., Eds.; American Chemical Society: Washington, DC, 1988, pp 408-419.
28. Hodinár, A.; Jyo, A. *Chem. Lett.* **1988**, 993-996.
29. Hodinár, A.; Jyo, A. *Anal. Chem.* **1989**, *61*, 1169-1171.
30. Ammann, D.; Huser, M.; Kräutler, B.; Rusterholz, B.; Schulthess, P.; Lindemann, B.; Halder, E.; Simon, W. *Helv. Chim. Acta* **1986**, *69*, 849-854.
31. Pressprich, K. A.; Maybury, S. G.; Thomas, R. E.; Linton, R. W.; Irene, E. A.; Murray, R. W. *J. Phys. Chem.* **1989**, *93*, 5568-5574.
32. Dong, S.; Sun, Z.; Lu, Z. *J. Chem. Soc., Chem. Commun.* **1988**, 993-995.

RECEIVED November 5, 1991

Chapter 15

Electropolymerized Films for the Construction of Ultramicrobiosensors and Electron-Mediated Amperometric Biosensors

Eugene R. Reynolds, Robert J. Geise[1], and Alexander M. Yacynych[2]

Department of Chemistry, Rutgers, The State University of New Jersey, New Brunswick, NJ 08903

Electropolymerized films of poly(1,3-diaminobenzene) were used in the construction of an ultramicrobiosensor for glucose and an electron-mediated glucose biosensor. The ultramicrobiosensor is based on 25 μm diameter platinum disk electrodes, and is capable of direct measurement. Glucose oxidase is immobilized by a combination of adsorption and glutaraldehyde cross-linking, and an electropolymerized, insulating film is used to prevent fouling and signals from interfering species. These sensors have small background currents and fast response times (15-30 s, 100% response).

An electron-mediated glucose biosensor was made by incorporating 1,1'-dimethylferrocene in electropolymerized 1,3-diaminobenzene, and immobilizing glucose oxidase. These biosensors had a linear response up to 76 mM glucose, showed little interference from oxygen, and they were stable for about three months.

Future directions in biosensors will require the miniaturization of individual sensors for *in-vivo* use, and for use in microsensor arrays (*1*). Ideally, these biosensors should be capable of direct, rather than differential measurement, and should have reasonable lifetimes.

Miniaturization of biosensors is important for a variety of reasons: 1) potential use in neurological and physiological studies. 2) improved biocompatibility. 3) measurement in very small volumes.

---

[1]Current address: GAF Corporation, 1361 Alps Road, Wayne, NJ 07470
[2]Corresponding author

4) determination of spatial concentration profiles.   5) multisensor arrays and probes.   6) biochips.   7) lower costs. *In-vivo* sensors are needed as biofeedback devices that provide information on a real-time basis, e.g., for artificial organs, clinical monitoring devices, etc.   As such, they are good candidates for use in physiological studies and neurological studies, such as *in-vivo* studies of neurotransmitters and other mammalian brain species.

Biosensors based on carbon fiber ultramicroelectrodes have been used to determine pyruvate (2) and glucose (3).   Glucose sensors using platinum ultramicroelectrodes have also been reported (4), including the entrapment of glucose oxidase in an electropolymerized film of polyaniline (5,6). Ikariyama and co-workers have used platinum ultramicroelectrodes modified with platinum black to construct very sensitive glucose sensors (7-13).

Most ultramicrobiosensors use differential measurement to overcome the problems of interferences and electrode fouling.   The practical use of these biosensors for direct measurement is limited by interferents, such as ascorbic acid, acetaminophen (paracetamol), uric acid, etc., which are present in complex matricies such as serum.   The specificity of the biochemical system is compromised by the partial selectivity of the electrode.   The electrode not only oxidizes the desired product (e.g., H_2O_2 formed in the enzymatic oxidation of glucose by glucose oxidase), but also any other species oxidizable at the working potential.   This produces a larger current response and a positive error.

Electrode fouling is the passivation of the electrode surface by the adsorption of nonelectroactive species.   This is a major problem in the analysis of biological samples.   High molecular weight species, such as proteins, are a major source of fouling, which results in a decreasing biosensor response over time.

Differential measurement has been used as a method for compensating for interferences and fouling.   This is particularly true where direct measurement is difficult or impossible, such as with miniaturized biosensors.   This method consists of subtracting the signal of a control electrode from the signal of the active working electrode.   The control electrode is modified in an identical fashion to the working sensor except that the biocomponent is eliminated or deactivated.   The major limitation is that the electrodes must be identical, otherwise the precision suffers, and as the electrodes become smaller, the more difficult it becomes to meet this criterion.

Electropolymerized films have been used to prevent interferences and fouling in biosensors constructed from reticulated vitreous carbon and platinum disk electrodes (14,15).   A biosensor constructed using electropolymerized films can have significantly improved diffusional properties due to the thinness of the film.   By engineering the components and properties of a biosensor on a microscopic scale, rather than using "bulk-technology" and physically assembling discrete macroscopic components, as is the conventional practice, an all-chemical method of construction can be achieved.   All-chemical methods of construction would be important for miniaturized sensors, lithographically made sensors (i.e., "sensors on a chip"), and for ease of manufacture in general. For this type of sensor, the sensitization layer thickness (this includes everything between the electrode surface and the solution, i.e., immobilized enzyme, polymer film, etc.) is approximately 10 nanometers, while for conventional sensors this thickness is about 10 micrometers.

Electropolymerized films are generally formed by the oxidation of phenols and/or aromatic amines, which results in an insulating polymer film that coats the electrode surface. These polymer films have permselectivities that allow the passage of H_2O_2, but prevent interferents from reaching the electrode surface. They are self-regulating films, with uniform thickness, that completely cover the electrode surface regardless of size or shape. The film maintains its uniformity because it only grows thick enough to become an insulator. Polymerization continues until the surface is completely covered, which is signaled by the current decreasing to a minimum, because the monomer cannot penetrate the film. Electropolymerized films offer a molecular approach to solving the problems associated with interferences and fouling. As this approach is applicable to all sizes and shapes of electrodes, it should prove useful in the construction of miniaturized biosensors.

Therefore, the first objective is to use electropolymerized films to construct an ultramicrobiosensor capable of direct measurement. The second objective is to construct a biosensor that is not dependent on oxygen concentration, and is stable enough for repeated use. This requires the use of a substitute for oxygen in the reaction of a flavo-enzyme, such as glucose oxidase. These substitutes are called electron mediators because they mediate electron transfer to regenerate the enzyme by oxidizing the prosthetic group at the active site. The reduced mediator is oxidized at the electrode surface giving a response proportional to the substrate.

Electron mediators are usually organic molecules that are redox active, such as ferrocene derivatives, benzoquinone, N-methylphenazium, and 2,6-dichlorophenolindophenol (DCPIP). Ferricyanide has also been used as an electron mediator. They offer the advantages of non-dependence on ambient oxygen, increased linear response, a lower working potential, less interference from other electroactive species, and perhaps an extended biosensor lifetime, because hydrogen peroxide is not being generated, which can contribute to the deactivation of the enzyme.

Much of the initial work on electron mediators involved adding the mediating species to the test solution. Ideally, the mediator should be confined to the sensing layer in order for the biosensor to act as a stand-alone device. The mediator must be mobile enough to interact with the active site of the enzyme, yet be confined to the surface of the biosensor for long-term stability. Accommodating these divergent requirements has proved to be a serious obstacle in obtaining a biosensor with long-term stability and capable of repeated use.

In 1984, Cass, et al. adsorbed ferrocene derivatives along with immobilized glucose oxidase onto graphite electrodes (*16*). A linear response was obtained over a 1-30 mM range. Jonsson and Gorton immobilized glucose oxidase on a graphite electrode with adsorbed N-methylphenazium ion as mediator (*17*). The response was linear from 0.5 to 150 μM and usable up to 2 mM glucose, but had the drawback of requiring the mediator to be renewed daily. In 1988, these researchers reported using a ferrocene derivative, which included an aromatic anchor to adsorb the mediator on a graphite electrode (*18*).

Foulds and Lowe synthesized ferrocene-modified pyrrole by amide linkage of N-(3-aminopropyl) pyrrole with ferrocene carboxylic acid (*19*). The ferrocene-modified pyrrole was then electropolymerized onto the electrode surface during which glucose oxidase entrapped in the film. The response was linear over 2-20 mM glucose and responded for two days in deoxygenated solution. Wang and co-workers incorporated various ferrocenes with glucose oxidase into carbon paste which was used successfully as a glucose sensor (*20*). Problems of fouling are overcome by polishing the electrode to remove the old layer exposing a fresh, sensing layer.

Another method to overcome oxygen dependence by confining a mediator at the electrode was developed by Heller using "electrical wires" (redox polymers) to connect the active site of the enzyme to the electrode surface (*21*). Hale, et al. used ferrocene-modified polysiloxanes incorporated into carbon paste electrodes as electron-mediators for a glucose biosensor (*22*). An electrochemical glucose sensor, using an electron mediator, is available commercially (Exactech, MediSense, Inc.), but because of mediator leeching it is limited to one use, and it is used as a disposable (*23*).

There are many problems to be overcome in mediator-based glucose biosensors, and no previously reported sensor solves all of these problems. Problems include interfering signals from electroactive species present in real samples, fouling of the the electrode surface by biomacromolecules, non-linear response in the normal clinical to diabetic range (2-20 mM glucose), lack of stability (short lifetime) and competition from oxygen. We have found that the use of an electropolymerized film, which incorporates an electron mediator and immobilized glucose oxidase results in a glucose sensor which solves many of these problems.

These sensors show little or no interference from oxygen, are protected from electroactive interferents and fouling by the film, exhibit linear range from below normal clinical level to well beyond diabetic range, and have a stability of greater than three months with little or no loss in response. This stability is due to the retention of the mediator in the sensing layer by the polymer film.

EXPERIMENTAL

Apparatus-Ultramicrobiosensor. All batch mode experiments were done in a Faraday cage, using a BAS CV-27 potentiostat with a low current module (Bioanalytical Systems, West Lafayette, IN). Electrochemical pretreatments were carried out using an EG&G PARC 264A potentiostat. A BAS silver/silver chloride reference electrode (3M NaCl) and platinum wire auxiliary electrode completed the cell. Compressed air was sparged through the cell to stir the solution.

Apparatus-Electron-Mediated Biosensor. All experiments were done with either an EG&G Princeton Applied Research (Princeton, NJ) Model 264A potentiostat or an EG&G PAR Model 174 potentiostat. A saturated calomel reference electrode (SCE) and a platinum mesh auxiliary electrode were used. Working graphite electrodes were constructed by sealing spectroscopic grade graphitic rods in glass tubing with epoxy. Copper wire was epoxyed to the graphite for electrical contact. The graphite was then sanded flush with the end of the glass rod, polished by normal methods, and sonicated in deionized/distilled water.

Materials- Ultramicrobiosensor. Platinum wire, 25 μm in diameter, was from Aesar-Johnson Matthey (Seabrook, NH). Phosphate buffer (0.1 M) was prepared using ACS certified phosphate salts (Fisher Scientific, Springfield, NJ). The pH was adjusted to 6.5 or 7.4 as needed, using potassium hydroxide or concentrated phosphoric acid. 1,3-Diaminobenzene (1,3-DAB) (Aldrich, Milwaukee, WI) was purified by sublimation. Solutions of 1,3-DAB were made with deoxygenated buffer, and blanketed with nitrogen during electropolymerization. Other chemicals used were ß-D(+)-glucose, and bovine serum albumin (Sigma, St. Louis, MO); 4-acetamidophenol (acetaminophen), L-ascorbic acid (vitamin C), uric acid, and 25% w/v glutaraldehyde (Aldrich). Stock solutions of L-ascorbic acid were made daily, using deoxygenated buffer. Glucose oxidase (EC 1.1.3.4) was Type II, from *Aspergillus Niger* (Sigma). Compressed air and nitrogen were from JWS Technologies (Piscataway, NJ). An Oxy-Trap (Alltech Associates, Deerfield, IL) was used to further purify the nitrogen.

Materials- Electron- Mediated Biosensor. 1,3- Diaminobenzene (Aldrich, Milwaukee, WI) was purified by sublimation. Phosphate buffer (0.1 M) was prepared with distilled/deionized water using ACS certified (Fisher, Springfield, NJ) phosphate salts. The pH was adjusted to 7.4 with concentrated phosphoric acid or potassium hydroxide. Other chemicals used were 1,1'-dimethylferrocene (Lancaster Synthesis, Windham, NH), L-ascorbic acid, 99 % (Aldrich), and ß-D(+)-glucose (Sigma, St. Louis, MO). Glucose oxidase was from Sigma (Type II from *Aspergillus Niger*) and glutaraldehyde, 25 % (wt %) was from Aldrich. Nitrogen and oxygen was of high purity grade and an oxy-trap was used in the nitrogen line.

Procedure-Ultramicrobiosensor. Ultramicroelectrodes were made by sealing approximately 1 cm of 25 μm diameter platinum wire in one end of an 8 cm length of 4 mm outer diameter soda-lime or soft glass tubing, using a glass-blowing torch.   Electrical contact to the wire is made using Wood's metal and a piece of insulated 22 gauge wire as a lead wire.   The electrode is then polished to a flat surface with successively finer grades of sandpaper and alumina, and sonicated in distilled/deionized water.   The resulting geometry is a microdisk.

After polishing and sonication, the ultramicroelectrodes (UMEs) are tested using cyclic voltammetry, E = +0.50 V to 0.00 V, vs Ag/AgCl, 50 mV/s, in a solution containing 5 mM ferricyanide, in pH = 7.4 buffer. Prior to enzyme immobilization, up to four UMEs are placed in a cell containing pH = 7.4 buffer, and attached to the working electrode lead (i.e., the multiple working electrodes are connected in parallel to the potentiostat).   The electrodes are then held at -1.5 V vs Ag/AgCl for 5 min, then +1.5 V for 5 min; then the entire process is repeated, for a total of 20 min.   This electrochemical pretreatment step enhances the sensitivity of the platinum electrodes to hydrogen peroxide, as well as the reproducibility of the sensitivity at a given electrode. It also decreases the background current at the operating potential of +0.58 V vs Ag/AgCl.

Glucose oxidase is immobilized via glutaraldehyde cross-linking.   A solution containing approximately 8500 units/mL is prepared in pH = 6.5 phosphate buffer containing 1.25% w/v glutaraldehyde.   The UMEs are inverted, a 5 μL drop is placed on each, and allowed to dry for 10 min. The resulting biosensors are then placed in pH = 7.4 buffer overnight (T = 4°C). All batch-mode determinations of glucose and interferences were performed at E = +0.58 V vs Ag/AgCl.

Cyclic voltammetry was used to electropolymerize poly(1,3-DAB) from a 3mM solution of the monomer in pH = 7.4 phosphate buffer.   The potential was cycled from 0.00 V to 0.80 V, vs Ag/AgCl, 2 mV/s for 18 h (81 scans total).

The effects of fouling were studied by obtaining glucose calibration curves for two ultramicrobiosensors, one without electropolymerized film, the other with poly(1,3- DAB).   Both sensors were then placed in a solution containing 3% w/v bovine serum albumin, at temperature of 4°C, for 6 h. After 6 h, both sensors were again calibrated.

Procedure-Electron-Mediated Biosensor.   1,3-DAB was electropolymerized onto the graphite from a 3 mM solution in phosphate buffer using cyclic voltammetry.   The potential was cycled from 0.00 V to +0.80 V and back to 0.00 V (vs SCE) at 5 mV/s for twelve scans.   1,1'-dimethylferrocene (1,1'- DMF) was adsorbed onto an inverted electrode.   Three successive drops of a saturated solution (10 μL each) were added to the surface and each was allowed to air dry before applying the next drop.   The electrode was rinsed with water after application of each drop.

Glucose oxidase was immobilized by crosslinking with glutaraldehyde. A solution containing approximately 5500 units/mL glucose oxidase was prepared by dissolving 0.22 g glucose oxidase (25,000 units/mL) in 950

μL phosphate buffer. Then, 50 μL, 25 % glutaraldehyde was added and the solution mixed. The glutaraldehyde concentration was 1.25 %. The enzyme solution (20 μL) was placed on the inverted electrode for approximately 25 min.

The experiments were performed in batch mode with stirring. The sensor was placed in 20.00 mL of buffer in a three electrode configuration. The potential was applied ($+150$ mV vs SCE) and the response was allowed to reach baseline. The required amount of analyte was added to measure the response to electroactive interferents and to generate a glucose calibration curve.

RESULTS AND DISCUSSION

Ultramicrobiosensor. Immobilization of glucose oxidase via glutaraldehyde cross-linking is a simple procedure that yields ultramicrobiosensors (UMBs) with good sensitivity (usually 8-12 pA/mM, over the human clinical range of 3-7 mM glucose) and fast response times (15-30 s for 100% response). These sensors can be used for direct measurement of glucose in simple matrices; however, without protection from interferences and fouling they will not work for a more complex sample, such as human serum.

For example, a typical UMB (enzyme only) has an 80 pA response to a sample of 5 mM glucose. The same sensor has a response of 900 pA to 0.44 mM uric acid, 180 pA response to 0.21 mM acetaminophen, and 140 pA to 0.11 mM ascorbic acid (these are maximum clinical amounts of each of the three major interferences that could be found in human serum). The total response due to interferences is more than fifteen times that of the response due to an average clinical amount of glucose.

Very thin films (ca. 10 nm) formed by electropolymerization can be used to effectively prevent signals due to interferences. Figure 1 shows a cyclic voltammogram for the electropolymerization of poly(1,3-DAB) onto a platinum ultramicroelectrode (25μm diameter). Note how the current decreases with each subsequent scan, indicating coverage of the electrode surface.

The current response of a completed UMB (enzyme and poly(1,3-DAB) film) is shown in Figure 2. Note how quickly the baseline is achieved after the potential (E$= +0.58$V vs Ag/AgCl) is applied, as well as the low background current. Note the enormous decrease in the response to each of the interferences (uric acid: 1.5 pA, $>99.8\%$ screened out; acetaminophen: 1.5 pA, $>99\%$ screened out; ascorbic acid: 0 pA). Now the total signal due to interferences is less than 5% of the signal due to 5 mM glucose.

For the electropolymerization of poly(1,3-DAB), four UMBs were placed in parallel to the working electrode lead, to insure that the electropolymerized film was formed under the same conditions and same length of time for all four UMBs, and to save time. Since the electropolymerization is a potentiostatically controlled process, all four

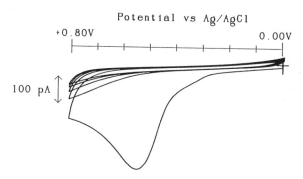

Figure 1. Cyclic voltammogram depicting the electropolymerization of 1,3-DAB onto a platinum ultramicroelectrode (25 μm diameter).

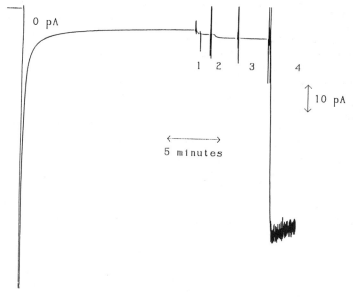

Figure 2. Batch mode response of an ultramicrobiosensor with poly(1,3-DAB) to maximum clinical amounts of interferences (1: 0.44 mM uric acid; 2: 0.21 mM acetaminophen; 3: 0.11 mM ascorbic acid), and 5 mM glucose (4) (E = +0.58 V vs Ag/AgCl).

UMBs act as working electrodes, and the recorded current would be the sum for all four. A sensor was considered useful as long as the total signal due to the three interferences (uric acid, acetaminophen, and ascorbic acid) was less than 5% of the signal for 5 mM glucose. One ultramicrobiosensor with poly(1,3-DAB) had a useful lifetime of up to four weeks.

A calibration curve for glucose at a UMB with poly(1,3-DAB) is shown in Figure 3. The calibration curve is piece-linear. The linearity over the normal human clinical range for glucose (3-7 mM) is good ($r=0.998$), and the sensitivity is 9.2 pA/mM. The calibration curve levels off at about 15-20 mM glucose, due to oxygen limitation.

An ultramicrobiosensor with poly(1,3-DAB) was used to determine the amount of glucose present in a simulated serum sample. The sample contained 5 mM glucose, 0.4 mM uric acid, 0.21 mM acetaminophen, in pH=7.4 buffer. Two glucose calibration curves were obtained, in the usual fashion. Then, 5 to 10 mL of each of the simulated serum samples was placed in the cell, +0.58 V was applied, and a reading was taken after allowing the current to come to a steady baseline. Then a third calibration curve was obtained.

Figure 4 shows the average calibration curve for these three trials. The response is linear over the range of 3-7 mM glucose, with a slope of 10.9 pA/mM, intercept of 28.1 pA, and $r=0.9945$. After correcting for background current, the samples are determined to have 4.73 mM glucose (5.4% error), and 5.10 mM glucose (2% error).

The effect of electrode fouling was tested by obtaining calibration curves at two UMBs, one with enzyme and poly(1,3-DAB), and one with enzyme only. Both UMBs were then placed in a 3% w/v BSA solution in pH=7.4 buffer, for 6 h, at T=4°C. After 6 h, the UMBs were removed, rinsed, and calibration curves were again obtained. The sensitivity of the UMB with polymer film decreased by 10%, some of which can be attributed to experimental error. The sensitivity of the UMB without polymer film decreased over 30%.

All chemical means of construction were used to construct an ultramicrobiosensor for glucose using a platinum ultramicroelectrode (25μm diameter). The sensors exhibit fast response time, and good sensitivity and linearity over the human clinical range for glucose. An electropolymerized film of poly(1,3-DAB) was used to prevent signals due to interfering species, and reduce the effects of fouling. The completed ultramicrobiosensor can be used for the direct measurement of glucose in complex matrices.

Electron-Mediated Biosensor. Electropolymerization of 1,3-diaminobenzene (1,3-DAB), followed by adsorption of 1,1'-dimethylferrocene (1,1'-DMF), and immobilization of glucose oxidase, results in an easily and quickly (<2 h) constructed glucose biosensor with excellent linearity and stability (>3 months). Figure 5 shows a proposed schematic of the sensing layer consisting of film/mediator/enzyme. The ferrocene is depicted as circles

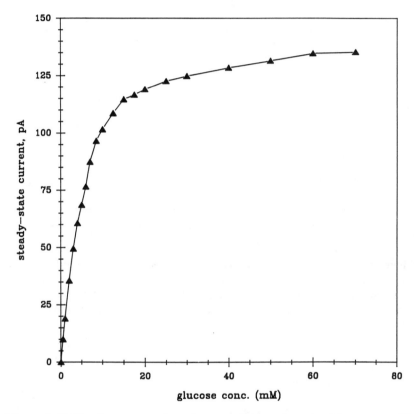

Figure 3. Calibration curve for glucose at an ultramicrobiosensor with poly(1,3-DAB) film, showing oxygen limitation of glucose response above 20 mM glucose (E = +0.58 V vs Ag/AgCl).

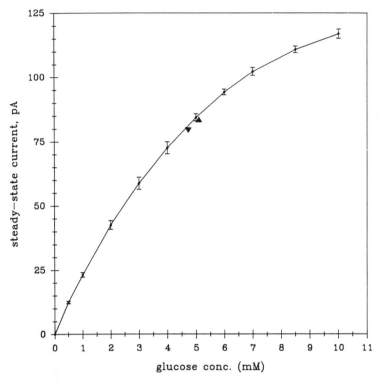

Figure 4. Calibration curve represents the average of three separate trials, including error bars, showing the reproducibility of the glucose response at an ultramicrobiosensor with poly(1,3-DAB) film. The points ▲, ▼ represent two simulated serum samples containing 5 mM glucose.

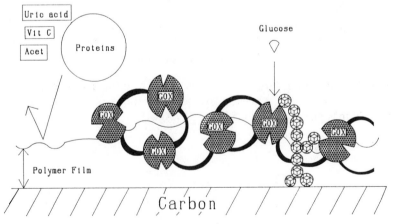

Figure 5. Schematic of sensing layer at electrode surface. Ferrocene mediator is indicated by circled cubic structures.

with hexagons. We feel that upon addition of the ethanol solution of mediator, the film swells (effect of ethanol), allowing ferrocene to pass into the polymer film. As the ethanol evaporates, the polymer film shrinks around the ferrocene, thus holding the mediator in place. There is now ferrocene through out the polymer film, this allows charge transfer from the surface to the electrode through a network of ferrocene molecules. When glucose oxidase is immobilized on top of this polymer film/mediator system, ferrocene can now, through this network, access both glucose oxidase and the electrode.

Response to Glucose. Figure 6 shows a typical strip chart recorder output for a poly(1,3-DAB)/1,1'-DMF/glucose oxidase carbon rod electrode at +0.15 V vs SCE in deoxygenated buffer. There is no response to 0.1 mM ascorbic acid, and the response to aliquots of glucose are fast (90 % response <20 s), are linear from 2-76 mM glucose (r=0.9950), and are noise-free. Typical outputs for H_2O_2-based glucose sensors are noisy, making measurement difficult and compromising precision and accuracy.

Ideally it is preferable to operate at ambient oxygen concentrations without interference from oxygen. This avoids purging all standards and samples of oxygen. In order for a mediator-based sensor to perform well under ambient O_2 conditions, the regeneration of the oxidized, active form of the enzyme by the mediator must be preferred over oxidation by O_2 because the latter case produces H_2O_2 which is not sensed at the working potential. The result, depending on the extent of interference by O_2 over mediator, is a decrease in response or no response to glucose.

Figure 7 shows that for poly(1,3-DAB)/1,1'-DMF/glucose oxidase-modified carbon biosensors, the mediator reaction, not the oxygen reaction, dominates almost completely. This is evidenced by the ambient O_2 and deoxygenated glucose curves being superimposed. Any competition from O_2 would produce H_2O_2 instead of reduced mediator and would lead to a decreased response (slope). Although there is competition from O_2 under saturated O_2 conditions at low glucose levels, it is the lack of competition from O_2 under ambient oxygen conditions that is important because the need to deoxygenate standards and samples is avoided.

Long Term Performance of Sensors. The sensors showed excellent long term stability stored in buffer at 4 °C. Table 1 shows the long term performance of four typical biosensors. The usual current response to 6 mM glucose for a new biosensor is about 2 μA. The precision of the absolute current response to 6 mM glucose remains within ±15 % for up to four months. This level of precision is unusual for amperometric biosensors. Normally, variations in current are compensated by calibration. The sensitivity of the biosensors, as indicated by the slope of the response, is also stable. This stability is due to the film which maintains the ferrocene within the sensing layer. Biosensors with adsorbed mediator and immobilized enzyme but without film are not stable for any period of

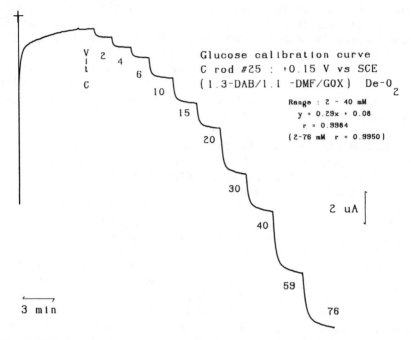

Glucose calibration curve
C rod #25 : +0.15 V vs SCE
(1.3-DAB/1.1 -DMF/GOX)   De-O_2

Range : 2 - 40 mM
y = 0.29x + 0.08
r = 0.9984
(2-76 mM r = 0.9950)

2 uA

3 min

Figure 6. Strip chart recorder output.   Glucose calibration curve: carbon electrode modified with poly(1,3-DAB)/1,1'-DMF/glucose oxidase at +0.15 V vs SCE; deoxygenated buffer.

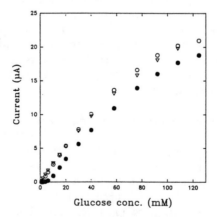

Figure 7. Glucose calibration curves under conditions of ∇, deoxygenation; ○, ambient oxygen; and ●, saturated oxygen. Carbon electrode modified same as Figure 2; +0.15 V vs SCE.

Table 1. Long Term performance of glucose biosensors. Response at +0.15 V vs SCE

Carbon rods modified with GOX

Film : 1,3-DAB   Mediator : 1,1'-DMF

Electrode	Resp. to 6 mM glucose (μA) n = 8	Slope (2-40 mM) (μA/mM) n = 8	# days
28	1.51 ± 0.27	0.22 ± 0.04	72
29	1.71 ± 0.23	0.24 ± 0.03	93
30	1.10 ± 0.12	0.18 ± 0.015	129
56	1.40 ± 0.18	0.20 ± 0.06	56

time due to leeching out of the ferrocene from the sensing layer (*16,17,23*).

In conclusion, a ferrocene-mediated glucose sensor was constructed by all-chemical means. They are easily and quickly constructed (<2 h) and show excellent linearity and stability. Also, this construction method is applicable to any graphite electrode, regardless of shape or size. Future work will focus on applying electron mediators to ultramicrobiosensors.

LITERATURE CITED

1. Reynolds, E.R.; Yacynych, A.M. *American Laboratory* **1991**, *March*, 19.
2. Suaud-Chagny, M.F.; Gonon, F.G. *Anal. Chem.* **1986**, *58*, 412.
3. Wang, J.; Lin, M. *Electroanalysis* **1989**, *1*, 151.
4. Kim, J.Y.; Lee, Y.H. *Biotechnol. Bioeng.* **1988**, *31*, 755.
5. Shinohara, H.; Chiba, T.; Aizawa, M. *Shokubai* **1987**, *29*, 70.
6. Shinohara, H.; Chiba, T.; Aizawa, M. *Sens. Actuators* **1987**, *13*, 79.
7. Ikariyama, Y.; Yamauchi, S.; Yukiashi, T.; Ushioda, H. *Anal. Lett.* **1987**, *20*, 1407.
8. Ikariyama, Y.; Yamauchi, S.; Yukiashi, T.; Ushioda, H. *Anal. Lett.* **1987**, *20*, 1791.
9. Ikariyama, Y.; Yamauchi, S.; Aizawa, M.; Yukiashi, T.; Ushioda, H. *Bull. Chem. Soc. Jpn.* **1988**, *61*, 325.
10. Ikariyama, Y.; Yamauchi, S.; Yukiashi, T.; Ushioda, H.; Aizawa, M. *Bull. Chem. Soc. Jpn.* **1989**, *62*, 1869.

11.   Ikariyama, S.; Shimeda, N.; Yamauchi, S.; Yukiashi, T.; Ushioda, H. *Anal. Lett.* **1988**, *21*, 953.

12.   Ikariyama, Y.; Shimeda, N.; Yukiashi, T.; Aizawa, M.; Yamauchi, S. *Bull. Chem. Soc. Jpn.* **1989**, *62*, 1864.

13.   Ikariyama, Y.; Yamauchi, S.; Yukiashi, T.; Ushioda, H. *J. Electroanal. Chem. Interfacial Electrochem.* **1988**, *251*, 267.

14.   Sasso, S.V.; Pierce, R.J.; Walla, R.; Yacynych, A.M. *Anal. Chem.* **1990**, *62*, 1111.

15.   Geise, R.J.; Adams, J.M.; Barone, N.J.; Yacynych, A.M. *Biosensors & Bioelectronics* **1991**, *6*, 151.

16.   Cass, A.E.G.; Davis, G.; Francis, G.D.; Hill, H.A.O.; Aston, W.J.; Higgins, I.J.; Plotkin, E.V.; Scott, L.D.L.; Turner, A.P.F. *Anal. Chem.* **1984**, *56*, 667.

17.   Jonsson, G.; Gorton, L. *Biosensors* **1985**, *1*, 355.

18.   Jonsson, G.; Gorton, L.; Petterson, L. *Electroanalysis* **1989**, *1*, 49.

19.   Foulds, N.C.; Lowe, C.R. *Anal. Chem.* **1988**, *60*, 2473.

20.   Wang, J.; Wu, L.; Lu, Z.; Li, R.; Sanchez, J. *Anal. Chim. Acta* **1990**, *228*, 251.

21.   Gregg, B.A.; Heller, A. *Anal. Chem.* **1990**, *62*, 258.

22.   Hale, P.D.; Boguslavsky, L.I.; Inagaki, T.; Karan, H.I.; Lee, H.S.; Skotheim, T.A.; Okamoto, Y. *Anal. Chem.* **1991**, *63*, 677.

23.   Higgins, I.J.; Alvarez-Icaza, M.; Hall, G.F.; Wilson, R.; Turner, A.P.F. *Pitt. Conf. Abst.* **1990**, 262.

RECEIVED November 5, 1991

Polymer Membranes on Planar Substrates

Chapter 16

Molecular Materials for the Transduction of Chemical Information into Electronic Signals by Chemical Field-Effect Transistors

D. N. Reinhoudt

Laboratory of Organic Chemistry, University of Twente, P.O. Box 217, 7500 AE Enschede, Netherlands

Synthetic receptor molecules such as crown ethers, hemispherands and calix[4]crown ethers complex selectively with cations. These complexation reactions are kinetically fast and reversible. This offers the possibility to use these processes to store chemical informaton. Field effect transistors (FETs) are attractive transducing devices because they are able to register and amplify chemical changes at the gate oxide surface of the semiconductor chip. Integration of synthetic receptor molecules and FETs into a defined chemical system allows the transduction of chemical information into an electronic signal. Surface silylation of the gate oxide of the FET and subsequent chemical attachment of polymer films reduces the original pH sensitivity of the FET and incorporation of receptor molecules in these films induces a specific sensitivity to a particular cationic species. The stability of the response of the FET is greatly enhanced by application of an intermediate polyhydroxyethyl methacrylate (polyHEMA) layer.

Different materials for the hydrophobic membrane in which the receptor is incorporated, have been investigated. Polysiloxanes that have the required glass transition temperature and dielectric constant provide a stable chemical system that transduces the complexation of cationic species into electronic signals. The material properties can be optimized by copolymerization of three building blocks viz. dimethyl-, (3-cyanopropyl)methyl-, and methacryloxypropylmethyl siloxane. CHEMFETs made with this terpolymer have fast response times ($\leq$ 1 sec.). With valinomycin and hemispherands (2) and (3) linear responses to changing K^+ concentrations are obtained in the range 10^{-5} - 1.0M (55-58 mV/decade) in a solution of 0.1M NaCl. Similar devices specific for Na^+ and Ca^{2+} have been obtained with other ionophores.

0097–6156/92/0487–0202$06.00/0
© 1992 American Chemical Society

Almost from the beginning (*1*) of "host-guest" (*2*) or "supramolecular" (*3*) chemistry, this field has been associated with possible technological applications. The fact that molecular recognition can be achieved by systematic variation of the structure of the receptor molecule offers almost unlimited possibilities to design selective receptors. However, the technological application not only requires the selective receptor but also the translation of a molecular into a macroscopic property. Our work involves both the synthesis of molecular receptors (*4-8*) and their applications e.g. in membrane transport (*9*), medicine (*10*), optical (*11*) and electronic (*12*) sensors.

In this chapter our work is described that deals with the development of chemically modified Field Effect Transistors (CHEMFETs) that are able to transduce chemical information from an aqueous solution directly into electronic signals. The emphasis of this part of our work will be on the materials that are required for the attachment of synthetic receptor molecules to the gate oxide surface of the Field Effect Transistor. In addition the integration of all individual components into one defined chemical system will be described. Finally, several examples of cation selective sensors that have resulted from our work will be presented.

Ion Sensitive Field Effect Transistors (ISFETs)

A field effect transistor (FET) measures the conductance of a semiconductor as a function of an electrical field perpendicular to the gate oxide surface (*13*). When the gate oxide contacts an aqueous solution a change of pH will change the SiO_2 surface potential ψ. A site-dissociation model describes the signal transduction, a function of the state of ionization of the amphoteric surface SiOH groups (*14*). Typical pH responses measured with SiO_2 ISFETs are 37-40 mV/pH unit (*15*). Our choice for an Ion Sensitive Field Effect Transistor (ISFET) as a transducing element was based on the fact that the SiO_2 surface contains reactive SiOH groups for the covalent attachment of organic molecules and polymers. In addition the FET has fast response times and can be made very small with existing planar IC technology.   FIGURE 1

Chemical Modification of the SiO_2 Oxide Surface

For chemical attachment of polymer films the gate SiO_2 surface was first silylated with either 3-(trimethoxysilyl)propyl methacrylate or 3-(triethoxysilyl)propylamine. The methacrylate modified surface was subsequently reacted with vinyl monomers (or prepolymers) and the amino groups, resulting from the reaction with 3-(triethoxysilyl)propylamine with isocyanato groups of polyurethanes.

The modified FETs were tested for their pH response and stability in contact with aqueous solutions. For comparison a physically attached polymer (VE) was included in these experiments (*16*). The results given in Figure 2 show that the pH sensitivity can be eliminated almost completely with acrylate polymers (ACE), polybutadiene (PBD) and also with the physically attached VE polymer. The latter, however, is not stable in time and a pH sensitivity similar to the untreated SiO_2 gate oxide returns after 5 days. The structures of the polymers are given in Scheme 1.

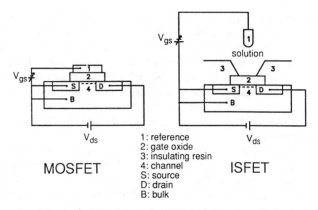

1: reference
2: gate oxide
3: insulating resin
4: channel
S: source
D: drain
B: bulk

MOSFET   ISFET

Figure 1. Schematic drawings of a MOSFET and an ISFET structure

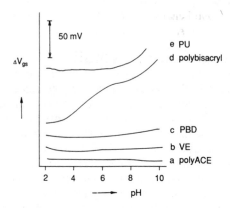

Figure 2. pH-dependence of modified ISFETs

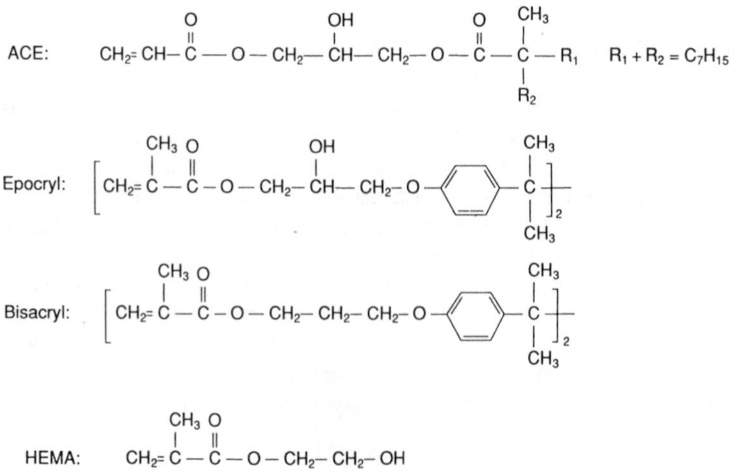

ACE:

$$CH_2= CH- \overset{\overset{O}{\|}}{C} - O - CH_2- \overset{\overset{OH}{|}}{CH} - CH_2- O - \overset{\overset{O}{\|}}{C} - \overset{\overset{CH_3}{|}}{\underset{\underset{R_2}{|}}{C}} - R_1 \qquad R_1 + R_2 = C_7H_{15}$$

Epocryl:

$$\left[CH_2= \overset{\overset{CH_3}{|}}{C} - \overset{\overset{O}{\|}}{C} - O - CH_2- \overset{\overset{OH}{|}}{CH} - CH_2- O - \underset{\underset{CH_3}{|}}{\overset{\overset{CH_3}{|}}{C}} \right]_2$$

Bisacryl:

$$\left[CH_2= \overset{\overset{CH_3}{|}}{C} - \overset{\overset{O}{\|}}{C} - O - CH_2- CH_2- CH_2- O - \underset{\underset{CH_3}{|}}{\overset{\overset{CH_3}{|}}{C}} \right]_2$$

HEMA:

$$CH_2= \overset{\overset{CH_3}{|}}{C} - \overset{\overset{O}{\|}}{C} - O - CH_2- CH_2- OH$$

Scheme 1. Structures of monomers used for polymerization

In principle such a chemically modified polyACE FET can be used as a reference FET (REFET) in differential measurements. This requires not only that the pH sensitivity is eliminated but also that a REFET should be insensitive to changes in ionic strength. In principle we can distinguish between two types of REFETs, that differ in the penetration to ions into the polymer layer. In a non-ion blocking REFET there will be ion exchange between the solution and the membrane and the resulting electrical potential is a membrane potential. In an ion-blocking REFET structure, there is a negligible ion exchange and the measured potential is a surface potential which results from reversible ion complexation reactions of the interface. We have measured the electrical relaxation by means of transconductance measurement, bulk modulation and light pulse response (*17*). These experiments revealed that polyACE REFETs is a non-ion blocking structure with electrical relaxation times of 0.37 ± 0.03 s (*18*). Because of this non-ion blocking nature it was not unexpected that polyACE REFETs exhibit an undesirable cation permselectivity (see Figure 3).

This permselective behaviour, also observed for most polymeric membranes that are used in ion selective electrodes e.g. PVC (*19*), may have two origins. A much lower mobility of anions compared with cations may give rise to a diffusion potential. Alternatively, immobile anionic sites in the membrane may influence the concentrations of exchangeable cations in the membrane, which would give rise to a boundary potential. We have been able to eliminate the cation permselectivity for a great deal by the addition of various amounts of lipophilic cations (didodecyldimethylammonium bromide, DDMAB). The initially high permselectivity of the poly ACE membrane (curve 1, Figure 3) changes upon the addition of 2.10^{-5} mol DDMAB. g^{-1} ACE) to give an almost complete insensitivity towards the K^+ concentration (curve 2). At higher DDMAB concentrations the REFET behaviour changes to that of an anion sensitive ISFET (curve 3).

Although our results show that a complete elimination of the sensitivity of such REFETs can be achieved, in a number of situations the combination of a pH sensitive ISFET and an ACE REFET can be used to measure pH values with an integrated FET device (Figure 4). An example of a differential measurement with a Ta_2O_5 ISFET/REFET combination is given in Figure 5.

Ion Sensitive FETs (CHEMFETs)

FETs that can transduce other than simple protonation/deprotonation reactions have been reported in the literature. In 1975 Moss et al (*20*) described an ISFET that has a PVC membrane *physically* attached to the gate oxide of an SiO_2 FET in which valinomycin was introduced to obtain a K^+ sensitive device. Although a response to variations in K^+ concentrations was observed there are some serious drawbacks that until now have prevented that such devices can compete with or even replace conventional Ion Sensitive Electrodes. Firstly, the physical

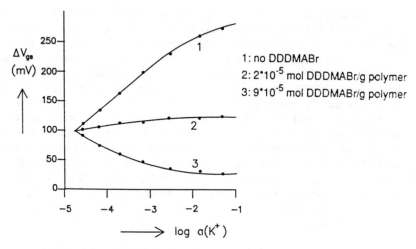

Figure 3. Permselectivity of an ACE modified FET

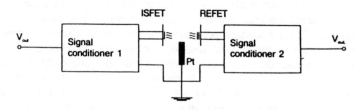

Figure 4. Differential measurement set-up

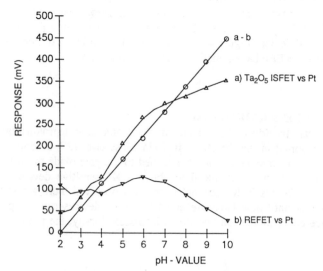

Figure 5. Differential measurement of a Ta_2O_5 ISFET/ACE REFET

attachment is not sufficiently stable in aqueous solutions and electroactive components leach out of the membrane. Secondly, the pH sensitivity is not completely eliminated, this leads among others to interference by CO_2 (*21*). However, the most serious fundamental problem is the fact that the interface between the gate oxide and the organic matrix is not defined in terms of common species. In the case of the pH sensitive ISFET this potential is thermodynamically defined by the (de)protonation equilibrium of the silanol surface groups.

In order to solve this problem we have introduced a chemically anchored polyhydroxyethyl methacrylate (polyHEMA) layer between the gate oxide and the polymer membrane that will contain the ion specific receptor molecules. Schematically this is depicted in Figure 6.

The polyHEMA layer can be buffered and the constant pH determines the potential at the gate oxide via the (de)protonation equilibrium *and keeps it at a constant* value. Besides it eliminates CO_2 interference as we could demonstrate experimentally.

The beneficial effect of the buffered polyHEMA layer on the stability of the response, even when the polymer membrane containing the valinomycine is only physically attached to this intermediate layer, is obvious. A comparison of the noise levels of systems with and without polyHEMA, given as peak-to-peak voltages (V_{pp}), is presented in Table 1.

New Materials for CHEMFETs

It is obvious that a PVC membrane can not fulfill the requirement of longterm stability because it can not be linked covalently to the polyHEMA layer and electroactive components or receptor molecules can not be covalently attached. From the work on Ion Selective Electrodes several general features for the design of membrane materials can be defined:

i)   The glass transition temperature must be low ($\leq$ o°C); current materials use additional external solvent mediators.

ii)   For cation permselectivity anionic sites e.g. tetraphenyl borates must be present. These compounds also reduce the membrane resistance.

iii)   The membrane must be hydrophobic.

If we want to apply these materials on ISFETs in addition the membrane materials should allow covalent attachment to polyHEMA, receptor molecules, and anionic sites.

We have investigated both (meth)acrylates and polysiloxanes. The first show excellent REFET properties and the latter can be synthesized in different compositions (vide infra).

All acrylate- and methacrylate-based membranes were synthesized by photopolymerization on top of the polyHEMA interlayer. The resulting FETs showed in the absence of an ionophore a cation response of 36-54 mV/decade and therefore we concluded that residual anionic groups must be present. Titration of the ACE monomer with KOH solution indicated the presence of 7.5×10^{-5} eq. acid.g^{-1} ACE monomer. As shown in Table 2 the ACE was chemically modified by reaction of the hydroxyl group. In this way acetyl, pentanoyl, and hydroxy

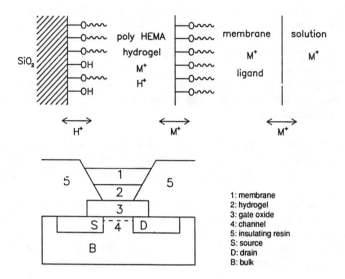

Figure 6. Schematic representation of a polyHEMA CHEMFET

Table 1. Noise level (V_{pp}) of modified CHEMFETS (0.1 M KCl)

Hydrophobic membrane	V_{pp} (mV)	
	Without polyHEMA/KCl (± 0.01)	With polyHEMA/KCl (± 0.005)
PVC[a]	0.10	0.030
ACE/Epocryl[b]	>10	0.020
Siloprene[c]	0.50	0.030
Silicone rubber[d]	0.20	0.040

[a] PVC-DOS/Valinomycin

[b] Polyacrylate/Valinomycin

[c] Siloprene (Fluka)/Valinomycin

[d] Dow Corning RTV3140/Valinomycin

hexanoyl ACE were synthesized. When valinomycin was introduced in these membranes we observed in all cases a sub-Nernstian response *and no* K^+ *selectivity*. We are unable to explain this unexpected behaviour.

Polysiloxanes (Scheme 2) can be obtained from commercially available starting materials. Three different types were investigated (*22*) viz. the addition type Wacker RTV-ME 625 and the condensation types Dow Corning 3140 RTV and Silopren (Bayer). These siloxanes form polymeric membranes by addition or

polycondensation. In contrast to the (meth)acrylates it was not necessary to introduce a polyHEMA interlayer to obtain electrically stable CHEMFETs, although such a layer has a positive effect on the reduction of the noise level of the output signal. The Wacker RTV-ME 625 is a two-component addition-type silicone rubber which is cured via the addition of SiH groups to vinyl groups under the influence of a platinum catalyst. Dow Corning 3140 RTV undergoes polycondensation by the uptake of water from the atmosphere catalyzed by a titanium catalyst. The Bayer Silopren material undergoes polycondensation in the presence of an organotin catalyst and its use in ISEs has been described (*23*). The CHEMFETs with these materials all exhibited K^+ selective responses. Optimal results were obtained with Silopren and either valinomycine (1) or hemispherands (2) and (3) as the ionophore (Figure 7).

Continuous immersion of these CHEMFETs in aqueous solution for 8 weeks showed only a minor decrease in the slope to 54 mV.(pK$^+$)$^{-1}$. This surprizing stability might be due to the bond formation of the membrane between the silanol groups and the gate oxide.

Photopolymerizable Siloxanes

The positive results with the polysiloxanes as membrane materials directly deposited on the gate oxide led us to investigate these materials for our ultimate chemical system for the transduction of complexation reactions by FETs. This work is comprised of two parts *viz.* the synthesis of photopolymerizable siloxanes that can be covalently attached to the polyHEMA intermediate layer and the synthesis of receptor molecules that are selective for K^+ which can be covalently incorporated in these membrane materials. For the deposition of the membranes on wafer scale via IC compatible technology, photochemical processes are a prerequisite.

In order to meet the requirements of Tg, dielectric constant, and covalent anchoring random terpolymers have been synthesized by emulsion polymerization of octamethylcyclotetrasiloxane, a mixture of trimer and tetramer of (3-cyanopropyl)methylcyclosiloxanes, and methacryloxypropylmethyldichlorosilane (Scheme 3). In order to optimize the properties of the resulting polysiloxane the ratio of the three reagents have been varied. An increasing amount of the cyanopropyl group will increase the dielectric constant of the membrane, and increasing the methacryl group percentage enhances the Tg. The optimal composition for a

Table 2. Effect of polymeric membrane composition on the K^+ sensitivity of CHEMFETs containing valinomycin

Polymer	Plasticizer[a]	Lipophilic salt[b]	K^+-sensitivity (mV/dec)
ACE	-	-	44
ACE	DBP (50 wt%)	-	48
ACE/Epocryl (1:1)	DBP (67 wt%)	-	36
ACE	Oct (50 wt%)	-	54
HydroxypentanoylACE	-	-	45
AcetylACE	-	-	45
PentanoylACE	-	-	45
ACE	-	$KB(PhCl)_4$	44
ACE	-	KDNNS	45

a) DBP: dibutylphtalate; Oct: Octanol

b) $KB(PhCl)_4$: potassium tetrakis(4-chlorophenyl)borate; KDNNS: potassium dinonylnaphtalene sulphonate

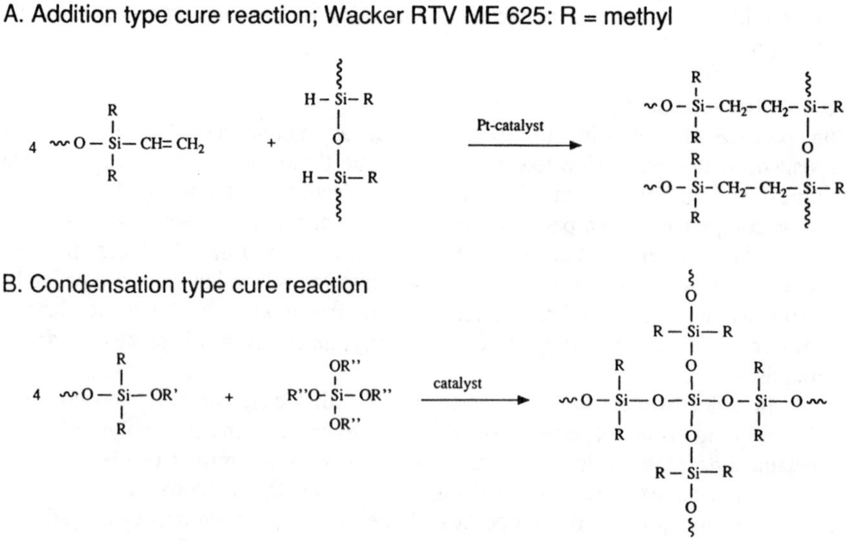
A. Addition type cure reaction; Wacker RTV ME 625: R = methyl
Pt-catalyst
B. Condensation type cure reaction
catalyst

i) Dow Corning 3140 RTV: R = R' = R'' = methyl; water from the air and a titanium catalyst are needed

ii) Bayer Silopren: R = methyl, R' = H, R'' = ethyl; an organotin catalyst is used

Scheme 2. Polysiloxane synthesis

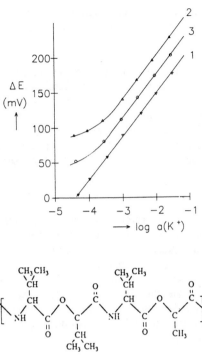

Figure 7. K⁺ response of silopren modified CHEMFETs

membrane that selectively detects monovalent cations has a ratio of 10:1:2. The materials were cross-linked by irradiation in the presence of 4 wt-% of 2,2-dimethoxy-2-phenylacetophenone.

For the synthesis of a photopolymerizable receptor molecule that has a high selectivity for K^+, we have modified the hemispherand synthesis according to Scheme 4. The final polymer membrane is represented by Figure 8. The response to a variation in K^+ concentration (10^{-5} - 10^{-1} M) in the presence of 0.1 M NaCl is given in Figure 9. The potentiometric selectivities (log K_{ij}) determined by the mixed solution method are -3.0 (Na^+), -3.5 (Ca^{2+}), -4.0 (Mg^{2+}), and -0.7 (NH_4^+). This renders this CHEMFET an excellent K^+-sensor with a longterm stability ($\geq$ 100 days).

Impedance Spectroscopy Measurements of Silicone Rubber Membranes

For the characterization of the polysiloxane membranes we have measured the impedance spectra in the presence of different concentrations of cations. Both valinomycin (1) and the two different hemispherands (2 and 3) have been incorporated in these membranes. Buck et al (24, 25) have shown that this technique allows to separate the surface exchange rates and the rates of bulk transport. The copolymer described above containing valinomycin initially has a very high resistance but after several days conditioning in a 0.1 M KCl solution the resistance is the same as that of a similar membrane to which $KB(4\text{-}ClC_6H_4)_4$ was added.

Also the final dielectric constants of both membranes are similar. This indicates that the final amount of charge carriers hardly differs and the contributions of the lipophilic anion to the membrane conductance is of the same order as of the chloride anion. Membranes with the hemispherand, either free or covalently attached, have about the same resistance. Because at the low frequency part of the impedance spectra a slight distortion was observed with the hemispherands, kinetic measurements were carried out with the valinomycin containing membrane.

When exposed to NaCl and KCl solutions the membrane shows an interesting behaviour. The impedance spectrum of a membrane conditioned in KCl shows an ideal semicircle, but when exposed to an NaCl solution the spectrum changes to a flattened semicircle, indicating an extra resistance term. In time this second contribution increased which means that the membrane resistance grows larger when the electrolyte is changed from KCl to NaCl. It was observed that the change in the spectrum is less pronounced when the membrane is exposed to low concentrations of NaCl (10^{-3} - 10^{-4} M). Obviously the ion exchange is faster at higher concentration.

The influence of the membrane selectivity on the resistance could nicely be demonstrated in a series of experiments with mixed electrolyte solutions. The NaCl concentration was maintained at 0.1 M and the KCl concentration was varied from 0.1 M to 10^{-6} M. In the concentration range 0.1-10^{-4} M KCl there is virtually no change in the spectra but at 10^{-5} M KCl there is a small change which becomes very prominent at 10^{-6} M KCl. Experiments starting from a low

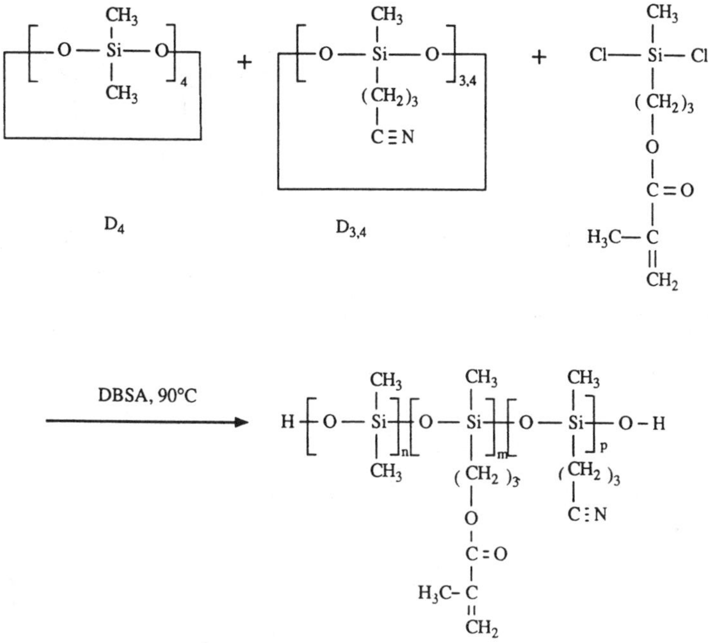

Scheme 3. Synthesis of a photopolymerizable polysiloxane

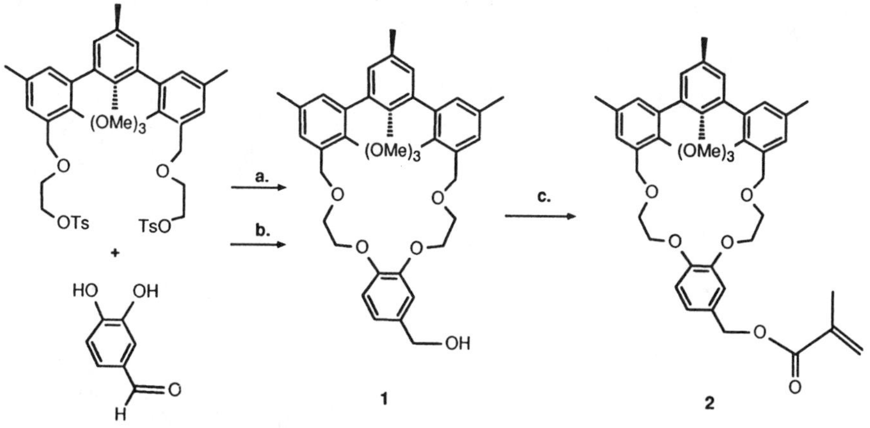

a. CsF, CH$_3$CN
b. NaBH$_4$, diglyme
c. DCC, DMAP, methacrylic acid, CH$_2$Cl$_2$

Scheme 4. Synthesis of a photopolymerizable ionophore

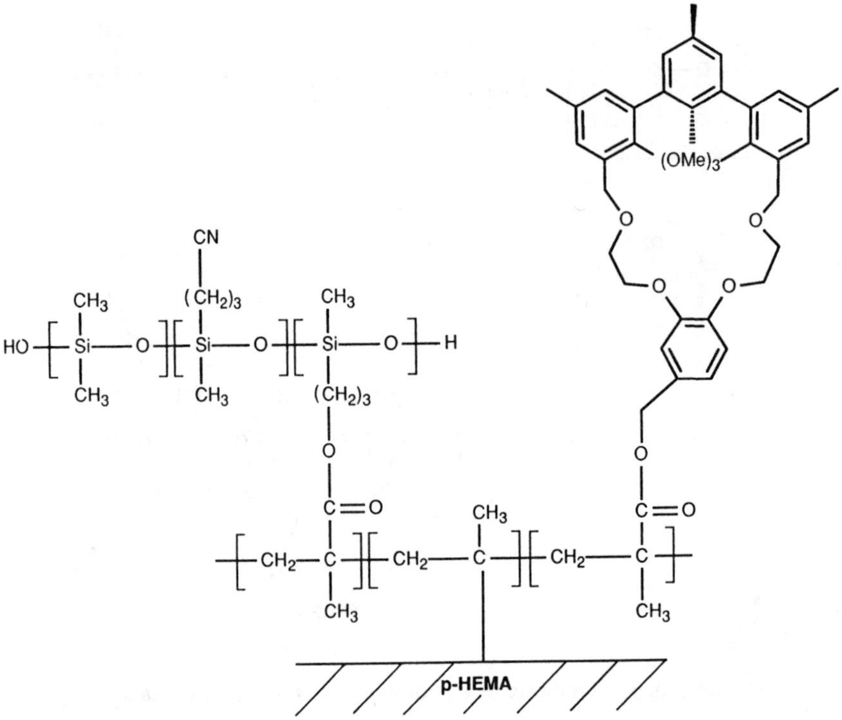

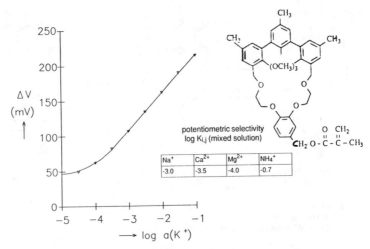

Figure 8. Schematic representation of the sensing membrane after photopolymerization

potentiometric selectivity log $K_{i,j}$ (mixed solution)			
Na$^+$	Ca^{2+}	Mg^{2+}	NH$_4^+$
-3.0	-3.5	-4.0	-0.7

Figure 9. K$^+$-response of a chemically bound ionophore

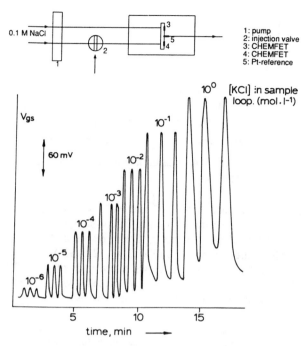

Figure 10. Flow injection analysis

concentration of KCl to 0.1 M KCl show the reversal of the events. These experiments show that the potassium/sodium ion exchange starts when the concentration ratio is of the same order as the potentiometric selectivity coefficient $\log K (K^+/Na^+) \approx -4.0$).

The Application of K^+-Selective CHEMFETs in Flow-Injection Analysis

Flow-injection analysis is a versatile technique to evaluate the performance of a detector system. CHEMFETs may have an advantage over ISEs because of their small size and fast response times. We have tested our K^+-sensitive CHEMFETs in a wall-jet cell with a platinum (pseudo-)reference electrode. One CHEMFET was continuously exposed to 0.1 M NaCl and the other to a carrier stream of 0.1 M NaCl in which various KCl concentrations in 0.1 M NaCl were injected. The linear response of 56 mV per decade was observed for concentrations of KCl above 5×10^{-5} M (Figure 9). When we used this FIA cell (Figure 10) for determination of K^+ activities in human serum and urine samples, excellent correlations between our results and activities determined by flame photometry were obtained (Figure 11).

CHEMFETs for Ca^{2+} and Na^+

Now we have developed the technology for making CHEMFETs we can extend our synthetic work on selective molecular receptors. The first targets are CHEM-

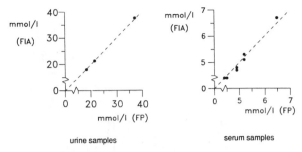

urine samples   serum samples

Figure 11. Correlation of FIA measurements with flame photometry (K^+ in human serum and urine samples)

FETS for the selective detection of Ca^{2+} and Na^+. The receptor molecules are a modified bisamide (4) and a tetra-0-alkylated calix[4]arene (5).

The Ca^{2+} CHEMFET exhibits a linear response (10^{-5} - 10^{-1} M) with potentiometric selectivities of $\leq$ - 2.6 (Na^+), $\leq$ - 2.6 (K^+), < - 3.7 (Mg^{2+}) and < -2.6 (NH_4^+). The Na^+ CHEMFET shows selectivities of $\leq$ -1.9 (K^+), $\leq$ -2.5 (Li^+), $\leq$ -3.0 (Pb^+) and $\leq$ -3.5 (Ca^{2+}). Currently we are investigating the selective detection of anions and transition metal cations.

Summary and Conclusions
We have shown that molecular recognition by synthetic receptor molecules can be transduced into electronic signals when these receptor molecules are incorporated in a polymer membrane that is covalently attached to the gate oxide of a Field Effect Transistor.

The properties of the membrane are critical and novel terpolymer siloxanes have been developed. High ion selectivities have been found for K^+, Na^+, and Ca^{2+} cation responses.

Acknowledgment: The author likes to thank all co-workers and colleagues, the names of whom are given in the literature cited. Financial support of Twente Technology Transfer BV, PRIVA BV and the Netherlands Foundations for Chemical Research (SON) with aid from the Netherlands Technology Foundation (STW), is gratefully acknowledged.

Literature Cited
1.   Pedersen, C. J. *Angew. Chem. Int. Ed. Engl.* **1988**, *27*, 1053.
2.   Cram, D. J. *Angew. Chem. Int. Ed. Engl.* **1988**, *27*, 1009.
3.   Lehn, J.-M. Angew. Chem. Int. Ed. Engl. **1988**, 27, 89.
4.   Reinhoudt, D. N.; Dijkstra, P. J. *Pure Appl. Chem.* **1988**, *60*, 477.
5.   van Staveren, C. J.; van Eerden, J.; van Veggel, F. C. J. M.; Harkema, S.; Reinhoudt, D. N. *J. Am. Chem. Soc.* **1988**, *110*, 4994.
6.   van Eerden, J.; Skowronska-Ptasinska, M.; Grootenhuis, P. D. J.; Harkema, S.; Reinhoudt, D. N. *J. Am. Chem. Soc.* **1989**, *111*, 700.
7.   van Veggel, F. C. J. M.; Bos, M.; Harkema, S.; Verboom, W.; Reinhoudt, D. N. *Angew. Chem. Int. Ed. Engl.* **1989**, *28*, 746.

8.   Ghidini, E.; Ugozzoli, F.; Ungaro, R.; Harkema, S.; Abu El-Fadl, A.; Reinhoudt, D. N. *J. Am. Chem. Soc.* **1990**, *112*, 6979.

9.   Nijenhuis, W. F.; van Doorn, A. R.; Reichwein, A. M.; de Jong, F.; Reinhoudt, D. N. *J. Am. Chem. Soc.* **1991**, *113*, 3607.

10.   Dijkstra, P. J.; Brunink, J. A. J.; Bugge, K. E.; Reinhoudt, D. N.; Harkema, S.; Ungaro, R.; Ugozzoli, F.; Ghidini, E. *J. Am. Chem. Soc.* **1989**, *111*, 7567.

11.   van Gent, J.; Lambeck, P. V.; Kreuwel, H. J. M.; Gerritsma, G. J.; Sudholter, E. J. R.; Reinhoudt, D. N. *Appl. Optics* **1990**, *29*, 2843.

12.   Reinhoudt, D. N.; Sudhölter, E. J. R.; *Adv. Mater.* **1990**, *2*, 23.

13.   Nadou, M. J.; Morrison, S. R., *Chemical Sensing with Solid State Devices*, Academic Press: Boston, 1989.

14.   Bousse, L.; Bergveld, P. *Sens. Actuators* **1984**, *6*, 65.

15.   van den Berg, A.; Bergveld, P.; Reinhoudt, D. N.; Sudhölter, E. J. R., *Sens. Actuators* **1985**, *8*, 129.

16.   Sudhölter, E. J. R.; van der Wal, P. D.; Skowronska-Ptasinska, M.; van den Berg, A.; Reinhoudt, D. N., *Sens. Actuators* **1989**, *17*, 189.

17.   Bergveld, P.; van den Berg, A.; van der Wal, P. D.; Skowronska-Ptasinska, M.; Sudhölter, E. J. R.; Reinhoudt, D. N. *Sens. Actuators* **1989**, *18*, 309.

18.   Skowronska-Ptasinska, M.; van der Wal, P.; van den Berg, A.; Reinhoudt, D. N. *Sens. Actuators* **1989**, *17*, 189.

19.   van den Berg, A.; van der Wal, P. D.; Skowronska-Ptasinska, M.; Sudhölter, E. J. R.; Reinhoudt, D. N.; Bergveld, P. *Anal. Chem.* **1987**, *59*, 2827.

20.   Moss, S. D.; Janata, J.; Johnson, C. C. *Anal. Chem.* **1975**, *47*, 2238.

21.   Fogt, E. J.; Untereker, D. F.; Norenberg, M. S.; Meyerhoff, M. E. *Anal. Chem.* **1985**, *57*, 1995.

22.   van der Wal, P. D.; Skowronska-Ptasinska, M.; van den Berg, A.; Bergveld, P.; Sudhölter, E. J. R.; Reinhoudt, D. N. *Anal. Chim. Acta* **1990**, *231*, 41.

23.   Mostert, I. A.; Anker, P.; Jenny, H.-B.; Oesch, U.; Morf, W. E.; Amman, D.; Simon, W. *Mikrochim. Acta Part 1* **1985**, 33.

24.   Lindner, E.; Niegreisz, Zs.; Toth, K.; Pungar, E.; Buck, R. P. *Anal. Chem.* **1988**, *60*, 295.

25.   Lindner, E.; Niegreisz, Zs.; Toth, K.; Pungar, E.; Berube, T. B.; Buck, R. P. *J. Electroanal. Chem.* **1989**, *259*, 67.

RECEIVED October 22, 1991

Chapter 17

Chemically Sensitive Microelectrochemical Devices

New Approaches to Sensors

Chad A. Mirkin[1], James R. Valentine[2], David Ofer[2], James J. Hickman[2], and Mark S. Wrighton[2]

[1]Department of Chemistry, Northwestern University, Evanston, IL 60208
[2]Department of Chemistry, Massachusetts Institute of Technology, Cambridge, MA 02139

New kinds of microelectrochemical sensors are described involving two redox active molecules immobilized onto a microelectrode. One redox species is chemically insensitive with respect to variation in $E_{1/2}$, e.g. a ferrocene derivative, and serves as an internal reference in a linear sweep voltammogram. The second species is chemically sensitive, e.g. a pH sensitive quinone or a CO sensitive ferraazetine derivative, which has an $E_{1/2}$ that varies with the changes in the chemical environment. A linear sweep voltammogram thus shows two waves, one for the reference molecule and one for the indicator molecule. The shift for the indicator wave along the potential or current axis provides a method for analyte detection. Surface derivitization, proof-of-structure, and proof-of-concept sensor functions are demonstrated.

Described herein are the proof-of-concept results demonstrating a new approach to electrochemical sensors based upon chemically sensitive microelectrochemical devices (1-2). A typical device consists of at least two individually addressable electrodes. The working electrode is a microelectrode and is derivatized with at least one molecule that has a chemically sensitive formal potential and serves as the indicator, and one molecule that has a chemically insensitive formal potential and serves as a reference. The indicator and reference molecules are confined to the electrodes either by monolayer self assembly techniques (3-10) or by dissolving in a thin film of solid electrolyte. Detection in these systems is accomplished by measuring either the potential difference, ΔE, associated with current peaks for oxidation (or reduction) of microelectrode-confined redox reagents, where the magnitude of ΔE can be related to the concentration of analyte, **Scheme Ia**; or in some systems, by measuring the change in current ΔI associated with the shifted peak, where ΔI can be related to the reaction of the indicator molecule with the analyte, **Scheme Ib**.

The Concept For A Two-terminal Microsensor.

Using macroelectrodes, Rubinstein was the first to demonstrate that the cyclic voltammetry of two jointly surface-confined electroactive species, one with a pH

0097–6156/92/0487–0218$06.00/0

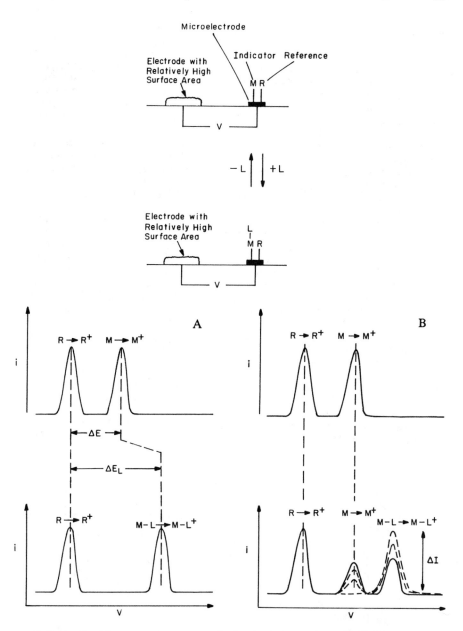

Scheme I. Concept of a two-terminal microsensor showing two possible idealized responses to a species L which binds to the indicator molecule M. A) The linear sweep voltammograms reveal a difference between the current peaks for oxidizing the reference molecule, R, and M or M-L. The position of the current peak along the potential axis, V, is variable and depends on the concentration of L. B) The linear sweep voltammograms reveal a decrease in amplitude for the current peak assigned to M and proportional growth of a new current peak assigned to M-L.

dependent redox potential and one with a pH independent redox potential, could be used to assay the pH of an electrolyte media (*11*). In the case of our two-terminal devices, current peaks for oxidation (or reduction) of the reference and indicator molecules are determined from two-terminal, linear sweep voltammograms using a counter electrode having an area much larger than the sensor electrode, **Scheme I**. A relatively large surface area counter electrode is needed so that the linear sweep of applied voltage produces only a change in the potential of the sensor electrode. For example, if the counter electrode is 10^3 times larger than the sensor electrode, the counter electrode potential will change ~1 mV upon application of a one volt potential difference. A key advantage of this device configuration is that it does not require a third reference electrode. A second advantage is that the footprint of the sensor can be quite small. And third, the system is self-assessing in that the linear sweep voltammetry yields an assessment of the surface coverage of the indicator and reference molecules. The sensor is viable as long as current peaks are measurable.

Realization of two-terminal microsensors like that represented in **Scheme I** depends on the discovery of viable reference and indicator molecules that can be confined to electrode surfaces. Described herein are three different microelectrochemical sensing systems. The first is a three-terminal microelectrochemical sensing system for CO based upon an indicator molecule, ferrocenyl ferraazetine, that selectively reacts with CO (*2*). The second microelectrochemical system is based upon a disulfide functionalized ferrocenyl ferraazetine that can be adsorbed onto Au or Pt via monolayer self-assembly techniques. Efforts to make a two-terminal CO sensor based upon the self-assembled ferraazetine will be discussed. Finally, a two-terminal system based upon the self-assembly of a pH sensitive hydroquinone and a pH insensitive ferrocene will be described (*1*).

A Solid State, Molecule-based Sensor for CO: The Solid-state Electrochemistry of Ferrocenyl Ferraazetine in MEEP/LiCF$_3$SO$_3$.

It has been reported that ferraazetine complexes **1a,b** show facile, reversible CO insertion to form ferrapyrrolinone complexes **2a,b**, equation 1 (*12-13*). We

$$\text{(1)}$$

1a, R= Bu^t
1b, R=Ph
1c, R=ferrocene

2

synthesized ferrocenyl ferraazetine **1c** with the aim of demonstrating a reversible redox active molecule which undergoes CO insertion to give a product with a different redox potential. Like **1a** and **1b**, **1c** reacts with CO to form a ferrapyrrolinone complex **2c**, in the dark. Significantly, while **1c** is photosensitive, **1c** at 298 K is chemically inert to 1 atm of the following gases: air (not containing CO), pure H_2, O_2, or CO_2. Using a solid state system similar to the ones pioneered by Murray (*14-16*), we have investigated the solid state electrochemistry of **1c** and **2c**. Our electrochemical system, diagrammed in **Scheme II**, is comprised of a microelectrode array (*17-19*), the solid electrolyte MEEP (poly[(2-(2'-methoxyethoxy)ethoxy phosphazene]) (*20-23*), and CO sensitive **1c**.

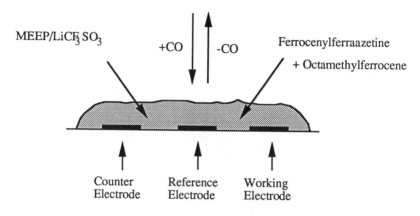

MEEP $\equiv$ Poly [bis(2-(2'-methoxyethoxy)ethoxy) phosphazene]

Scheme II. Solid-state microelectrochemical system for detecting CO.

Synthesis of a CO Sensing Molecule. Complex **1c** was isolated as a microcrystalline solid from the reaction of ferrocenyl phosphinimine, $FcN=PPh_3$ and $Fe_2(\mu\text{-}CH_2)(CO)_8$,(*24*) and has spectral features similar to **1a** and **1b**, equation (2). (*12-13*) FTIR shows that ~1 mM **1c** is converted rapidly ($t_{1/2} < 1$ min) and quantitatively to **2c** upon exposure to 1 atm of CO in THF solution, and it also confirms that the same chemistry occurs in a thin film of MEEP/Li[CF_3SO_3] at 298 K, Figures 1a and 1b respectively.

CO Dependent, Solid State Electrochemistry. The solid-state electrochemistry of **1c** and **2c** was investigated at Pt microelectrodes (144 μm long x 2 μm wide x 0.1 μm thick), **Scheme II**, and compared to the electrochemical behavior in THF/[n-Bu$_4$N]PF$_6$, Figure 2. Octamethylferrocene (*25*) has been used as an internal reference in the solid electrolyte medium because the formal potential of octamethylferrocene is ~400 mV negative of the formal potential of sensor molecule **1c**. To functionalize the electrodes, 6 mg (0.012 mmol) of **1c** and 3 mg (0.010 mmol) of octamethylferrocene were dissolved in a 50% solution of 4:1 MEEP/Li[CF_3SO_3] in THF and a film was cast on the microelectrode array. A cyclic voltammogram of a 1:1 mixture of octamethylferrocene and **1c** in a THF-saturated Ar atmosphere is shown in Figure 2a. When the atmosphere is changed to a THF-saturated CO atm, a ~100 mV shift in $E_{1/2}$ relative to the

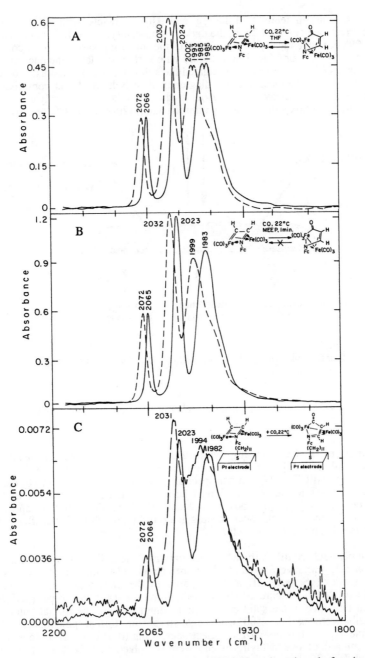

Figure 1. A) FTIR of ferraazetine, **1c**, in THF before (———) and after (-----) exposure to CO. B) FTIR of ferraazetine, **1c**, in 4:1 MEEP/[LiCF₃SO₃] before (———) and after (-----) exposure to CO. C) Specular reflectance FTIR of ferraazetine, **1d**, adsorbed onto a Pt electrode before (———) and after (-----) exposure to CO.

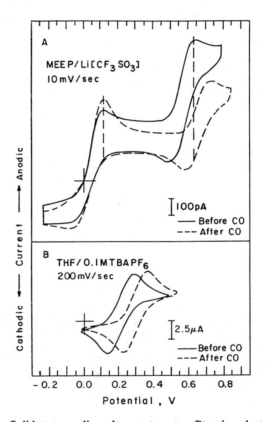

Figure 2. A, Solid-state cyclic voltammetry at a Pt microelectrode of a 1:1 mixture of octamethylferrocene (wave at more negative potential) and ferrocenyl ferraazetine (waves at more positive potentials) dissolved in a 4:1 mixture of MEEP/Li[CF$_3$SO$_3$] before and after exposure to CO. B, Cyclic voltammetry (vs AgNO$_3$/Ag) at a Pt disk (1-mm diameter) of 0.2 mM **1a** in THF/0.1 M [n-Bu$_4$N]PF$_6$ before and after the addition of CO. (Reproduced from ref. 2. Copyright 1990 American Chemical Society.)

octamethylferrocene is observed in less than 1 min. A similar effect is observed for the electrochemistry of compound **1a** in the liquid electrolyte, Figure 2b. At intermediate conversion of **1c** to **2c**, the cyclic voltammogram shows signals for each complex. Therefore, in this system sensing is accomplished by monitoring the growth of the electrochemical response for **2c**.

Irreversible CO Insertion In The Solid State. In the THF/[n-Bu$_4$N]PF$_6$ medium, the conversion of **2c** back to **1c** can be effected by purging the system of CO using an Ar stream with $t_{1/2}$< 1h at 25 °C. In the solid state-state system, reversion was atempted by pulling a vacuum for 20 min on the MEEP/Li[CF$_3$SO$_3$]/**2c** mixture. $E_{1/2}$ of the ferrocenyl unit remained constant, indicating that loss of CO does not occur. In addition, FTIR of a similarly prepared sample showed that **2c** remained in the mixture with no apparent reformation of **1c** after vacuum treatment for 20 min or upon standing for 8 h in air at 25 °C. Li$^+$ coordination to the O atom of **2c** to give **3c** may be responsible for the irreversibilty observed in the solid-state experiment, equation (2). ^{1}H NMR supports this conclusion where 7.5

$$ \textbf{2c} \qquad\qquad\qquad \textbf{3c} $$

(C(O)CHC̲H̲) and 27.5 Hz (C(O)C̲H̲CH) shifts in the resonances for the matallacyclic ring H resonances can be observed upon the addition of 0.1 M Li[CF$_3$SO$_3$] to a THF-d$_8$ solution of **2c**. Similarly, alkyl cations and H$^+$ are known to attack the O atom of the metallacyclic ring carbonyl of **2b** and **2c**, and the product formed from the methylation of **2b** has been crystallographically characterized (*13*). Furthermore, alkali cations, through an interaction similar to that proposed in equation (2), are known to inhibit the decarbonylation of transition metal acyl complexes (26).

Device Applications. Because of the irreversible reaction between **1c** and CO to form **2c** and the concomitant irreversible 100 mV shift in $E_{1/2}$ in the solid-state, accumulation of CO may be monitored by measuring the growth of the electrochemical response for **2c**. Such a device could be used to measure an accumulated exposure to CO, or simply as a disposable detector that measures a change in CO concentration in a given environment by monitoring the ratio of the currents associated with **2c** and **1c**. One of the device limitations is that because of slow diffusion in the solid electrolyte, slow scan rates must be used, and if the indicator and reference molecules of interest are unstable in one redox form, the system may not be durable. An advantage of a molecule based approach to sensing systems is that the selectivity of the system can be designed by choice of indicator molecule. In fact **1c** does not react with the usual atmospheric gases or H$_2$ demonstrating that the solid-state microelectrochemical system in **Scheme II** is selective for CO.

Towards a Two-terminal Solid-state Sensor for CO: The CO Dependent Electrochemistry of Self-assembled Ferrocenyl Ferraazetine and 11-Ferrocenylundecyl Thiol.

A CO Sensor Based Upon Self-assembled Ferrocenyl Ferraazetine. Having demonstrated the CO dependent solid-state electrochemistry of ferrocenyl ferraazetine, we synthesized a ferrocenyl ferraazetine molecule with disulfide functionality (**1d**), **Scheme III**. The specific aim was to design a CO sensitive molecule that could be confined to the working electrode of a two-terminal device via monolayer self-assembly techniques. Disulfides have been shown to irreversibly adsorb to Au and Pt surfaces (*3-10*). ^{1}H NMR and mass spectrometry are consistent with the proposed structure for compound **1d**. The FTIR spectrum of **1d** in THF exhibits metal carbonyl bands at 2067, 2024, 1989, 1985 cm^{-1} similar to the spectra for other ferraazetine derivatives **1a-c** (*2, 5-6*). Like derivatives **1a-c**, **1d** reacts with CO (1 atm) at 298 K in CH_2Cl_2 to form a ferrapyrrolinone complex **2d**, equation (3).

$$\qquad \qquad \text{(3)}$$

1d   **2d**

Electrode Modification. Au or Pt macroelectrodes (~0.02 cm^2) may be modified with **1d** by soaking them in a 0.1 mM hexane solution of **1d** at 298 K for 24 h. The electrochemical response is consistent with about one monolayer of molecules, ~2 x 10^{-10} mol/cm^2. An internal reference molecule, bis(1, 1', 11-undecylthiol)ferrocene **4a**, may be incorporated into the film by soaking the **1d** modified electrode for 30 min in a ~1 mM hexane solution of **4a**, equation (4). The electrochemical response is almost ideal ($\Delta E_p \cong 0$ and $i_p \propto$ scan rate) and shows two waves consistent with binding **4a** onto the Au. Total coverages determined from integration of the current-voltage curves are 8-9 x 10^{-10} mol/cm^2. Figure 3 shows the cyclic voltammetry of a similarly treated 0.02 cm^2 electrode before and after exposure to CO. The wave at the more negative potential is assigned to the internal reference molecule and does not shift upon exposure to CO. The wave at more positive potential is assigned to sensor molecule **1d** and it shifts 120 mV upon exposure to CO indicating formation of **2d**, equation (3). The conversion of **1d** to **2d** on Au or Pt surfaces may also be monitored by specular reflectance IR. Shown in Figure 1c, is the FTIR spectrum of a **1d** treated Pt electrode (~1 cm^2) before and after exposure to CO. The terminal CO region (2200-1800 cm^{-1}) is displayed in Figure 1c. Note the shift to higher energy for the metal carbonyl bands after the electrode is exposed to CO, and note the similarity of the spectra to those obtained for **1c** in THF solution and a MEEP/$LiCF_3SO_3$ film, Figures 1a and 1b, respectively. Similar results are obtained with Au microelectrodes (200 μm^2). The solid state electrochemistry of **1d** adsorbed onto Au and Pt microelectrodes is currently being investigated.

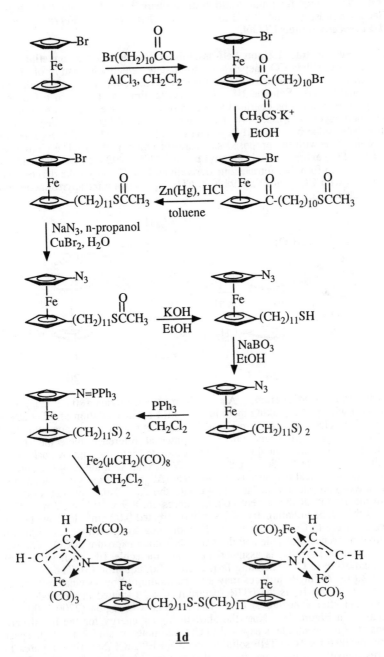

Scheme III. The synthesis of surface-confineable, CO sensing molecule, **1d**.

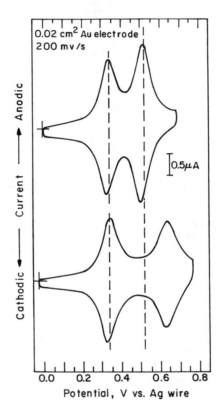

Figure 3. Cyclic voltammetry at a 0.02 cm^2 Au electrode treated with ferraazetine, **1d**, and bis(1,1', 11-undecylthiol)ferrocene, **4a**, in CH$_2$Cl$_2$/0.1 M [n-Bu$_4$N]PF$_6$ before (top) and after (bottom) exposure to CO.

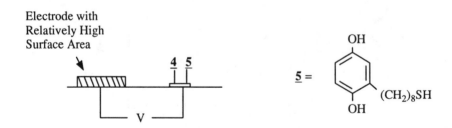

$$\xrightarrow[\substack{\text{hexane}\\24\text{ h}}]{\underline{\mathbf{1d}}} \qquad \xrightarrow[\substack{\text{hexane}\\30\text{ min}}]{\underline{\mathbf{4a}}} \qquad (4)$$

Au or Pt electrode

4a, R^1= (CH$_2$)$_{11}$SH, R^2= (CH$_2$)$_{11}$SH

4b, R^1=H, R^2= (CH$_2$)$_{11}$SH

4c, R^1=H, R^2= $\overset{\overset{\displaystyle O}{\|}}{C}$(CH$_2$)$_{10}$SH

Two-terminal, Voltammetric Microsensor with an Internal Reference For Measuring H$^+$ Activity.

This system is prepared by the self-assembly of a hydroquinone alkane thiol **5**, which has a pH dependent redox center (27), and the chemically insensitive

Electrode with
Relatively High
Surface Area

5 =

(CH$_2$)$_8$SH

Scheme IV. Cross-sectional view of a two-terminal voltammetric microsensor based on the self-assembly of a hydroquinone thiol **5** and a ferrocenyl thiol **4** serving as reference and indicator respectively.

ferrocenyl alkane thiol **4b** or acyl alkane thiol, **4c** (28-29), **Scheme IV**. **4b**, **4c**, and **5** have been previously attached separately to Au electrodes (30-33). Au microelectrodes ($\sim$10^3 μm^2) or macroelectrodes ($\sim$1 cm^2) can be modified with **5** and **4c** by dipping the Au into solutions containing one or both of the thiol reagents. Figure 4 shows cyclic voltammograms for Au macroelectrodes modified with pure **4c**, pure **5**, and a combination of the two. The electrochemical response

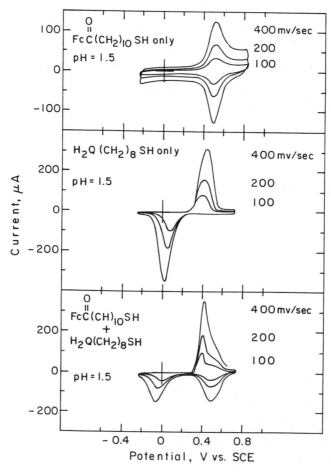

Figure 4. Cyclic voltammetry at three scan rates for Au macroelectrodes in 1.0 M NaClO$_4$ at pH 1.5 buffer (phosphate) derivatized with only acyl ferrocene thiol, **4c**, (5.2 x 10^{-10} mol/cm^2) (top); only quinone thiol, **5**, (5.6 x 10^{-10} mol/cm^2) (middle); and a mixture of **4c** and **5** at 2.8 x 10^{-10} mol/cm^2 and 2.8 x 10^{-10} mol/cm^2, respectively (bottom). Reproduced with permission from ref. 1. Copyright 1991 American Association for the Advancment of Science.

is persistent and consistent with about one monolayer of redox active molecules. Coverages determined from integration of the current-voltage curves are 3-5 x 10^{-10} mol/cm^2. The cyclic voltammetry for a Au microelectrode derivatized with **4c** and **5** at two values of pH, Figure 5, indicates that the redox potential for the Fc$^+$/Fc system, equation (5), is pH insensitive, whereas the redox response for the Q/QH$_2$ system, equation (6), depends on pH. The microelectrodes may be run vs. a macroscopic Pt counter electrode in a two-terminal configuration, and the cyclic voltammograms are superimposable on the curves shown in Figure 5 where a saturated calomel reference electrode (SCE) was used. Figure 6 shows, for pH 0-10, the lack of an effect of pH on $E_{1/2}$ of the surface confined Fc$^+$/Fc system and also shows the pH dependence of the potential difference between the cathodic current peaks, ΔE_{pc}, for the processes shown in equations (5) and (6). The linear response to the solution pH comprises the basis for a pH sensor system where the Fc$^+$/Fc serves as the reference and the Q/QH$_2$ serves as the indicator, **Scheme I**.

$$\text{(5)}$$

4

$$\text{(6)}$$

5

    The surface-confined Fc$^+$/Fc system behaves ideally (*34*) at all values of pH investigated (0-10). The surface-confined Q/QH$_2$ is not ideal, in that there is a large difference in the potential for the anodic and cathodic current peaks. Such behavior is well documented for other quinones (*35-36*). Despite the non-ideality, the effect of pH on the electrochemical response of Q/QH$_2$ is reproducible. Both the anodic and cathodic current peaks for the quinone system shift to more positive potentials at lower pH.

    A pH Sensor For Highly Acidic Media. Study of electrodes modified with **4b** and **5**, and subsequently examined in highly acidic media establishes a possible application of this device. Figure 7 illustrates the electrochemical response in aqueous media containing different concentrations of HClO$_4$. Amazingly, the electrochemical response of the redox molecules persists even in 10 M HClO$_4$. Note that the response for the Q/QH$_2$ system moves from ~0.5 V negative (pH = 11) of **5** to ~0.5 V positive (10 M HClO$_4$) of **5** for the media used. Measuring H$^+$ activity in highly acidic media is thus possible with a **4b/5**-modified electrode. Earlier work has established the constancy of the redox potential of a surface-confined ferrocene at very high H$^+$ activity (*36*), justifying our use of ferrocene as an internal reference in such media.

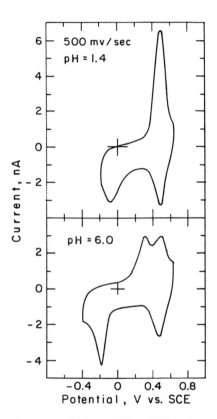

Figure 5. Cyclic voltammetry (500 mV/s) of Au microelectrodes derivatized with a mixture of **4c** and **5** at pH 1.4 and pH 6.0. The solutions used were phosphate buffers in 1.0 M NaClO$_4$ base electrolyte; an SCE reference electrode was used. Reproduced with permission from ref. 1. Copyright 1991 American Association for the Advancment of Science.

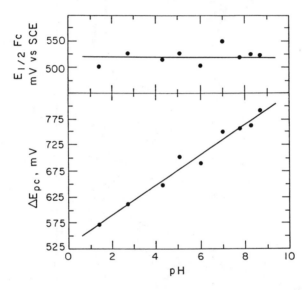

Figure 6. Plot of $E_{1/2}$ vs. SCE for the surface-confined acyl ferrocene thiol vs. pH (top) and plot of difference in cathodic current peak for surface-confined acyl ferrocenium and quinone vs. pH from two-terminal, voltammetric scans. All data are from voltammograms recorded at 500 mV/s in 1.0 M $NaClO_4$ in buffered solution. Reproduced with permission from ref. 1.   Copyright 1991 American Association for the Advancment of Science.

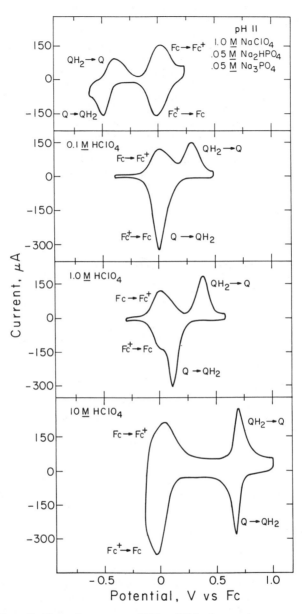

Figure 7. Cyclic voltammetry (500 mV/s) for a Au macroelectrode derivatized with alkyl ferrocene thiol, **4b**, and quinone thiol, **5**, in (from top to bottom): 1.0 M NaClO₄ buffered to pH 11 with phosphate; 0.1 M HClO₄; 1.0 M HClO₄; 10 M HClO₄. Note that the reference is taken to be the average position of the oxidation and reduction waves for the ferrocene system. Reproduced with permission from ref. 1. Copyright 1991 American Association for the Advancment of Science.

Summary. The proof-of-concept microsensor systems described herein can be easily extended by use of other specific indicator molecules. It should be clear that monolayer redox reagents and microelectrodes are not required, but their use has the advantage that very small amounts of charge are involved in detection. The self-assembly of thiol and disulfide reagents provides a reproducible method, applicable to many chemical functionalities (3-10, 30-33), for assembly of device-active materials, but there are a large number of other electrode modification techniques that can be useful (34). Au electrodes derivatized with thiol reagents are quite robust, but obviously long term durability is an issue in many sensor applications. The pH sensing system can be used intermittently over a period of several weeks with reproducible response to variation in pH. The two CO sensing systems are not nearly as durable as the pH sensing system and the signal is lost after prolonged cycling. Nevertheless the systems illustrate some of the important requirements necessary to make useful two-terminal microelectrochemical sensors. Experiments are in progress to demonstrate multiple-response, two-terminal microsensors by incorporating more than one indicator molecule.

Experimental

Instruments and equipment: IR spectra were recorded using either a Nicolet 60SX or 170SX Fourier transform infrared spectrometer. ^{1}H NMR were recorded using a Bruker WM250 or AC250 Fourier transform spectrometer and mass spectra were recorded using a Finnegan MAT System 8200 mass spectrometer.

Materials: All solvents were dried by stirring over Na/benzophenone (pentane, tetrahydrofuran, toluene) or CaH_2 (CH_2Cl_2) and were freshly distilled prior to use. All chemicals were commercially available and used as received unless otherwise specified. All reactions and procedures were performed under Ar atmosphere unless otherwise specified. Bu_4NPF_6 was recrystallized from CH_3CH_2OH prior to use. MEEP was donated by Professor Harry Allcock at the Pennsylvania State University. Octamethylferrocene and compounds **1c** (2), **4b**, **4c**, and **5** (2) were prepared by literature methods. $[N_3CpFeCp-(CH_2)_{11}S]_2$ was prepared by a modification of the Nesmayonov method for the preparation of triazoferrocene from bromoferrocene (37). Experiments involving **1c** and **5** have been previously described (1-2).

Synthesis of [(PPh)$_3$P=N-CpFeCp-(CH$_2$)$_{11}$S]. To a CH_2Cl_2 solution of $[N_3CpFeCp-(CH_2)_{11}S]_2$ (100 mg, 0.082 mmol) was added 25 mg (0.092 mmol) of PPh_3. The color changed from orange to red and after 2 h, the solvent was evaporated. Chromatography on alumina with 1:1 CH_2Cl_2/pentane as the eluent gave unreacted $[N_3CpFeCp-(CH_2)_{11}S]_2$. When the eluent was changed to 1% diethyl ether in CH_2Cl_2, a dark orange band of $[(PPh)_3P=N-CpFeCp-(CH_2)_{11}S]_2$ came off the column. The solvent was removed via rotary evaporation yielding 95 mg (0.073 mmol) of an oily yellow product, $[(PPh)_3P=N-CpFeCp-(CH_2)_{11}S]_2$.
$[(PPh)_3P=N(Fc)(CH_2)_{11}S]_2$: Yield = 90% MS (FAB+, High Res.) Cald: 1293.4798; Found: 1293.4793. ^{1}H NMR: δ 7.81 (m, 6H), 7.02 (m, 9H), 4.09 (m, 2H), 3.92 (m, 2H), 3.89 (m, 4H), 2.62 (t, 2H, J_{HH}= 7.5 Hz), 2.56 (t, 2H, J_{HH}= 7.6 Hz), 1.66 (m, 2H), 1.37-1.18 (m, 16H).

Synthesis of CO sensor molecule 1d. 95 mg (0.073 mmol) of $[(PPh)_3P=N(Fc)(CH_2)_{11}S]_2$ was dissolved in CH_2Cl_2 and added to a flask containing 28 mg (0.08 mmol) of $Fe_2(\mu-CH_2)(CO)_8$ dissolved in 20 mL of CH_2Cl_2. The mixture was stirred for 24 h, and the color changed from yellow to dark red.

The solvent was removed via rotary evaporation and the residue was chromatographed on silica gel. 1:3 CH_2Cl_2:pentane as an eluent gave a light yellow band of unreacted PPh_3. 1:1 CH_2Cl_2:pentane as eluent brought down a red band of **1d**. Solvent was removed via rotary evaporation yielding 69 mg (0.05 mmol) of a microcrystalline solid, **1d**.

1d: Yield = 68 % **MS** (FAB+, High Res.) Cald: 1381.0075; Found: 1381.0080. **IR** (CH_2Cl_2) 2066 (m), 2026 (vs), 1989 (s), 1995 (s) cm^{-1}. ^1H **NMR** (C_6D_6) δ 6.04 (d, J_{HH}= 1.9 Hz), 5.53 (d, J_{HH}= 1.9 Hz), 3.97 (m), 3.93 (m), 3.74 (m), 3.59 (m), 3.74 (m), 3.59 (m), 2.57 (t, 2H, J_{HH}= 7.25 Hz), 2.19 (t, 2H, J_{HH}= 8.0 Hz), 1.66 (m, 2H), 1.31 (m, 16 H).

Electrode surface derivatization with 1d and 4a. In a typical experiment, Au or Pt electrodes of the appropriate size were soaked in a 0.1 mM hexane solution of **1d** for 24 h. Cyclic voltammetry in CH_2Cl_2/Bu_4NPF_6 solution indicated monolayer coverage (4-8 x 10^{-10}) of **1d** as did specular reflectance FTIR spectroscopy. The **1d** treated electrodes were then soaked in a 0.1 M hexane solution of **4a** for 15 min. Cyclic voltammetry in CH_2Cl_2/Bu_4NPF_6 indicated that the monolayer formed in this manner consisted of approximately 1:1 **1d:4a**.

Exposure of a 1d and a 4a treated electrode to CO. An electrode treated with **1d** and **4a** in the aforementioned manner was placed in a 50 mL 3-neck flask containing 30 mL of 0.1 M CH_2Cl_2 solution of Bu_4NPF_6. CO was then bubbled into the solution for 10 min and a cyclic voltammogram was subsequently recorded. The wave assigned to **1d** had shifted +120 mV relative to the **4a** wave. Similarly, when a 1x1 cm^2 Au electrode treated with **1d** and **4a** was placed in a vial and purged with CO for 10 min, specular reflectance IR indicated surface conversion of **1d** to **2d**.

Acknowledgments

This work has been supported by the National Science Foundation under grant CHE-9002006 awarded in 1990 to C.A.M. in the form of an NSF Postdoctoral Research Fellowship and in part on grant support to M.S.W. from the National Science Foundation, the Office of Naval Research, and the Defense Advanced Research Projects Agency.

Literature Cited

1. Preliminary results have been reported in Hickman, J.J.; Ofer, D.; Laibinis, P.E.; Whitesides, G. M.; Wrighton, M. S. *Science* **1991**, *252*, 688.
2. Preliminary results have been reported in Mirkin, C. A.; Wrighton, M. S. *J. Am. Chem. Soc.* **1990**, *112*, 8596.
3. Nuzzo, R. G.; Allara, D. L. *J. Am. Chem. Soc.* **1983**, *105*, 4481.
4. Brown, A. P.; Koval, C.; Anson, F. C. *J. Electroanal. Chem.* **1976**, *72*, 379.
5. Bain, C. D.; Whitesides, G. M. *J. Am. Chem. Soc.* **1988**, *110*, 5897.
6. Porter, M. D.; Bright, T. B.; Allara, D. L.; Chidsey, C. E. D. *J. Am. Chem. Soc.* **1986**, *108*, 3559
7. Bain, C. D.; Troughton, E. B.; Tao, Y. T.; Evall, J.; Whitesides, G. M.; Nuzzo, R. G. *J. Am. Chem. Soc.* **1989**, *111*, 321.
8. Bain, C. D.; Whitesides, G. M. *Science* **1988**, *240*, 62.
9. Whitesides, G. M.; Laibinis, P. E. *Langmuir* **1990**, *6*, 87.
10. Bain, C. D., Whitesides, G. M. *Angew. Chem.* **1989**, *101*, 522.
11. Rubinstein, I. *Anal. Chem.* **1984**, *56*, 1135.

12. Mirkin, C. A.; Lu, K-L; Geoffroy, G. L.; Rheingold, A. L.; Staley, D. L. *J. Am. Chem. Soc.* **1989**, *111*, 7279.
13. Mirkin, C. A.; Lu, K-L; Snead, T. E.; Young, B.; Geoffroy, G. L.; Rheingold, A. L. *J. Am. Chem. Soc.* **1991**, *113*, 3800.
14. Pinkerton, M. J.; Mest, L. E.; Zhang, H.; Watanabe, M.; Murray, R. W. *J. Am. Chem. Soc.* **1990**, *112*, 3730.
15. Geng, L.; Longmire, M. L.; Reed, R. A.; Parker, J. F.; Barbour, C. J.; Murray, R. W. *Chem. Mater.* **1989**, *1*, 58.
16. Geng, L.; Longmire, M. L.; Reed, R. A.; Kim, M-. H.; Wooster, T. T.; Oliver, B. N.; Egekeze, J.; Kennedy, J. W.; Jorgenson, J. W.; Parker, J. F.; Murray, R. W. *J. Am. Chem. Soc.* **1989**, *111*, 1614.
17. White, H. S.; Kittlesen, G. P.; Wrighton, M. S. *J. Am. Chem. Soc.* **1984**, *106*, 5375.
18. Paul, E. W.; Ricco, A. J.; Wrighton, M. S. *J. Phys. Chem.* **1985**, *89*, 1441.
19. Talham, D. R.; Crooks, R. M.; Cammarata, V.; Leventis, N.; Schloh, M. O.; Wrighton, M. S. Proceedings of NATO ASI "Lower-Dimensional Systems and Molecular Electronics" Spetses, Greece, 1989, R. M. Metzger, P. Day, G. Papavassiliou, eds., Plenum Press: New York, 1991, p. 657-664.
20. Allcock, H. R.; Austin, P. E.; Neenan, T. X.; Sisko, J. T.; Blonsky, P. M.; Shriver, D. F. *Macromolecules* **1986**, *19*, 1508.
21. Austin, P. E.; Riding, G. H.; Allcock, H. R. *Macromolecules* **1983**, *16*, 719.
22. Blonsky, P. M.; Shriver, D. F.; Austin, P. E.; Allcock, H. R. *J. Am. Chem. Soc.* **1984**, *106*, 6854.
23. Blonsky, P. M.; Shriver, D. F.; Austin, P. E.; Allcock, H. R. *Solid State Ionics*, **1986**, *18*, 258.
24. Sumner, C. E., Jr.; Collier, J. A.; Pettit, R. *Organometallics* **1982**, *1*, 1350.
25. Schmit, V. G.; Ozman, S. *Chem. Zeit.* **1976**, *100*, 143.
26. Collman, J. P.; Hegedus, L. S.; Norton, J. R.; Finke, R. G. *Principles and Applications of Organotransition Metal Chemistry*; University Science Books: Mill Valley, CA, 1987; pp 355-399 and references therein.
27. "Introduction to Organic Chemistry", A. Streitwiser and C. Heathcock, 2nd Ed., Macmillan, New York (1981), pg 1014-1020.
28. Peerce, P.J.; Bard, A. J. *J. Electroanal. Chem.* **1980**, *108*, 121.
29. Modro, T. A.; Yates, K.; Janata, J. *J. Am. Chem. Soc.* **1975**, *97*, 1492.
30. Stern, D. A.; Wellner, E.; Salaita, G. N.; Laguren-Davidson, L.; Lu, F.; Batina, N.; Frank, D. G.; Zapien, D. C.; Walton, N.; Hubbard, A. T. *J. Am. Chem. Soc.* **1988**, *110*, 4885.
31. Hubbard, A. T. *Chem. Rev.* **1988**, *88*, 633.
32. Chidsey, C. E. D.; Bertozzi, C. R.; Putvinski, T. M.; Mujsce, A. M. *J. Am. Chem. Soc.* **1990**, *112*, 4301.
33. Hickman, J.J., Ofer, D.; Laibinis, P. E.; Whitesides, G. M.; Wrighton, M. S. *J. Am. Chem. Soc.* **1991**, *113*, 1128.
34. Murray, R. W. in *Electroanaltical Chemistry*, A. J. Bard, Ed. (Dekke, New York, 1984), Vol 13, p 191 and references therein.
35. Laviron, E.; *J. Electroanal. Chem.* **1983**, *146*, 15.
36. Laviron, E.; *J. Electroanal. Chem.* **1984**, *164*, 213.
37. Nesmayanov, A. N.; Drozd, V. N.; Sazonova, V. A. *Dokl. Akad. Nauk SSSR*, **1963**, *150*, 321.

RECEIVED October 22, 1991

Chapter 18

Macro- to Microelectrodes for In Vivo Cardiovascular Measurements

Richard P. Buck[1], Vasile V. Coşofreţ[1], Tal M. Nahir[1],
Timothy A. Johnson[2], Robert P. Kusy[3], Kirk A. Reinbold[3],
Michelle A. Simon[3], Michael R. Neuman[4], R. Bruce Ash[5], and
H. Troy Nagle[5]

[1]Department of Chemistry, [2]Department of Medicine, Division
of Cardiology, and [3]Biomedical Engineering, University of North Carolina,
Chapel Hill, NC 27599
[4]Department of Obstetrics and Gynecology, MetroHealth Medical Center,
Cleveland, OH 44109
[5]Department of Electrical and Computer Engineering, North Carolina State
University, Raleigh, NC 27695

Chemical sensors for cardiology include passive voltage measuring probes (e.g. inert metals or $Ag/AgCl,Cl^-$), passive membrane sensors for ions (most importantly H^+, K^+, Na^+ and Ca^{2+}) and current passing electrodes for defibrillation studies (and related pacemaker applications) and for simple monitoring (e.g. dissolved oxygen). These categories are important for three main lines of research: ischemic events, arrhythmias and sudden cardiac failure.

For the first studies, three general classes of cardiovascular changes have been identified, and our electrodes are important to each: electrical (e.g. decreased resting membrane potential, ST-TQ segment changes, fall in upstroke velocity), metabolic (e.g. loss of ATP and intracellular O_2, gain in CO_2 and lactic acid) and ionic (e.g. fall in intracellular and extracellular pH, concomitant rise in extracellular K^+ concentration). Ischemic events have ionic markers, specifically hydrogen and potassium ions, with characteristic and reproducible time courses of change after brief periods of ischemia. Crucial questions remain unanswered on irreversible and irreproducible ionic concentration changes after long term ischemia or multiple occlusions. How these are related to the chemistry and energetics can be probed with effective sensors.

In arrhythmia studies, understanding the pattern of fields and voltages during normal heartbeat, during fibrillation, and following defibrillation is a principle goal. To this end, micro voltage sensors, with long term stability and ability to resolve local potentials in space, are required. Chemical consequences of fibrillation and defibrillation require in vivo monitoring of chemical species, especially H^+, K^+, Na^+ and Ca^{2+}. Also required are assays for neutral charge species such as oxygen and various isozymes.

0097–6156/92/0487–0237$06.00/0

A major challenge is development of materials, manufacturing technology and sensor products for acute in vivo monitoring and for chronic applications in cardiology and in vascular diagnostics and intervention. Implantable macrosensors serve as a guide to design of microsensors of flexible (Kapton polyimide) and stiff (silicon "stiletto") formats for spatial resolution of chemical species concentrations in myocardium and epicardium. Materials synthesis, selection, characterization and testing are followed by prototype macrosensor fabrication, testing in vitro, in suspensions, in tissues and in animals. Parallel studies of implanted materials and implanted electrodes give the history of rejection or biocompatibilty and the directions for new materials synthesis and selection.

RESULTS AND DISCUSSION

Our purpose is to produce practical, chemically characterized and physically tested chemical sensors for in vivo applications in cardiology and, ultimately, other biomedical fields. The present main focus is on macro membrane-based pH sensors, reduction to micro size, in vivo testing and preliminary application by cardiologists.

IMPROVED pH and pK$^+$ SOLVENT-POLYMER MEMBRANE SENSORS
        Failure sources and analysis have pointed out previous design inadequacies (1,2). Selection and synthesis of new materials for optimized polymer electrode strength (resistance to puncture) have uncovered better materials, e.g. new and improved poly(vinyl chloride), PVC, of very high molecular weight (3). The new PVC's, made by Occidental Chemical Company, Pottstown, PA, include a product 410 and experimental products made by suspension polymerization. Fractionation of an experimental PVC and two very high molecular weight (VHMW) commercial PVC's by Polysciences and Aldrich has been achieved using gel permeation chromatography (GPC) with tetrahydrofuran solvent. The molecular weight distributions in Figure 1 were obtained. Various fractions can be combined to develop polymer blends with controlled physical characteristics.
        In Fig. 2 is shown a determination of the relationship between relative viscosity of PVC solutions in cyclohexanone and GPC - determined weight average molecular weights. A series of Occidental Chemical Co products were used. The data were fit to the Mark-Houwink equation:

$$\eta = KM^{\alpha} = 1.23x10^{-4} \; M^{0.80} \tag{1}$$

The predicted solid line is simply a linear regression of the experimental and commercial values.
        Membrane puncture strength measurements were performed on plasticized samples of Aldrich VHMW and Occidental 410 PVC's. Samples were 33.3 wt% PVC, and 66.7 wt%. dioctylsebacate (DOS) plasticizer. Thicknesses were calculated from densities using weights and areas. Results are given in Figure 3. Membranes were hydrated for 24 hours to duplicate conditions of use. Smaller membranes are stronger per unit thickness.

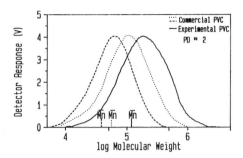

Fig. 1 Fractionation of poly(vinylchlorides) using HPLC/GPC; --- Polyscience PVC; VHMW Aldrich PVC; solid line: experimental Occidental Chemical Co. Suspension-polymerized PVC; PD = polydispersity index = 2 (random distribution).

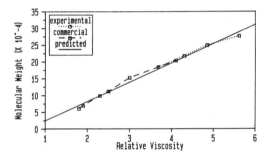

Fig. 2 Molecular weight vs. relative viscosity; Polymers were "off-the-shelf" Occidental Chemical Co. commercial products, as well as our experimental high molecular weight PVC produced in small batches at Occidental Chemical Co.

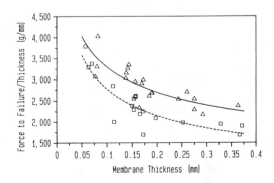

Fig. 3 Strength-Thickness relationship for Occidental Chemical Co. PVC.

Ultimate tensile strength vs. molecular weights for PVC samples are shown in Figure 4. Tensile strengths were determined from 4-inch dumbell-shaped PVC film samples.

Carrier based pH membranes (4-7) have traditionally required the addition of trapped, hydrophobic negative sites, typically tetraphenylborate (TPB) and p-chlorotetraphenyl borate (p-ClTPB). In comparative studies we have frequently noticed the improved pH response of the membranes containing additional sites compared with those with only naturally occurring fixed sites, found in all the PVCs we have tested. Specifically, there is a distinctive deterioration in accuracy in the latter sensors at low pH. In addition, membranes prepared from aminated PVC with TPB have previously shown a good pH response (8). However, our preliminary impedance studies have shown that undoped aminated PVC membranes have a relatively low conductivity when compared with the neutral carrier designs above.

Earlier sensors were made of plasticized PVC with neutral carriers: TDDA (tridodecylamine) for H^+ and valinomycin for K^+. Accelerated deterioration tests of the proton sensor have been performed by heating TDDA in nitrogen, air and oxygen (9). Partially destroyed TDDA carriers have then been incorporated into the electrodes, and their responses were tested. Results demonstrated a deleterious effect of air oxidation on the response slopes.

A major concern in neutral carrier ion selective electrodes is leakage of the ionophore from the membrane into the test solution. Under the suggested in vivo use, this problem must be eliminated. Recently, we have tested the pH response of a plasticized PVC membrane containing a new, exceedingly oil-soluble carrier, which contains a lipophilic imino chain {9-(diethylamino)-5-octadecanoylimino-5-benzo phenoxazine}, or DOBP (Figure 5). The results are comparable to or better than those obtained with TDDA membranes, in both the near-Nernstian slope and the strong rejection of interfering cations. Furthermore, the intrinsic noise level of the new carrier electrode was about 1/10 the value for the glass electrode in the same solutions. Some of the noise reduction may stem from the lower resistance of the plastic film electrode.

To interpret the mechanisms related to the proper response at different pH levels, impedance studies were conducted on TDDA and DOBP membranes in various solutions. In both types of membranes, with and without TPB, the bulk semicircle decreased in size when the ion-selective membranes were soaked in low pH bathing solutions (Figure 6). This can be related to the previous observation (5) that at very acidic media, i.e. at pH less than 3, amine compounds, such as TDDA, tend to be protonated in the membrane. Consequently, chloride ions from contacting test solution enter the membrane to satisfy electroneutrality. This failure of the Donnan exclusion principle implies the presence of more ions in the membrane, and therefore a higher conductivity or a reduced real impedance ("size" of the semicircle). Obviously, under such circumstances the membranes cease to be selective to H^+.

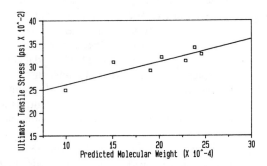

Fig. 4 Ultimate Tensile Strength vs. molecular weights of PVC samples predicted from relative viscosity measurements for Occidental Chemical Co. experimental products.

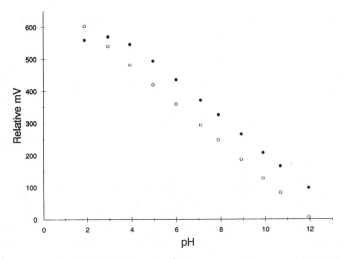

Fig. 5 pH response of a membrane containing DOBP (solid circles), compared with a glass electrode response (open circles).

MACRO TO MICROSENSOR DESIGNS

A macrosensor design for cardiology has been described previously. These devices, while considered "macro", are quite small, but do not require microelectronics type of processing (10,11). As noted, during ischemic events local changes in ion distribution, oxygen, lactate and isozymes are expected. Therefore, microsensors with good spatial resolution and rapid response times are needed. Application of microsensor technology to the fabrication of the potentiometric pH sensors allows reproducible geometries at the tens of microns spacing as one dimensional arrays. Presently, micro versions of the same devices are approached using the "stiletto" or dagger design with multiple sensing sites, typically squares 40 x 40 microns or, alternatively, circles 50 microns in diameter. The overall length of a dagger, shown in Figure 7, is approximately 1 cm, and the bonding pad width is about 0.8 cm. Each of the sensing sites contains several or all of the components shown in Figure 8.

To date, dagger designs have been used for voltage measurements by coating the silver sites with silver chloride. These electrodes function as well as similar devices based on tungsten wires as voltage probes. Also, the design features of a sensor shown in Figure 8 were used in a prototype pH sensor approximately 2 mm in diameter. The responses to different pH buffer solutions were as good as a reference glass electrode.

ACUTE IMPLANTATION

Our target is a reliable pH sensor for acute response applications in vivo to supplement voltage measuring electrodes and K^+ sensors. Using the design of a sensor on a Kapton substrate, we have prepared a relatively large (2 mm diameter) K^+ ion selective electrode and tested it in vivo in a pig heart. The results are shown in Figure 9, and are similar to those obtained from a silver wire based potassium ion sensor. Since the difference in design between this and a H^+ sensor is only the neutral carrier selected and incorporated into the PVC membrane, we expect a good performance from a similar design of a pH sensor. Work towards this end is now in progress.

CHRONIC IMPLANTATION

Chronic in vivo implantation of passive and active sensors is among the most important and challenging aspects in the sensor field. The collection of sensor designs described above may prove completely or partially inadequate for the in vivo context. For example, in a series of tests conducted on microelectrodes implanted in rats, a significant deterioration of bare AgCl sites was evident after 14 days. In some cases, even the underlying silver was gone. A possible solution which is under investigation now, is a biocompatible polymer coating, which will double as an ion selective membrane, or a biocompatible coating of porous character that permits easy access to the chemically sensitive device. Many coating have been considered, polyHEMA, polyethylene oxides and polyurethanes.

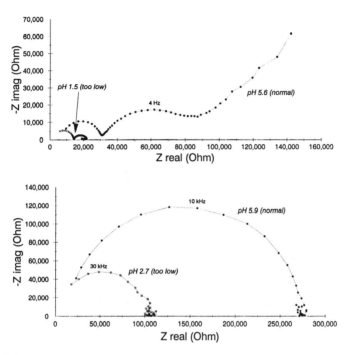

Fig. 6 Impedance plane plots for TDDA membranes with TPB (top) and without TPB (bottom) as a function of bathing solution pH.

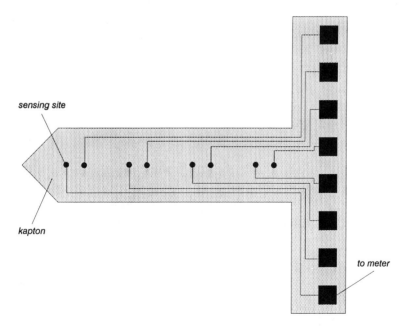

Fig. 7 "Stiletto" type micro "plunge" electrode.

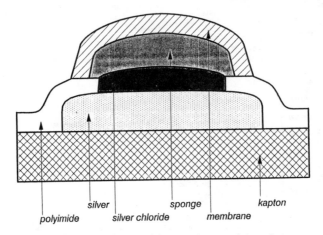

Fig. 8 Cross section of an ion selective site.

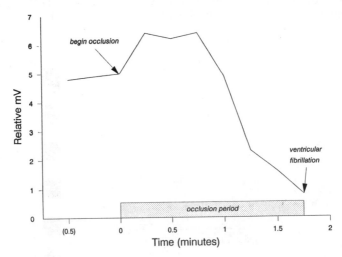

Fig. 9 Subepicardial recording using a potassium sensor on Kapton.

CONCLUSIONS

The design of microelectrodes for cardiovascular measurements involves many challenges, both in science and engineering. The composition and size of the sensors and the connections to them must be guided by the goal of in vivo testing. The problems of design of microsensors, their calibration and stability are particularly great, since these devices need to be implanted for long periods of time. They must be robust enough to withstand the cardiovascular dynamics. Local hostile environment reaction to a sensor is almost inevitable, especially when the implantation is for long periods of time.

To confront these difficult tasks, we have incorporated old and new solutions. For example, strong yet flexible Kapton polyimide substrates are promising, and the neutral-carrier type ion-selective membranes offer the advantage of using similar fabrication methods for sensors of different ions of interest. We are also working on the biocompatibility problems, and all new designs are subject to in vivo tests.

ACKNOWLEDGEMENT

The authors wish to thank the ERC (Engineering Research Center) funded by National Science Foundation, North Carolina Biotechnology Center, and industrial associates, and the Occidental Chemical Company.

LITERATURE CITED

1. Kusy, R.P. and Buchanan, J.W., Microbeam Analysis, (Russell, P.E. ed.), San Francisco Press, 1989, pp 63-64.
2. Kusy, R.P., Whitley, J.Q. and Buchanan, J.W., A. I. Ch. E., National Meeting on Materials of Biomedical Applications, Philadelphia, PA, August 1989.
3. Simon, M.A., Department of Biomedical Engineering, University of North Carolina, Chapel Hill, unpublished results.
4. Erne, D., Amman, D., and Simon, W., Chimia, 33, 88 (1979).
5. Amman, D., Lanter, F., Steiner, R.A., Schulthess, P., Shijo, Y. and Simon, W., Analyt. Chem., 53, 2267 (1981).
6. Oesch, U., Brzozka, Z., Xu, A., Rusterholz, B., Suter, G., Pham, H.V., Welti, D.H., Amman, D., Pretsch, E. and Simon, W., 58, 2285 (1986).
7. Chao, P., Amman, D., Oesch, U., Simon, W. and Lang, F., European J. of Physiology (Pflugers Arch.), 411, 216 (1988).
8. Ma, S.C., Chaniotakis, N.A. and Meyerhoff, M.E., Analyt. Chem., 60, 2293 (1988).
9. Reinbold, K. A., Master's thesis, Curriculum of Biomedical Engineering, University of North Carolina, Chapel Hill, August 1990.
10. Johnson, T.A., Engle, C.L., Kusy, R.P., Graebner, C.A. and Gettes, L.S., The Electrochemical Society Spring Meeting, Montreal, Canada, May 1990.
11. Johnson, T.A., Engle, C.L., Kusy, R.P., Knisley, S.B., Graebner, C.A. and Gettes, L.S., Am. J. Physiology, 258, H1224 (1990).

RECEIVED October 22, 1991

Chapter 19

Polymeric Matrix Membrane Field-Effect Transistors

Sodium Ion Sensors for Medical Applications

S. Wakida

Material Chemistry Department, Government Industrial Research Institute, Osaka, Midorigaoka 1–8–31, Ikeda, Osaka 563, Japan

Sodium ion-selective field-effect transistors (Na^+ ISFETs) were prepared by using three different types of polymeric matrix materials, such as polyvinyl chloride, bio-compatible polymer (polyurethane) and Urushi (natural oriental lacquer). Their electrochemical characteristics were discussed in connection with their characteristics of polymeric matrix membranes.

Ion-selective field-effect transistors (ISFETs) are ion sensors that combine the electric properties of gate-insulator field-effect transistors and the electrochemical properties of ion-selective electrodes (ISEs). ISFETs have attracted much attention for clinical and biomedical fields because they could contain miniaturized multiple sensors and could be routinely used for continuous *in vivo* monitoring of biological fluid electrolytes (*e.g.*, Na^+, K^+, Ca^{2+}, Cl^-, *etc.*) during surgical procedures or at the bedside of the patients in clinical care unit (*1*).

Two types of Na^+ ISFET have been reported so far. One was an inorganic sodium-aluminum-silicate (NAS) glass ISFET, which was fabricated by the hydrolysis of a mixed solution of metal alcoholates, followed by thermal treatment (*2*), or by the ion implantation technique (*3,4*). The other type of Na^+ ISFET was prepared by coating with so-called solvent polymeric membrane, such as polyvinyl chloride (PVC) membrane.

In the present paper, the electrochemical characteristics of Na^+ISFETs with PVC, Urushi and bio-compatible polymer (KP-13, polyurethane) are discussed in connection with their characteristics of polymeric matrix membranes.

Experimental

Preparation of KP-13. A pre-polymer of KP-13, which has isocyanate groups at it's both terminal group, was synthesized with polyethylene oxide, polydimethyl-siloxane, polyethylene oxide, polytetramethylene glycol and 4,4'-diphenylmethane diisocyanate in the presence of diazobicycloundecene catalyst at 50 °C for 1 hour in the

0097–6156/92/0487–0246$06.00/0

mixed solvent of dioxane and N,N-dimethylacetamide. The KP-13 was obtained by chain-lengthening reaction of the prepolymer with ethylene glycol at 50 °C for 2 hours. The synthesis of KP-13 is outlined in Scheme 1. The synthetic method was reported in detail elsewhere (5). The antithrombogenicity of the KP-13 shows better results with Lee White test and implantation test in canine vein, compared with Cardiothane[TM] and Biomer[TM] (5).

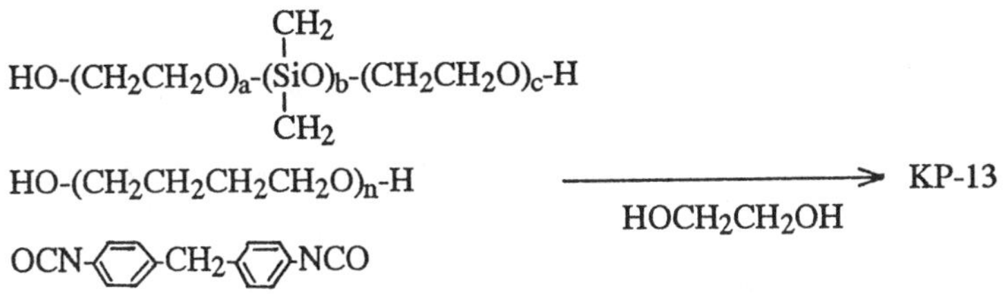

$$CH_2$$
$$|$$
$$HO-(CH_2CH_2O)_a-(SiO)_b-(CH_2CH_2O)_c-H$$
$$|$$
$$CH_2$$

$$HO-(CH_2CH_2CH_2CH_2O)_n-H \qquad\qquad\longrightarrow KP\text{-}13$$
$$HOCH_2CH_2OH$$

$$OCN{-}\langle\bigcirc\rangle{-}CH_2{-}\langle\bigcirc\rangle{-}NCO$$

Scheme 1   Preparation of KP-13; (a+c):b = 68:32.

Preparation of sodium PVC/ISFET.   Na[+] PVC/ISFETs were obtained by dip-coating with the tetrahydrofuran (THF) solution composed of the mixture of 7.1 *wt.* % of Na ionophore   (1,1,1-tris(1'-(2'-oxa-4'-oxo-5'-aza-5'-methyl)dodecanyl)propane; ETH 227, Fluka AG), 63.9 *wt.* % of 2-nitrophenyloctylether (NPOE; Dojin Research Laboratories Co. Ltd.), 0.4 *wt.* % of sodium tetraphenylborate (NaBPh$_4$, Dojin Research Laboratories Co. Ltd.) and 28.6 *wt.* % of PVC (Dojin Research Laboratories Co. Ltd.) onto the Si$_3$N$_4$ gate of the ISFET devices (0.5 mm x 5.5 mm x 0.2 mm; catheter type ISFET donated by Shindengen Electric Mfg. Co. Ltd.) at several times to avoid pin-holes. The resulting Na[+] PVC/ISFETs were allowed to dry overnight. The thickness of the membrane was approximately 0.1 mm.

Preparation of sodium Urushi/ISFET.   A mixture of 5 *wt.* % of ETH 227, 45 *wt.* % of di-2-ethylhexylphthalate (DOP; Kishida Chemical Co. Ltd.) containing 0.5 *wt.* % of potassium tetrakis(4-chlorophenyl)borate (Dojin Research Laboratories Co. Ltd.) and 50 *wt.* % of Urushi (Saito Urushi Co. Ltd.) was coated on the FET devices and then the resulting Na+ Urushi membranes were hardened for 10 days at 30 °C and 90% relative humidity. The thickness of the membrane was approximately 0.1 mm. The surface of the Urushi matrix membrane was lustrous, smooth and adhesive to the gate of the device. The hardening mechanisms were discussed in detail elsewhere (6).

Preparation of sodium   KP-13/ISFET.   Na[+] KP-13/ISFETs were obtained by dip-coating with the THF mixture of 4 *wt.* % of ETH 227, 36 *wt.* % of NPOE containing 0.5 *wt.* % of NaBPh$_4$ and 60 *wt.* % of KP-13 onto the device in the same manner as PVC/ISFETs. The thickness of the membrane was approximately 0.1 mm.

Measurements with sodium ISFETs.   After the prepared ISFETs were conditioned in 10^{-3} M NaCl solution for a few hours to stabilize the potential response, their potential response was measured *vs.* Ag/AgCl reference electrode with a source-follower circuit (ISFET mV/pH meter; Shindengen Electric Mfg. Co. Ltd.) at 25 °C in the dark. The frozen horse serum (Working Certified Reference Serum for ISEs;

Chemical Inspection and Testing Institute, Tokyo) was used as the standard blood serum.

Results and discussion

Sensitivity of sodium ISFETs. The Na$^+$ ISFETs with PVC, Urushi and KP-13 showed almost the same linear response range in the Na$^+$ activity range from $10^{-3.5}$ M to 10^0 M and showed almost the same sensitivity (slope of potential change per decade) of 55, 53 and 50 mV per decade change of Na$^+$ activity, respectively as shown in Figure 1.

Response times of sodium ISFETs.

The response times of the three kinds of Na$^+$ ISFETs are also within seconds of one another. The above-mentioned characteristics, such as linear response range, sensitivity and response time are almost the same in PVC, Urushi and KP-13 matrix ISFETs.

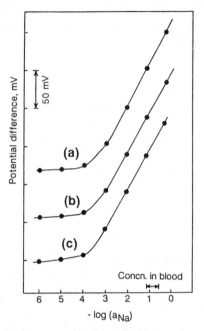

Figure 1.   Calibration curves of Na$^+$ ISFETs (a) PVC, (b) Urushi, (c) KP-13.

Selectivity of sodium ISFETs.   The selectivity of the Na$^+$ ISFETs were evaluated with the selectivity coefficients obtained by the mixed solution method. The selectivity coefficients are summarized in Table I. The selectivity of the PVC and Urushi matrix ISFETs are almost the same, however, KP-13 matrix ISFETs have   relatively poor selectivity against K$^+$, as compared to the PVC and Urushi matrix ISFETs. Since the concentrations of Na$^+$ and K$^+$ of the blood are 138-146 mM and 3.7-4.8 mM respectively, it is considered that Na$^+$ KP-13/ISFETs have some interference of K$^+$ in measuring blood serum.

Table I.   Logarithms of selectivity coefficients of Na$^+$ ISFETs

Interfering *ion*(M)	*Concn. of* *M*(M)	$\log K_{NaM}^{pot1)}$	$\log K_{NaM}^{pot2)}$	$\log K_{NaM}^{pot3)}$
K$^+$	10^{-1}	-1.9	-1.5	0.1
NH$_4^+$	10^{-1}	-1.6	-1.6	-0.5
Ca^{2+}	10^{-1}	0.1	0.0	-0.7
Mg^{2+}	10^{-1}	-1.8	-2.4	-2.3

1) PVC/ISFET; 2) Urushi/ISFET; 3) KP-13/ISFET

It has been reported that the polypropylene glycol (PPG) membrane responds to alkali metals (7). Since KP-13 (a kind of polyurethane) has the same polyether parts as PPG, it is considered that the reason of the different selectivity may be the direct interaction of the polyether part of the KP-13 matrix material.

Matrix mechanisms of sodium Urushi and PVC/ISFETs.   The electro-chemical characteristics, such as linear response range, sensitivity, selectivity and response time of the Urushi matrix ISFETs are similar to those of the PVC matrix ISFETs. The reason of the same characteristics is discussed from the standpoint of matrix mechanisms as follows. The obtained results indicate that these characteristics are mainly determined not by polymeric matrix materials but by sodium-sensing materials, including the membrane solvent (NPOE *etc.*). Therefore, it is considered that the polymeric matrix materials, such as PVC and Urushi only act as a hydrophobic support polymer and that the major part of surface of the matrix membrane should be covered with the membrane solvent containing the Na ionophore.

In the Urushi matrix membrane, it is supposed that the solvent at the surface would be supplied from the porous bulk phase of the mechanically hard Urushi matrix membrane by capillary action, which was discussed in detail elsewhere (6), while in the PVC matrix membrane, the solvent of the surface would be supplied by the leaching process from the mechanically soft PVC matrix membrane, because the electron micrograph of the PVC matrix membrane indicates structural damage (8).

Stability of sodium ISFETs.   The PVC and KP-13 matrix ISFETs have some drift characteristics of a few mV per hour. The lifetimes of the both ISFETs are about 1 week. It is considered that the drift and durability is caused by the poor membrane adhesion to the ISFET device (9). The Urushi matrix ISFETs exhibited a drift < 0.1 mV per hour and durability > 1 month because of the strong adhesion of the Na^+ sensing membrane to the ISFET device.

Johnson *et al.* reported adhesion studies of some polymeric matrix membrane by using PVC, Urushi and copolymer (vinylchloride/vinyl alcohol copolymer matrix membrane with treatment of tetrachlorosilane) by using an ultrasonic bath (10). They also reported the use of Urushi gives improved adhesion lasting over 5 hours and that in some cases the membrane lasted for over 20 hours. The stability of the Urushi matrix ISFETs was discussed in detail elsewhere (6).

Application to standard blood serum.   The potential responses of the Na^+ ISFETs in standard blood serum are shown in Figure 2. In the PVC/ISFETs, the potential value decreased vigorously with time. It is supposed that this behavior is caused by coating with components, such as protein *etc.*, in blood serum over the PVC matrix membrane. The KP-13 and Urushi matrix ISFETs gave a stable response corresponding to the Na^+ activity, though the response time is about 1 minute respectively. The slow response may be due to the adsorption of serum protein. The fairly stable response of the Urushi matrix ISFETs compared with the PVC matrix ISFETs might be brought about by the difference of the solvent supply mechanism. Since the Urushi matrix ISFETs have some drift problems with regard to long term stability of the continuous measurements of blood serum, the studies on the improved bio-compatibility of the Urushi matrix ISFETs are now under way.

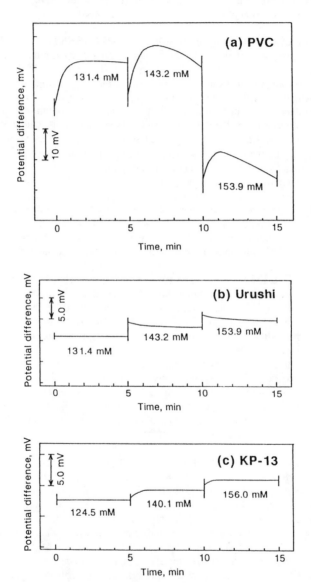

Figure 2.   Potential *vs.* time responses of Na⁺ ISFETs in standard blood serum;
(a) PVC, (b) Urushi, (c) KP-13.

Acknowledgments

The author wishes to thank Dr. Higashi, Dr. Ujihira and Dr. Hiiro for valuable comments and discussions.

Literature Cited

1   *Analytical and Biomedical Applications of Ion-Selective Field-Effect Transistors;* Bergveld, P.; Sibbald, A., Eds.; Comprehensive Analytical Chemistry 23, Elsevier, Amsterdam, 1988.
2   Abe, H.; Esashi, M.; Matsuo, T., *IEEE Trans. Electron Devices,* **1979**, *ED-26,* pp.1939.
3   Sanada, Y.; Akiyama, T.; Ujihira, Y.; Niki, E., *Fresenius' Z. Anal. Chem.,* **1982**, *312,* pp.526.
4   Ito, T.; Inagaki, H.; Igarashi, I., *IEEE Trans. Electron Devices,* **1988**, *ED-35,* pp.56.
5   Kira, K.; Minokami, T.; Yamamoto, N.; Hayashi, K.: Yamashita, I., *Seitai Zairyo,* **1983**, *1,* pp.646.
6   Wakida, S.; Yamane, M.; Higashi, K.; Hiiro, K.; Ujihira, Y., *Sens. and Actuators,* **1990**, *B1,* pp.412.
7   Jaber, A. M. Y.; Moody, G. J.; Thomas, J. D. R., *Analyst,* **1977**, *102,* pp.943.
8   Harrison, D. J.; Cunningham, L. L.; Li, X.; Teclemariam, A.; Permann, D., *J. Electrochem. Soc.,* **1988**, *135,* pp.2473.
9   Janata, J., *Chem. Rev.,* **1990**, *90,* pp.691.
10   Johnson, S.; Moody, G. J.; Thomas, J. D. R., *Anal. Proc.,* **1990**, *27,* pp.79.

RECEIVED October 23, 1991

Chapter 20

Polymeric Membranes for Miniature Biosensors and Chemical Sensors

M. Koudelka-Hep, A. van den Berg, N. F. de Rooij

Institute of Microtechnology, University of Neuchâtel, Breguet 2, 2000 Neuchâtel, Switzerland

On-wafer membrane deposition and patterning is an important aspect of the fabrication of planar, silicon based (bio)chemical sensors. Three examples are presented in this paper : amperometric glucose and free chlorine sensors and a potentiometric ISFET based calcium sensitive device. For the membrane modified ISFET, photolithographic definition of both inner hydrogel-type membrane (polyHEMA) and outer siloxane-based ion sensitive membrane, of total thickness of 80 μm, has been performed. An identical approach has been used for the polyHEMA deposition on the free chlorine sensor. On the other hand, the enzymatic membrane deposition for a glucose electrode has been performed by either a lift-off technique or by an on-chip casting.

Conceptually most electrochemical (bio)sensors rely on a combination of a solid-state transducer and one or more organic membranes. These membranes impart the sensor sensitivity, selectivity, interference free detection, diffusion limitation and outer protection. In the case of miniature devices, the transducer part is frequently realized using silicon microfabrication technology which allows a highly uniform production of small transducers for both, potentiometric and amperometric type sensors. However, the advantages of the well established microfabrication technology are somewhat mitigated by the second step of the sensor realization, i.e. the membrane deposition and adhesion onto planar transducers, that often limits the sensor reproducibility, performance and life-time. In consequence, the membrane deposition technology together with the development of new membrane materials are likely to be, in the near future, the most important means to effectively design new devices and to improve existing ones.

In our earlier work we have studied the realization of an amperometric transducer using thin film and photolithography techniques (1) as well as that of an Al_2O_3 type pH-ISFET (2). The use of the former in an enzymatic glucose electrode (3) and of the latter in for example a K^+ sensitive device (4) has been reported. In both cases the control of membrane thicknesses and adhesion was the most critical parameter of the sensor realization. In this paper we will focus on different on-wafer and on-chip

0097–6156/92/0487–0252$06.00/0
© 1992 American Chemical Society

membrane deposition techniques investigated with a view to improve the overall control of the sensor fabrication process. Deposition of four different membranes - enzymatic and polyurethane membranes for the glucose sensor, polyHEMA membrane for a free chlorine sensor and polyHEMA plus polysiloxane membrane for the Ca^{2+} sensitive device are described and results obtained using different deposition techniques are reported.

Experimental and Results

Transducers : Described in detail elsewhere (*1,3*), the realization of the amperometric transducer consists basically of the photolithographic patterning of two thin-film metallic layers i.e. Pt and Ag, which were deposited by electron beam evaporation onto the $Si/SiO_2/Al_2O_3$ substrate. The resulting planar electrochemical microcell (overall dimensions of 0.8 mm x 3 mm x 0.38 mm) comprises three electrodes - Pt working and counter electrodes and a Ag/AgCl reference electrode. The surface area of the working electrode is 0.1 mm^2.
The ISFET transducer, described in a previous paper (2), is an n-Si, p-well type device with the sensitive gate consisting of a thermally deposited SiO_2 (1000 Å) and an APCVD deposited Al_2O_3 (600 Å) layers. The gate area is 20 μm x 500 μm and the overall chip dimensions are 0.75 mm x 3 mm x 0.38 mm.

Enzymatic Membrane : The enzymatic membranes of the glucose sensor are formed using the chemical co-cross-linking of glucose oxidase (50 mg/ml) and bovine serum albumin (80 mg/ml) by glutaraldehyde. The membranes are deposited either on each individual transducer by casting or on-wafer by spin coating followed by lift-off. Individual casting allows rather thick membranes of 5 - 30 μm to be obtained using different cast volumes. The reproducibility of membrane thickness is typically ± 20 % within a series of 10 electrodes. The spin-coated membranes, on the other hand, present the advantage of better thickness control (± 5 % on one wafer), but the attainable thickness of membranes is currently limited to 2 μm. This results from the low viscosity of the enzymatic mixture. Although it can be increased within a certain limit by increasing the overall proteinic concentration this was not yet sufficient to obtain the thicker membranes of about 10 μm which are required for *in vivo* glucose sensors. It has been found experimentally that a membrane of this thickness when covered with an outer polyurethane membrane gives an optimal compromise between sensor response (diffusion-controlled) and mechanical stability (thicker membrane are less stable upon hydration). The sensors with spin-coated enzymatic membranes of 1 μm thickness have a very short response time (t $_{90}$ = 10 s) and linear range up to 1.5 mM glucose (5) and can be used in, for example, flow injection analysis of diluted glucose samples.
The sensors having cast enzymatic membranes allow measurements of glucose in a concentration range up to 6 mM, but with high dependence on dissolved oxygen concentrations. For glucose measurements *in vivo*, the linear range has to be extended (the physiological range of interest lies between 3 and 16 mM) and the pO_2 dependence minimized. This is achieved by depositing an additional membrane of polyurethane, dip coated from a 4 % polyurethane solution in tetrahydrofuran and dimethylformamid (9 : 1 v/v). The resulting effect is the extension of the linear range, covering now the clinically important glucose range and to lower the dependency of the sensor signal on changing $pO_2 \geq 5\%$ (3).
An example of the response of the glucose sensor implanted subcutaneously in a rat is shown in Figure 1. The glycaemia modification was performed by an intramuscular administration of glucagon at time zero followed by insulin at time 30'. The so called "apparent subcutaneous glycaemia" (*6,7*), calculated using one point *in vivo* calibration matched quite closely the concomittantly measured plasma glycaemia.

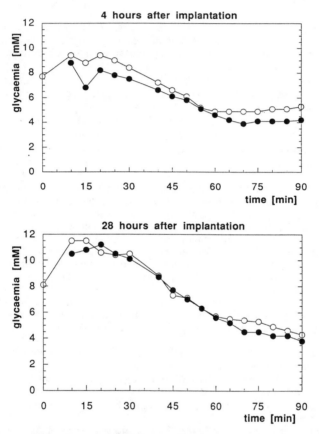

Figure 1. Apparent subcutaneous glycaemia (closed symbols) calculated using one point calibration compared to the actual plasma glycaemia (open symbols), measured by conventional method. Upper pannel shows the sensor response after 4 hours of implantation and the lower one after 28 hours of implantation. Conditions : intramuscular administration of glucagon (2.6 μg) at time zero followed by insulin (100 mU) at time 30'.

The same kind of glucagon-insulin test is also used for longer-term experiments. The results of the first series of tests (n = 15) showed that, when periodically calibrated, 90% of implanted sensors followed correctly the blood glycaemia changes for three days and 35% up to seven days. The problem of various interferences, especially that of paracetamol, is now under investigation.

Hydrogel Membranes : The adhesion of enzymatic membranes was satisfactory without any prior pretreatment of the surface. The case of the hydrogel membranes of polyHEMA type is different, since in order to ensure good membrane adhesion, a chemical pretreatment has to be performed prior to the membrane deposition. Since the membrane material used contains methacrylic groups that are crosslinked during polymerization, a silanization with a methacrylic functional silane was chosen to modify the transducer surface. Although the treatment of oxide surfaces with silanes is well-documented (8), the application of such treatment to the surface of Pt electrodes was investigated in our group. A special procedure using an oxygen plasma to form an appropriate platinum oxide on the electrode surface has been developed (9). The Pt plasma oxidation, unlike the commonly used electrochemical oxidation, has the advantage of being applicable on-wafer level and so the next step, the wafer functionalization with 3-(trimethoxysilyl)propyl methacrylate (Aldrich) can be carried out on whole wafers (10). The functionalization can be easily verified by cyclic voltammetry measurements of individual electrodes (9).

A certain volume (depending on the required membrane thickness) of the mixture composed of 57.5 % wt. hydroxyethyl methacrylate (HEMA) (Fluka), 38 % wt. ethyleneglycol (EG) (Merck), 1 % wt. dimethoxyphenylacetophenone (DMAP) (Aldrich), 2.5 % wt. polyvinylpyrrolidone K 90 (PVP) (Aldrich) and 1 % wt. tetraethyleneglycol dimethacrylate (TEGDMA) (Fluka) is cast on the wafer by micropipette. A Mylar® sheet is then pressed onto the wafer in order to avoid the oxygen quenching of the photopolymerization reaction. The monomer mixture is then photopolymerized (UV-light) using a photomask. Exposure times of 30 s to 3 min were used and the development was carried out in ethanol.

The membrane thickness reproducibility was tested over the wafer on 32 evenly distributed membranes. A mean thickness of 61.8 μm (s.d. = 7.3 μm) was found across the whole wafer, while a lower ($\pm$ 4 μm) typical standard deviation was found on more closely located membranes. Prior to measurements, the membranes are preconditioned in 0.1 M KCl. No degradation of the membrane adhesion was observed during an immersion of 2 hours in an ultrasonic bath containing 0.1 M KCl.

Free Chlorine Sensor : Figure 2 shows cyclic voltammograms of a bare amperometric transducer carried out in order to determine the optimum value of the polarization voltage for free chlorine detection. As can be seen, the oxygen interference on the chlorine detection starts to be prohibitive at potentials E < 0 V vs. SCE. This was also confirmed by chronoamperometric measurements performed at 0.15V, 0.05 V and - 0.05 V (Figure 3). Clearly at the lowest potential an offset current due to the oxygen reduction is observed. At higher potentials this offset is suppressed with the highest sensitivity corresponding to a polarization voltage of 0.05 V. This voltage was chosen to perform the calibration of two sensors with membrane thicknesses of 10 and 50 μm respectively. The calibration curves (Figure 4) were obtained by adding appropriate amounts of 10^{-2} M NaClO solution to 2.10^{-3} M NaCl (pH 7) solution. A small sensitivity difference at low and high free chlorine concentration was observed in both cases (2.2 and 1.8 nA/mg/l for the 10 μm membrane, 0.45 and 0.3 nA/mg/l for the 50 μm thick membrane).

The same transducers covered with the polyHEMA membranes can serve as a basis for several other amperometric sensors. The optimum trade-off between rapid response (thin membrane) and stirring independency (thicker membranes) should be found for each particular application (11). In many cases, the hydrogel membrane

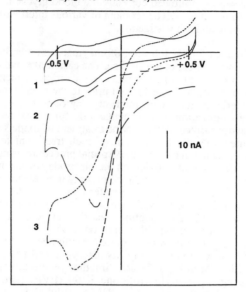

■ 10^{-1} M KCl, pH 7, 50 mV/s

■   1)N_2 2)N_2 +10^{-3} M HClO   3)ambient air

Figure 2. Cyclic voltammetry showing oxygen interference on chlorine detection. Bare amperometric sensor in a 0.1 M KCl solution, 1) N_2, 2) N_2 + 10^{-3} M HClO and 3) air saturated. Electrode : thin film Pt, scan range from 0.55 V to - 0.6 V vs SCE, scan rate 50 mV/s. (Reproduced with permission from ref. 9. Copyright 1991 IEEE.)

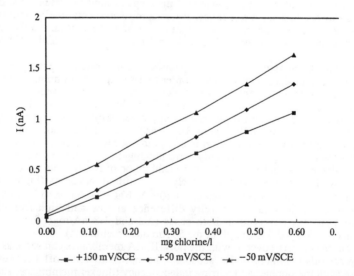

Figure 3. Calibration curve of the chlorine sensor with a 10 μm polyHEMA membrane at three different potentials : lower: 150 mV, middle 50 mV and upper -50 mV vs SCE. (Reproduced with permission from ref. 9. Copyright 1991, IEEE.)

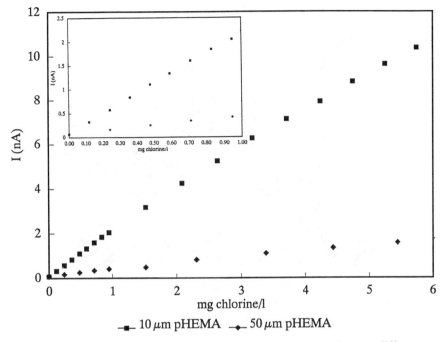

Figure 4. Chlorine sensor calibration curves at E = 50 mV for two different thicknesses of the polyHEMA membrane : 10 μm (upper) and 50 μm (lower). Insert shows in detail the lower concentration part of the calibration graph. (Reproduced with permission from ref. 9. Copyright 1991, IEEE.)

constitutes only the inner part of the sensor construction, that is they are often covered with an additional, usually hydrophobic, membrane. The technology of the deposition and patterning of double membrane layers will be presented here for the case of a Ca^{2+} sensitive ISFET.

Ca^{2+} Sensitive Polysiloxane Based Membranes : The necessity of using an inner hydrogel membrane in the case of membrane modified ISFETs has been demonstrated previously (*10, 12, 13*). The outer membrane that was used here is a polysiloxane-type ion sensitive membrane. This type of membrane has the advantage over the conventionally used plasticized PVC membrane of being intrinsically plastic and possessing good adhesion properties to several substrates (*14*). Moreover, the methacrylic functionality present in the chosen polysiloxane enables not only a UV-initiated polymerization and cross-linking but also ensures a covalent bond to the previously silanized surface. Following the silanization procedure, polyHEMA membranes of varying thicknesses (20 - 50 μm) were deposited and patterned on the pH-sensitive ISFET-gates using the same procedure as described for the amperometric transducers. The membranes were preconditioned in a 1 : 1 mixture of a pH 4 buffer and 0.1 M $CaCl_2$.

The ion sensitive membrane mixture consisted of a methacrylic functional polysiloxane containing 1 % wt. of the DMAP, 0.7 % wt. Ca ionophore (ETH 129, Fluka) and 0.4 % wt. potassium tetrakis(4-chlorophenyl)borate (Fluka). The UV-photopolymerization is carried out under a Mylar ® film during 2 - 5 min depending on the membrane thickness (*15*). The two step development in xylene was completed by a thorough water rinse and drying procedure. An SEM photograph of the complete Ca^{2+} sensor is shown in Figure 5. The underlaying 25 μm thick polyHEMA

Figure 5. SEM photograph of a part of a wafer showing ISFET structures with a 25 μm thick polyHEMA membranes and a 80 μm thick double membrane (pHEMA and polysiloxane). (Reproduced with permission from ref. 15. Copyright 1991 Sens. and Actuators.)

membrane was covered with a 55 μm thick Ca^{2+} sensitive polysiloxane based membrane. It can be seen that the membrane definition is better than 20 μm, which is quite satisfactory considering the membrane thickness patterned.

The good adhesion of membranes onto the surface of Al_2O_3 is illustrated by the fact that they withstand an ultrasonic bath treatment for several hours without any loss of adhesion. Without the silanization pretreatment, on the other hand, all membranes lift-off within 5 minutes of ultrasonic agitation.

Ca^{2+} **Sensitive ISFET** : Figure 6 shows the reproducibility of the response of three ISFETs covered with double photopolymerized membranes tested by titrating demineralized water ($\rho > 12$ MΩ.cm) with either 10^{-2} M $CaCl_2$ (for low concentration range i.e. $< 10^{-3}$ M) or 1 M $CaCl_2$ solutions (for the higher concentrations). The measured sensitivity is 28.7 mV/decade and the extrapolated value of the detection limit is $10^{-6.4}$ M Ca^{2+}. A relatively high detection limit is experienced because no calcium buffer was used in the background solution. The selectivities towards Mg^{2+}, Na^+ and K^+ were measured by the separate solution method using 0.1 M unbuffered metal chloride solutions and the measured sensitivities : log K_{CaMg}^{Pot} = -4.9, log K_{CaNa}^{pot} = -4.5 and log K_{CaK}^{pot} = -4.1 respectively. These values are in good agreement with the values reported in the literature for conventional PVC membranes using the same ionophore (16).

However, it should be noted that due to the apolar nature of the siloxane membrane used, the concentration of dissociated salts inside the membrane is low. This leads to high membrane resistances resulting in a relatively high signal instability (few mV). Therefore, a modification of the siloxane based membrane to increase its dielectric constant is currently investigated.

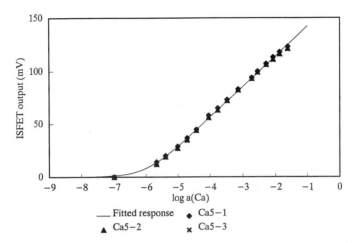

Figure 6. Response of three different calcium sensitive ISFETs covered with photopolymerized polyHEMA and polysiloxane membranes upon titration of demineralized water with $CaCl_2$. T = 25 °C. (Reproduced with permission from ref. 15. Copyright 1991 Sens. and Actuators.)

Conclusions

In order to maintain the advantage of the microfabrication approach which is intended for a reproducible production of multiple devices, parallel development of membrane deposition technology is of importance. Using modified on-wafer membrane deposition techniques and commercially available compounds an improvement of the membrane thickness control as well as the membrane adhesion can be achieved. This has been presented here for three electrochemical sensors - an enzymatic glucose electrode, an amperometric free chlorine sensor and a potentiometric Ca^{2+} sensitive device based on a membrane modified ISFET. Unfortunately, the on-wafer membrane deposition technique could not yet be applied in the preparation of the glucose sensors for *in vivo* applications, since this particular application requires relatively thick enzymatic membranes, whilst the lift-off technique is usable only for the patterning of relatively thin membranes.

The membrane deposition technology has to be adapted for each particular application. This is also true of membrane materials, where the additional requirement of new polymeric materials seems to be almost of more importance.

Acknowledgments : This work was supported by Swiss National Science Foundation, Committee for the Promotion of Applied Scientific Research, Fondation Lord Michelham of Hellingly and GIE of the Groupe Lyonnaise des Eaux (France).

Literature cited

1. Gernet S., Koudelka M., de Rooij N.F., *Sens. and Actuators*, **1989**, 18, 59-70.
2. van den Vlekkert H., de Rooij N.F., *Analusis*, **1988**, 16, 110.
3. Koudelka M., Gernet S. de Rooij N.F., *Sens. and Actuators*, **1989**, 18, 157-165.
4. van den Vlekkert H., Francis C.G., Grisel A., de Rooij N.F., *Analyst,* **1988**, 16, 1029-32.

5. Gernet S., Koudelka M., de Rooij N.F., *Sens. and Actuators*, **1989**, 17, 537-40.
6. Velho G., Frogue Ph., Thévenot D., Reach G., *Diabetes, Nutrition & Metabolism*, **1988**, 3, 227-33.
7. Koudelka M., Rohner-Jeanrenaud F., Terretaz J., Bobbioni-Harsch E., de Rooij N.F., Jeanrenaud B., *Biosensors & Bioelectronics*, **1991**, 6, 31-36.
8. Pluedemann E., *Silane Coupling Agents*, Plenum Press, New York, **1982**.
9. van den Berg A., Koudelka M., van der Schoot B.H., Verney-Norberg, Krebs Ph., Grisel A., de Rooij N.F., *Proceedings Transducers' 91*, San Francisco, **1991**, 233-6.
10. Sudhölter E.J.R., van der Wal P.D., Skowronska-Ptasinska M., van den Berg A., Bergveld P., Reinhoudt D.N., *Anal. Chim. Acta,* **1990**, 230, 59-65.
11. van den Berg A., Koudelka M., van der Schoot, Grisel A., *Proceedings of the 3rd Int. Meeting on Chemical Sensors,* Cleveland, OH, **1990**, 140-3.
12. Janata J., *Sens. and Actuators*, **1983**, 4, 255.
13. Fogt E.J., Untereker D.F., Norenberg M.S., Meyerhoff M.E., *Anal. Chem.*, **1985,** 57, 1995.
14. van der Wal P.D., Skowronska-Ptasinska M., van den Berg A., Bergveld P., Sudhölter E.J.R., Reinhoudt D.N., *Anal. Chim. Acta,* **1990**, 231, 41-52.
15. van den Berg A., Grisel A., Verney-Norberg E., *Sens. and Actuators*, **1991**, 4, 235-8.
16. Schefer U., Ammann D., Pretsch E., Oesch U., Simon W., *Anal. Chem.*, **1986**, 58, 2282-5.

RECEIVED December 11, 1991

Chapter 21

Thick-Film Multilayer Ion Sensors for Biomedical Applications

Salvatore J. Pace and James D. Hamerslag

Medical Products, Glasgow Site, Box 509, P.O. Box 6101, E. I. du Pont de Nemours and Company, Newark, DE 19714–6101

Planar format thick film ion sensors have been designed for biomedical applications. The multilayer sensor structure consists of layers sequentially deposited on a ceramic substrate using a screen printing process. All layers are of solid composition with the uppermost chemically active layer consisting of polyvinyl chloride (PVC). Because no aqueous layers are used in the construction, the resulting devices are stable and robust. The PVC layer serves as a common membrane vehicle for a multitude of ion specific ligands that may be patterned on a substrate. This paper describes the chemical and electrochemical principles of design, construction and measurement of ion sensing devices for blood electrolytes; K^+, Na^+, Cl^-, HCO_3^-, pH and the potential for many other clinically significant tests. The more pragmatic issues of packaging design and materials of construction are treated in the context of device performance (i.e., analytical efficacy, reliability and stability) for the biological application intended.

There exists a need for reliable, low cost chemical sensors for clinical diagnostic applications and, in particular for, the management of patients undergoing a medical crisis. The demands for cost containment and more effective delivery of health care has added a premium on innovative technology. What is needed is the ability to respond quickly to the diagnostic demands of physicians at the patient's bedside. Among the most frequently requested tests are blood electrolytes (Na^+, K^+, Cl^- and HCO_3^-) for both routine clinical profiling of blood serum and for the more critical emergency situations. We have already reported on the design of blood gas sensors based on similar design[1,2] and discussed strategies for enzyme and immunosensor structures encompassing the principles of design and construction[3,4] described in this article. This report focuses on a planar format multilayer ion sensor structure, assembled by a thick film printing process. All materials of construction are of solid composition and no aqueous layers are employed.

The chemically active, ion specific layer consists of a polyvinyl chloride (PVC) film in direct contact with the sample. The PVC membrane is tailored for

0097–6156/92/0487–0261$06.00/0

selective ion detection and may be patterned to achieve a multitude of tests within a single device. A unique feature of this electrochemical device is the absence of aqueous gel layers conventionally applied to stabilize interfacial potentials. Attempts to stabilize the interface between the solid conductor and the polymer membrane has proven illusive for Ion Sensing Field Effect Transistors (ISFET's), yet it is well recognized in the art that the key to technological success rests on the ability to effectively transfer electrical charge across an electrical/chemical (or heterogeneous) interface without effectively altering its potential. The sensors described below are stable chemical to electrical transducers, they are robust, amenable to large scale manufacture and with appropriate patterning of membranes can enable chemical profiling of biological fluids for a variety of medical applications.

Experimental

The multilayer sensor structure consists of cermet and polymer based layers sequentially deposited on a 96% alumina ceramic substrate using a thick film screen printing process. The cermet layers are of ceramic-metal composition which require firing at a temperature of 850°C and the polymer layers are cured at temperatures below 100°C. Layout of this multilayer sensor structure is shown in Figure 1.

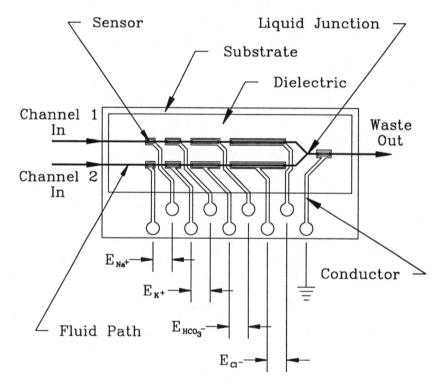

Figure 1. Layout of planar multilayer sensor

The sensors were fabricated using a manual thick film screen printer, a convection oven and a box furnace. The order in which the layers were deposited on the ceramic substrate is illustrated in Figure 2 and their composition characterized in Table I. The fabrication sequence is described follows: 1) The conductor layer was printed and dried at 150°C for 15 minutes. 2) The dielectric layer, made by two print/dry steps, was printed and dried at 150°C for 15 minutes. 3) The substrates were then fired at 850°C for 10 minutes. 4) The interfacial (carbon) layer was printed and dried at 100°C for 60 minutes. 5) The membrane (PVC) layer, made by two print/dry steps, was printed and dried at 50°C for 60 minutes. This last step was repeated for each of the different membrane formulations.

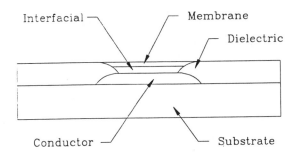

Figure 2. Cross section of sensor

The PVC membrane compositions consist of a neutral carrier and ion exchange ionophores as indicated in Table I. The K^+ membrane includes valinomycin as the ionophore and is plasticized by 2-ethyl-hexyladipate (Fluka). The Na^+ membrane contains methyl monensin and tris(2-ethylhexyl)phthalate. Tri-n-dodeclyamine is the ionophore for pH and tridodecylmethyammonium chloride is the ion exchanger for Cl^-. HCO_3^- utilizes 4-butyltrifluorophenone for CO_3^- detection as well as trioctylammonium chloride ion exchanger. Both anion membranes are plasticized with dioctyladipate.

A test system, controlled by personal computer (PC), was developed to evaluate the performance of the sensors. A schematic of this system is shown in Figure 3. The signals from the sensors were amplified by a multi-channel electrometer and acquired by a 16 bit analog to digital data acquisition board at a resolution of 0.0145 mV/bit. The test fixture provided the electrical and fluid interface to the sensor substrate. It contained channels which directed the sample, reference and calibrator solutions over the sensors. These channels combined down stream of the sensors to form the liquid junction as shown in Figure 1. Contact probes were used to make electrical connection to the substrate. Fluids were drawn through the test fixture by a peristaltic pump driven by a stepper motor and flow of the different fluids was controlled by the pinch valves.

The assay protocol consisted of a single two point calibration to determine the slope of each sensor and a one point or offset calibration prior to each assay measurement. The two point calibration was conducted as follows. Reference solution was pumped into channel 1 (Figure 1), calibrator 1 into channel 2 and the response of the sensors was measured. Calibrator 2 was then pumped into channel

Table I. Sensor composition

LAYER	DESCRIPTION	THICK [um]
1 Conductor	Ag DuPont QS175	17
2 Dielectric	Glass DuPont QS482	34
3 Interfacial	Carbon DuPont 7861D	10

4 Membrane	PVC (Plasticized)
	 $K^{+(12,13)}$ Valinomycin
	 $Na^{+(14,15)}$ Methyl Monensin
	 $Cl^{-(10,11)}$ Tridodecylmethyl−
	                              ammonium chloride
	
	 $HCO_3^{-(8,9)}$ 4−butyltrifluorophenone
	                              Trioctylammonium chloride
	
	 $H^{+(6,7)}$ Tri−n−dodecylamine
		10

2, displacing calibrator 1 and a measurement taken. The slope of each sensor was then calculated. The offset calibration was accomplished by pumping reference and calibrator 1 solutions into channels 1 and 2 respectively, and then measuring the response of the sensors. The sample was then pumped into channel 2, displacing calibrator 1 and a measurement taken. Air bubbles were incorporated into the beginning of the fluid streams to minimize contamination of the solutions, a major source of assay error.

Theory

Aside from the selectivity criterion that is essential to all ion specific electrodes, the principal objective of applied design is to physically and chemically control the phase and interphase boundaries across the multiple layers that comprise the electrode structure. The conduction path of electrical charge across all the phases including the solid conductors and external measurement circuitry, as well as the chemical charge across polymeric and solution phases, may be represented by the schematic illustrated in Figure 4.

The sources of measurement error are easily identifiable from Figure 4 and are represented in electrochemical cell notation as:

$$Cu\ /\ Ag\ /\ C(PVA)\ /\ PVC\ /\ a_i[s]\ //\ a_i[r]\ /PVC\ /\ C(PVA)\ /\ Ag\ /\ Cu$$

The potential at the interphases Cu/Ag and Ag/C are easily controllable and are submicrovolt in magnitude. The C(PVA)/PVC interphase is designed to sustain small charge transfer rates less than 10^{-12} amperes, required by the electrometer amplifier to accomplish the measurement. The net chemical changes occurring at the two interphases, Figure 4, are inconsequential, rendering stable and matching potentials. The more uniform the interphase chemistry of the sensor pair, the more effective the

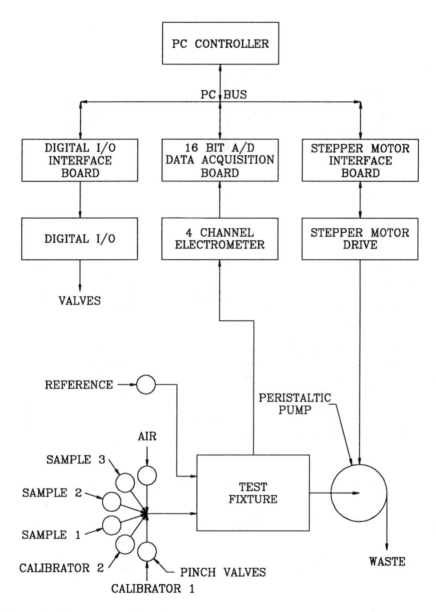

Figure 3. Test system schematic

compensation for offsets. Transient responses caused by water uptake in the PVC phase and the time required to establish the space charge within the PVC, are equally minimized by the differential measurement. An effective interface is one that achieves inter-layer adhesion and is electrically non-polarizable. The polyvinyl acetate (PVA) binder for the carbon (C) promotes the stability of the C(PVA)/PVC interface. The interphase potentials ultimately responsible for the ion activity

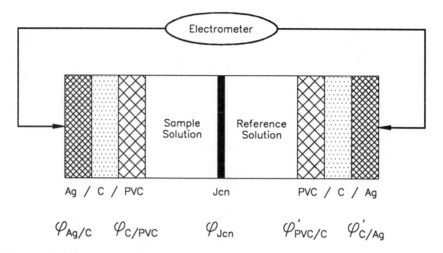

Figure 4. Schematic of electrochemical cell

measurement are those for the PVC/Solution interphases, controlled by the sample and reference solution ion activities $a_i[s]$ and $a_i[r]$ respectively. The analytical signal is a measure of the ion activities of the reference and sample solutions, all other factors of the fabrication process being equal.

The measurement of ion activities assumes chemical equilibrium between the PVC membrane and the electrolyte bearing solutions. The time domain chemical and dielectric space charge changes that occur are minimized by membrane composition and sensor design and are considered negligible during the measurement period. Hence, the potential dependence of the ion activity is characterized by the Nernst equation. The following thermodynamic expressions describe the potentials of the individual electrodes and the electrode pairs:

$$\mathcal{E}_s = \mathcal{E}_s^\circ + \frac{RT}{nF}\ln a_s + \varphi_{Ag/C} + \varphi_{C/PVC} + \varphi_{Jcn} \tag{1}$$

$$\mathcal{E}_R = \mathcal{E}_R^\circ + \frac{RT}{nF}\ln a_R + \varphi_{PVC/C}' + \varphi_{C/Ag}' \tag{2}$$

$$\Delta\mathcal{E}_{s/R} = \Delta\mathcal{E}_{s/R}^\circ + \frac{RT}{nF}\ln\frac{a_s}{a_R} + \Delta\varphi_{Ag/C} + \Delta\varphi_{C/PVC} + \varphi_{Jcn} \tag{3}$$

$$\Delta\mathcal{E}_{s/R} = \xi + \frac{RT}{nF}\ln\frac{a_s}{a_R} \tag{4}$$

$$\text{where } \xi = \Delta\mathcal{E}_{s/R}^\circ + \Delta\varphi_{Ag/C} + \Delta\varphi_{C/PVC} + \varphi_{Jcn} \tag{5}$$

Equations 1 and 2 are expressions for half cell potentials. $\mathcal{E}_s$ and $\mathcal{E}_r$ expresses the potential for the Solution/PVC membrane interfaces and correspond to the sample and reference solutions, respectively. The comparative measurement is described by the difference expression in equation 3. If the interfacial potential differences are insignificant in magnitude, then only ion activity impacts on the differential signal. Any offset potentials (ξ's) or mismatch of the half-cell pairs is compensated by exposing both electrodes to a solution of constant composition. Because the logarithmic term in equation 3 drops out at an activity ratio of 1, the remaining terms constitute the offset potential. The offset compensation accomplishes more than potential mismatch corrections of the polymeric layers, it also compensates for activity coefficients variations and the junction potential variation as expressed by equations 4 and 5. Sample assay is then accurately performed by the comparative measurement of the sample solution relative to a reference solution of constant and known composition.

Junction potentials caused by charge separation at liquid to liquid interfaces are minimized by design, Figure 1, and control of ionic strength variances. Streaming potentials are a measurement artifact caused by electrolyte flow and easily minimized by the use of a solution electrical ground. The former error is effectively compensated by the offset correction while the latter is virtually eliminated by accomplishing the measurement on static test solutions. An added benefit of the comparative measurement is the virtual elimination of signal drift because of the inherent subtraction of transient responses.

The space-charge definition within the dielectric (PVC) membrane changes with time. Such changes are a direct consequence of membrane exposure to sample solution and are controlled by membrane formulation. The two factors of consequence to pragmatic design (not accounted for by equations 1-5) are the transient response of the membrane and signal drift due to membrane dielectric constant and interfacial potential changes.

Results And Discussion

The measurement of sodium ion concentration in human blood specimens is the most demanding of the electrolyte tests. The clinical range in humans is quite narrow (i.e., 136 to 142 mEq/L), placing acute signal resolution requirements on a method that must discern 1 mEq/L at a level of 150 mEq/L of sodium. Less than 0.2 mV signal must be resolved, making this a benchmark performance test. Our discussion of results will focus on sodium, primarily for this reason, but also for the sake of clarity. Finally, the performance of all four electrolyte sensors (i.e., Na^+, K^+, Cl^- and HCO_3^-) are given.

The transient response of the sodium sensor was measured from the dry state as a half-cell versus a Calomel Reference Electrode and also differentially between identical half cells. The responses, shown in Figure 5, are the consequence of the sensors in contact with solutions of equal composition. Traces [a] and [b] are the responses of each of the half-cells comprising a sensor pair versus an external Calomel Reference Electrode and trace [c] corresponds to the differential signal between half-cell pairs. The initial ± 400 mV signal corresponds to amplifier saturation and quickly settles to a steady state. Each of the two electrodes stabilize to a steady-state signal in less than a second to a potential level biased to the reference electrode. The corresponding differential signal, trace [c], removes the bias and settles to a zero volt signal within less than 300 ms. The differential signal achieves steady state faster and is drift free, a characteristic of such differential measurements. Trace [d] represents the measurement of the sensor pair closest to the conduit outlet and the time difference between [c] and [d] of approximately 150 ms reflects the time delay in fluid exposure between the first sensor pair and the last sensor pair in the fluid channel.

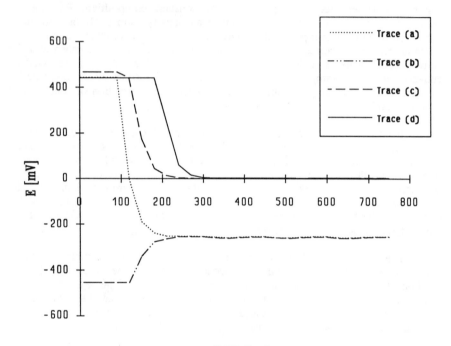

Figure 5. Transient response of sodium sensor

The response of sensor pairs to varying electrolyte concentrations is shown in Figure 6. The four sensors are all sodium selective and differ only in geometric area. Each trace represents a sensor response in time as the sodium solution concentration varies from 70 to 150 mEq/L. Each of the four sensor responses is stable and reproduces the same signal levels as the exposure alternates to the three concentration levels of the test. A signal offset is observed that is in part geometry related since the sensors differ in surface area from 1.0 to 8.0 mm^2. The responses to concentration changes suggest that sensor response time follows precisely the actual concentrations of the fluids in contact with the ion selective membrane. The rate limiting step of the signal response is related to fluid flow of sample delivery within the fluid channel and not any chemistry or ion transport dynamics within the membrane. Note the reproducibility of the signal in time for any given concentration.

A logarithmic plot of potential as a function ion concentration results in a linear relationship that extends well beyond the clinical range. A linearity study that compares a set of Na$^+$ sensors of varying geometries is plotted in Figure 7, the data corresponds to multiple parallel lines of nearly identical 59 mV slopes and varying

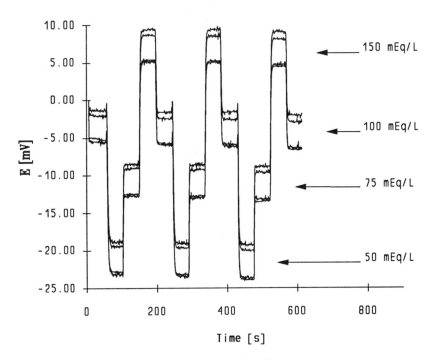

Figure 6. Real time concentration response of sodium sensor

intercepts that reflect signal offsets of several millivolts. Although the geometric variation is a correctable parameter, not all forms of offsets are (e.g., junction potentials), and thus, an offset compensation/calibration is appropriate to achieve accurate assays.

The effect of electrode geometry was further studied to establish the importance of this parameter on design and performance. The data on Figure 8. indicates that as the geometric area increases from 1.0 to 8.0 mm², the signal to noise characteristic improves and so does the drift, slope response and assay precision. In a separate study addressing the design of pH sensors[4], 44 such devices of identical surface area were compared in offset potentials using a pH 7.40 phosphate buffer as a test solution. The measured potential distribution resulted in a mean offset potential of 0.27 mV and standard deviation of 0.86 mV. This offset potential data is indicative of the uniformity achievable for matched sensor pairs.

The precision and accuracy performance was tested by correlating the measured assays of horse serum specimens to assay values obtained by a standard method. The precision was measured by replicating assays at three sodium serum levels that extend well beyond the clinical range. The correlation plot is shown in Figure 9 and corresponds to a set of samples encompassing a sodium range of 100 to 170 mEq/L.

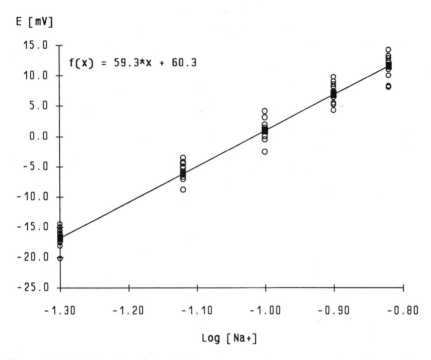

Figure 7. Linearity of sodium sensor

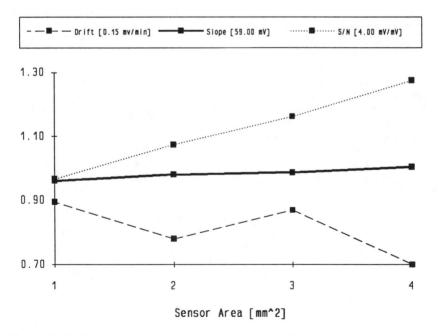

Figure 8. Sodium sensor performance as a function of area

Assay accuracy was evaluated by comparing the analytical results of the subject sensors to those analyzed by the Dimension Clinical Chemistry System[*]. The analyses of serum samples were correlated and assays plotted in linear relationship as shown in Figure 9. Error flags accompany the variances in the assays for replicate measurements for both methods and the composite coefficient of variation (CV) for all the sodium assays for the test method is 0.34%. This compares favorably with the clinical performance requirement of CV ≤ 1.0%. Linear regression analysis of the correlation data results in a slope of very nearly one and an intercept indicative of very little bias. A correlation coefficient of very nearly one is a good measure of assay accuracy. The sensors were repeatedly subjected to serum samples over an extended time period and the slope response intermittently measured and recorded as a function of time. Figure 10 graphs such slope response versus time. The actual longevity relative to time exposure is mediated in part by the membrane formulation and geometric parameters, but the above data indicates that such sensors command a practical and useful lifetime when in contact with serum specimens.

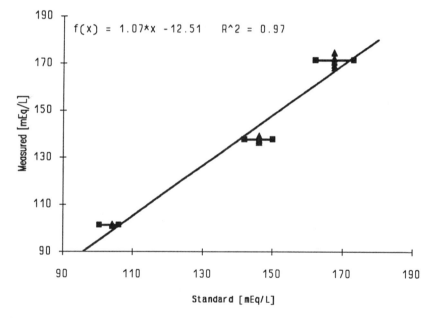

Figure 9. Correlation study of sodium sensor assay performance

Sodium, potassium and chloride sensors were evaluated using undiluted serum specimens with no sample pretreatment of any kind. The performance data set, summarized by Table II, comprises a minimum of 8 sensors and 60 samples. The bicarbonate assays were performed separately on serum specimens buffered at pH 9.0. Clinical efficacy is normally judged by the response linearity, precision and

---

[*] Dimension is a product of the Dupont Company, Medical Products, Wilmington, DE

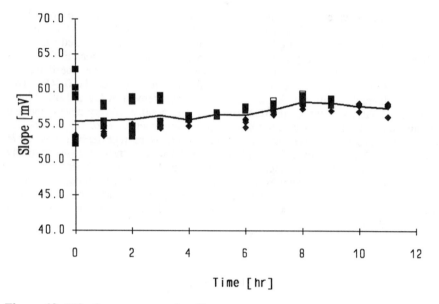

Figure 10. Life time response of sodium sensor

Table II. Performance summary

		Na	K	Cl	HCO3*
Slope	mV	57.20	57.90	-44.48	-24.00
Precision	%CV	0.34	0.30	0.74	4.50
Correlation					
	slope	1.07	1.03	0.89	1.15
	intercept	-12.51	-0.21	11.70	-2.60
	R^2	0.97	0.99	0.99	0.95

* Serum samples buffered at pH of 9.0

accuracy parameters that are detailed in the data table. The results are well within the clinical analytical "goal" standards sought by clinicians during routine medical practice.

Conclusions

The planar format thick-film ion sensors described herein lead to stable and robust devices. The fabrication process is amenable to volume manufacturing and integration of multiple sensors on the same substrate so that the entire electrolytes profile may be obtained from as little as 50 microliters of undiluted serum sample in a single sample operation. The fluid system is simple to implement and does not contribute to assay error.

The elimination of aqueous gel layers afford practical benefits. The sensors do not require refrigeration or special handling and enjoy extended shelf life as well. The migration of polymer layer ingredients are confined by design, the principal attribute to extended stability. The one point calibration in actual usage compensates for many inadequacies of the potentiometric measurement. Although, the results discussed might lead us to believe that calibration could be eliminated with the improvement of layer uniformity, long term instability due to inter-layer migration remains a formidable challenge.

This basic design is applicable to other clinically significant tests and a variety of biosensors designs are achievable by modification of the PVC membrane. Many strategies to achieve such ends is a subject of intense research activity which are only now beginning to yield application dividends.

Acknowledgements

The authors acknowledge the contribution of our DuPont colleagues; Jay Bass, Dick Speakman, Dave Thompson and Mark Walters.

References

1.   S. J. Pace, P. P. Zarzychi, R. T. McKeever and L. Pelosi, Proc. of 3rd. International Conference on Solid-State Sensors & Actuators, 1985, Philadelphia, USA
2.   S. J. Pace & M. J. Jensen, Proc. of The 2nd Irtn'l Meeting on Chem. Sensors, 1986, Bordeaux, France
3.   S. J. Pace, Medical Instrumentation, 1985, 40, no.4, pp 168, July-August
4.   S. J. Pace, Chemically Sensitive Electronic Devices, 1980, Elsevier Sequoia
5.   D. P. Hamblen, C. P. Glover and S. H. Kim, U.S. Patent #4,053,381, 1977
6.   P. Shulthess, Y. Shijo, H. V. Pham, E. Pretsch, D. Amman and W. Simon, Anal. Chem. Acta., 1981, 131, pp 111
7.   D. Amman, F. Lanter, R. Steiner, P. Schulthess, Y. Shijo and W. Simon, Anal. Chem., 1981, 53, pp 2267
8.   H. B. Herman and G. A. Rechnitz, Anal. Chem. Acta., 1975, 76, pp 155
9.   H. B. Herman, and G. A. Rechnitz, Science, 1974, 184, pp 1074
10. C. J. Coetzee, H. Freiser, Anal. Chem., 1968, 40, 2071
11. C. J. Coetzee, H. Freiser, Anal. Chem., 1969, 41, 1128
12. L. A. R. Pioda, V. Stankova, and W. Simon, Anal. Lett., 1969, 2, 665
13. W. Simon, Swiss Patent #479,870, 1989
14. O. Kadem, E. Luebel and M. Furmansky, Ger. Offen. #2,027,128, 1970
15. K. Tohda, et. al., Anal. Sci., 1990, 6, 227

RECEIVED October 22, 1991

HYDRATION-DEPENDENT POLYMER APPLICATIONS

Chapter 22

Environmentally Sensitive Polymers as Biosensors

The Glucose-Sensitive Membrane

Lisa A. Klumb, Stephen R. Miller, Gary W. Albin, and Thomas A. Horbett

Department of Chemical Engineering and Center for Bioengineering, Mail Stop BF–10, University of Washington, Seattle, WA 98195

A glucose sensitive membrane consisting of a poly(acrylamide) hydrogel with covalently bound phenol red dye and immobilized glucose oxidase is described as an example of a polymer-based biosensor. Other polymers that are environmentally sensitive and change their swelling, permeability, or hydraulic flow rate are also reviewed. The color change of the phenol red dye in response to the pH decrease induced by the enzyme catalyzed turnover of glucose to gluconic acid was detected spectrophotometrically. Using a flow cell mounted in a diode array spectrophotometer, steady state and transient spectra of the membrane in response to glucose were obtained and analyzed. The glucose oxidase loading, the membrane thickness, and the flow rate of the buffer solution in the flow cell were varied. Typical steady state pH decreases were 0.1-0.4 of a pH unit, and the response time to reach half of the maximum response was 3-10 minutes, depending primarily upon the membrane thickness.

A biosensor utilizes a biological component to translate the concentration of a specific analyte of interest into a signal detectable by some chemical or physical means (1). Successful operation of a biosensor requires that the biological component and the signal it transduces be localized to and concentrated within a region in close proximity to the detection system. Immobilization of enzymes within a polymeric matrix ensures concentration and localization of the enzymic reaction, and creates a convective barrier, thus preventing dilution and convective removal of the product species before it is detected. Enzymes immobilized on or near the detection system are frequently used as the

0097–6156/92/0487–0276$06.00/0

biological component because of their catalytic activity and high specificity (2). Enzymes are capable of quickly and selectively converting an undetectable analyte into a product detectable by a variety of means including, but not restricted to, the use of electrochemical, thermal, and optical techniques (3). If the polymeric matrix itself exhibits measurable changes induced by the enzyme catalyzed reaction, the detection system and the immobilization matrix become one and the same.

Environmentally sensitive polymers properties' change in response to changes in temperature, solvent composition, pH, and other variables (*see reviews: 4, 5*). The purpose of this report is to examine the use of environmentally sensitive polymers containing immobilized enzymes as biosensors, using the specific examples of the glucose sensitive polymers developed in our labs (*6-11*). The glucose sensitive polymers we have studied are pH sensitive polymers containing immobilized glucose oxidase. This enzyme converts glucose to gluconic acid which lowers the pH within the microenvironment of the polymer. Three types of pH sensitive polymers currently studied in our labs are: 1) those polymers which swell as the pH is lowered due to the protonation of weakly basic groups within the polymer matrix; 2) a composite system which changes its hydraulic permeability in response to pH by changing the pore size of a porous substrate; and 3) a polymer whose pH sensitivity is conferred by the covalent incorporation of a dye which changes color in response to pH. The first two types are described here in brief and the last one will be the example used to demonstrate the glucose sensing capabilities of such systems.

Glucose Sensitive Membranes

An example of a glucose sensitive polymer which swells in response to glucose is a copolymer of hydroxyethylmethacrylate (HEMA) and dimethylaminoethylmethacrylate (DMAEMA). The glucose oxidase is immobilized within this hydrogel by entrapment during the polymerization procedure using either a frozen irradiation technique or chemically initiated polymerization (*6, 7*). The DMAEMA monomer contributes weakly basic ($pK_a \sim 7.0$) amine groups to the copolymer which, as the pH is lowered, become charged due to amine protonation. The increasingly protonated hydrogels swell as a result of the charge repulsion forces. These copolymers are being developed for the controlled delivery of insulin (*6, 7, 9-11*). As these polymers swell, the water content increases, thereby increasing the permeability of water soluble solutes (such as insulin) through the hydrogel matrix.

We have demonstrated that these types of polymers swell and increase their permeability to insulin in response to pH changes at physiologic ionic strength and pH (*6-10*). For example, a polymer with

20 monomer % (v/v) DMAEMA when equilibrated in buffer at pH 7.4 and then at pH 6.8, exhibited a 25% increase in its water content (water content at 7.4 was 60%) (*10*). It has also been shown that the size as well as the water content increases in response to glucose. Almost a 15% increase in diameter and a 25% increase in water content in response to a 100 mg% glucose challenge was observed, although these experiments were performed in distilled water with a slightly different polymer formulation (*7*). Finally, when these polymers are made macroporous via the inclusion of more than 55% (v/v) solvent (usually ethylene glycol in water) in the monomer solution, they increase their permeability to insulin in response to pH changes and glucose (*6, 11*). Insulin permeabilities were 8 to 12 times higher at pH 4.0 than pH 7.4, and 2.4 to 5.5 times higher after introducing 400 mg% glucose.

A similar system is being developed by Siegel, *et al.* (*12*) using a hydrophobic matrix such as an alkyl acrylate instead of the hydrophilic HEMA. They have demonstrated large increases in the swelling of such polymers, even over a narrow pH range. They conceptualize a mechanochemical insulin pump design in which the polymer swelling action, due to the glucose oxidase conversion of glucose to gluconic acid, acts as a pumping system yielding glucose controlled insulin delivery. A glucose sensor based on Siegel's polymers, would require the incorporation of glucose oxidase, and the measurement of the forces induced by such a polymer as it swells in response to glucose.

Another system being developed in our labs consists of grafting a weakly acidic polymer onto a porous substrate such as an ultrafiltration membrane (*10*). This composite system has been shown to change its hydraulic permeability, and also the bulk flow of insulin in the presence of a pressure gradient as a function of the solution pH. An acrylic acid or methacrylic acid monomer solution (5-10% v/v in water) was graft polymerized onto the surface of a polyvinylidene difluoride substrate (0.22 micron pore size) using a radio frequency glow discharge plasma initiated polymerization technique (*10, 13, 14*). At physiologic pH (7.40) the acidic groups are charged and it is thought that the polymer chains are expanded thus effectively closing the pores of the substrate. As the pH is lowered in this system, the acidic groups are neutralized and the polymer chains shrink thereby opening the pores of the substrate. Hydraulic flowrates under a pressure of 16 psi, have been shown to increase from 0.26 mL/min at pH 7.4 to 7.2 mL/min at pH 4.0, a 28-fold change (Horbett, T. A. , University of Washington, Seattle, unpublished data). Insulin flowrates under pressure have been shown to be 10 to 13 times greater at pH 2.4 than pH 7.8 (*10*). Iwata, *et al.* developed a similar system and showed that the hydraulic flow rates change the greatest in the pH range between 3 and 4 for poly(acrylic acid) grafted onto porous PVdF (*13*) which agrees with data from our lab.

Immobilization of glucose oxidase via carbodiimide chemistry onto a poly(acrylic acid) grafted porous cellulose substrate was performed by Ito, *et al.* (*14*). They showed that this system's permeability to insulin in response to a 3600 mg% glucose challenge was 1.7 times greater than the permeability prior to the addition of glucose.

Since the above mentioned systems all depend on the pH decrease generated within the polymers by the enzymatic conversion of glucose to gluconic acid, it was useful to determine the magnitude and kinetics of the induced internal pH changes (*8, 9, 11*). This was done by covalently incorporating phenol red, a pH sensitive dye, into poly(acrylamide) gels with immobilized glucose oxidase and measuring the pH-induced color changes spectrophotometrically. The rest of this paper is devoted to describing in greater detail this type of polymer and its response to pH and glucose induced changes.

Materials and Methods

Membrane Preparation. The poly(acrylamide) (pAAm) polymers with covalently bound phenol red and immobilized glucose oxidase (GO) were prepared by redox-initiated solution polymerization with N,N'-methylene-bis-acrylamide as the cross linking monomer, N,N,N',N'-tetramethyl-ethylenediamine (TEMED) as an accelerator, and ammonium persulfate as the redox initiator (*8, 9*). Membrane sheets were cast between glass plates separated by stainless steel shims (0.125 or 0.305 mm thick). Due to the high amount of initiator necessary for the covalent binding of the phenol red, pre-cooling of all solutions prior to mixing and casting, cooling of the casting plates and shims, and pre-cooling an aluminum block used to remove heat generated during the polymerization reaction, were all required to protect the enzyme from heat denaturation (*8*). It was necessary to bind phenol red covalently within the polymer in this manner (*8, 15*) to prevent the water soluble dye from freely diffusing out of the hydrogel during an experiment.

Polymerization was allowed to proceed for at least 24 hours after which time the assembly was soaked in buffer at room temperature and the hydrogel membrane carefully removed. The membrane was allowed to swell in buffer for several days, with multiple changes of buffer, before its use in an experiment.

Mechanical support was found necessary to prevent rupture of the membranes when mounted in the flow cell during an experiment (see next section). Dacron mesh (0.060 mm thick, 150 grids/inch) was used as a support either mounted in the flow cell between the membrane and the flow, or incorporated directly into the membrane during the polymerization.

Determination of Membrane pH. Measurements of the membrane pH were made while solutions of air-saturated CBS (citrate buffered saline: 0.0195 M citrate and 0.154 M ionic strength) with or without glucose, were pumped with a peristaltic pump past one side of the membrane as it was mounted in a plexiglass flow cell (8, 9). The pH induced color changes were measured spectrophotometrically with the plexiglass flow cell mounted in the cuvette holder of a Perkin-Elmer Lambda Array 3840 spectrophotometer and complete spectra were recorded as a function of time on a Perkin-Elmer PR-210 chart recorder. The diode array spectrophotometer allows the collection of spectra every few seconds. It has been shown that the pH induced changes are linearly proportional to the ratio of the % light transmission (%T) at wavelengths from 570-510 nm to the %T at the pH-independent reference wavelength of 700 nm (8, 9, 15). The spectral change at a particular wavelength was defined as this ratio (for example $\%T_{510}/\%T_{700}$) and was used to calibrate the system as CBS solutions of known pH were pumped past the membrane. The calibration and steady state operation of the system has been described previously (8, 9). The %T was recorded as a function of time to obtain the kinetics of the system and then digitized using a SAC "Graf/Pen" sonic digitizer. These data were then analyzed to obtain the time required to reach half of the maximal change in %T, $t_{1/2}$, where the maximal change is defined as the difference between %T at time = 0 and time = ∞. Time = 0 is defined as the moment when the composition within the flow cell (not in the membrane) first begins to change, and is equal to the time when the pump is restarted after insertion of the pump chamber into a new solution, plus the space time of the pump chamber. The space time of the pump chamber is its volume divided by the volumetric flowrate.

The flow cell mixing characteristics were investigated by alternating solutions of the polysaccharide blue dextran with CBS alone and measuring the light transmission through the cell as a function of time. Also the volume of the pump chamber was determined to be 1.78 cm^3 using this technique.

Results

Figure 1 shows a series of spectra recorded at steady state for a poly(acrylamide)/glucose oxidase/phenol red (pAAm/GO/PR) membrane as the pH (Figure 1a) or glucose concentration (Figure 1b) of the solution being continuously pumped past the membrane is changed. Figure 1a shows the membrane responsiveness to pH demonstrating the pH dependent region of the spectra between 500 and 600 nm, and the pH independent region near 700 nm. The glucose sensing capability is shown in the spectra of Figure 1b. As the glucose concentration was

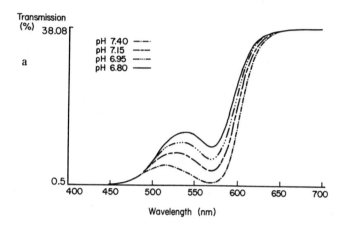

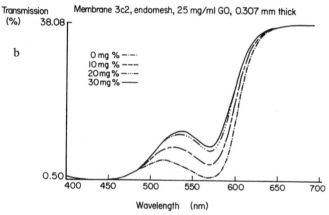

Figure 1. Transmission spectra of a pAAm/GO/PR membrane (25 mg/mL, 0.307 mm-thick) at four different pH values (a) and four different glucose concentrations (b). The spectra were obtained in flowing CBS buffer (0.4 mL/min). Figure 1a: pH 7.4 (—·—), pH 7.15 (— — —), pH 6.95 (—····—), and pH 6.80 (—). Figure 1b: 0 mg% (—·—), 10 mg% (— — —), 20 mg% (—····—), and 30 mg% (—).

increased, the spectra changed in the same way as the pH induced changes. This demonstrates that the enzymatic reaction in the presence of glucose was generating an internal membrane pH drop, and that the internal pH decrease was maintained in the presence of the buffer, due to the steady state turnover of glucose.

The steady state pH of the membrane calculated from calibration curves is shown as a function of the inlet glucose concentration in Figure 2 for various enzyme loadings (9). The plateau in membrane pH observed near 40 mg% glucose for the enzyme loadings greater than 1.0 mg/mL is typical of systems which use the glucose oxidase enzyme. This phenomenon has been described elsewhere and is the combined result of the enzymatic requirement for molecular oxygen and the low solubility of oxygen in aqueous solution for systems which are substrate diffusion limited (9, 16) (see Discussion).

Typical changes in the % light transmission with time in response to the blue dextran in the flow cell are shown in Figure 3. With a flowrate of 2.3 cm^3/min., approximately 3 minutes were required for complete clearance of the flow cell's contents, although it is nearly 90% complete in 1 minute. Figure 4 shows the kinetics for a 0.307 mm-thick pAAm/GO/PR membrane in the flow cell as the inlet glucose concentration was changed from 0-10 mg% and then from 10-20 mg% for a flowrate of 0.4 mL/min. The response times for this particular membrane are seen to be on the order of 30 to 40 minutes to reach steady state, though less than 10 minutes were required to reach the half maximum point ($t_{1/2}$).

Tables I-III contain a summary of the response time data as a function of the membrane thickness, the glucose oxidase loading, and the flowrate of buffer solution through the flow cell. The reproducibility of the experiments is indicated by the replicated data points in Tables I-III. The deviations in some cases are large and may be due to unduplicated folds in the membrane when assembled in the flow cell and possible membrane motion or fluttering, especially at higher flowrates during an experiment (8).

Table I shows the effect of membrane thickness on the response times for the two different types of Dacron mesh support (internal and external) used for the membranes in the flow cell. Thickness increases the response time for both the pH and glucose concentration changes. Use of the internally incorporated Dacron mesh may possibly decrease the response time due to an increase in the porosity of the gel allowing greater solute transport within the membranes.

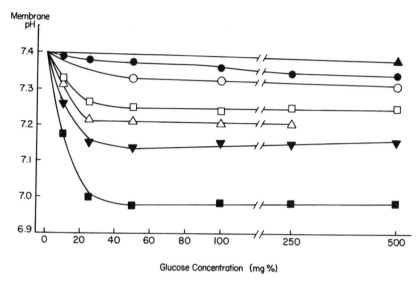

Figure 2. Effect of glucose oxidase concentration on the steady state pH in 0.129 mm-thick pAAm/GO/PR membranes in response to glucose concentration in flowing CBS buffer (2.0 mL/min). Glucose oxidase concentrations (mg/mL) were: ▲ : 0.05; ● : 0.125; O: 1.0; □ : 4.0; △ : 10; ▼ : 25; ■ : 50. (Reproduced with permission from ref. 9. Copyright 1987 Elsevier Science Publishers)

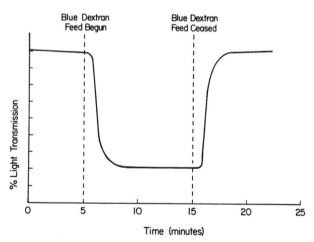

Figure 3. Light transmission through flow cell (no membrane in place) during wash-in and wash-out of a blue dextran solution at 2.24 mL/min.

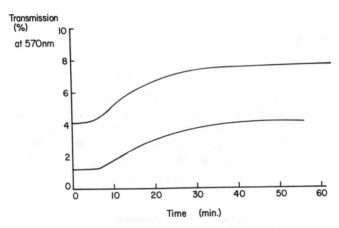

Figure 4.   Light transmission through a 0.307 mm-thick pAAm/GO/PR membrane with internal Dacron mesh and 25 mg/mL glucose oxidase, as a function of time for changes in glucose concentration from 0-10 mg% (lower curve) and 10-20 mg% (upper curve). Flowrate of buffer was 2.0 mL/min.

Table I.   Response Times as a Function of Membrane Thickness

External Change (in Glucose or pH)	Thickness (mm)	$t_{1/2}$ (min.)
Internal Mesh pH 6.8 - 7.4	0.080	3.1
	0.307	5.4
0 - 30 mg%	0.080	2.4
	0.307	7.8
External Mesh pH 7.4 - 7.1	0.129	2.5
	0.307	8.5, 9.8, 10.1
100 - 0 mg%	0.129	5.5
	0.307	5.4, 8.5, 13.3

Each membrane had 25 mg/mL GO in monomer solution. Flowrate was 2.0 mL/min.

The response time of a single membrane as the flowrate of buffer solution into the flow cell was varied is shown in Table II. Flowrate seems to significantly influence the response time of both the pH changes and the glucose concentration changes which might suggest that the cell mixing kinetics are on the same order of magnitude as the membrane kinetics. However, since the cell clearance time is inversely proportional to the flowrate, the role of clearance time in the observed kinetics should be much greater than what was actually seen for the large flowrate changes shown in Table II. Additionally the data for the 100 mg% glucose concentration increase at 2.0 mL/min. were analyzed using a deconvolution procedure to separate the cell mixing kinetics, as measured in the blue dextran experiments, from the membrane kinetics. This procedure resulted in a $t_{1/2}$ calculated for the membrane response alone to be 5.0 minutes which compares favorably to the 5.7 minutes reported in Table II. Based upon this example for an intermediate flowrate used in these studies, it is concluded that the cell mixing time is much faster than the membrane response in most of the experiments and therefore the $t_{1/2}$ values reported here largely reflect the membrane response alone. The flowrate effect observed and shown in Table II must be due to changes other than in cell mixing times (see Discussion).

Table II. Response Times as a Function of Flowrate

External Change (in Glucose or pH)	Flowrate (mL/min.)	$t_{1/2}$ (min.)
pH 7.4 - 7.1	1.8	10.5
	10.5	8.5
0 - 30 mg%	0.236	9.2
	1.8	9.8
	2.0	7.8
	11.0	6.2
0 - 100 mg%	1.8	6.8
	2.0	5.7
	10.5	3.9

Membrane was 0.307 mm thick, 25 mg/mL GO in monomer solution, and internal Dacron mesh.

The effect of the glucose oxidase concentration in the monomer solution is shown in Table III for two different membrane thicknesses using the external Dacron mesh support. The glucose oxidase concentration has no effect on the response time to a change in external pH for either the thin or thick membrane. As was seen in Figure 2, the enzyme loading strongly affects the steady state performance of the system in response to glucose, but it is seen not to effect the transient response time significantly in the 0.129 mm-thick membranes. In the 0.307 mm-thick membranes, however, the response times appear to be faster at the GO loading of 25 mg/mL than at 2.5 mg/mL.

Discussion

The steady state performance as a function of glucose concentration (Figure 2) showed the plateau in response near 40 mg% glucose for enzyme loadings greater than 1.0 mg/mL, which is typical for systems that use glucose oxidase and oxygen as the electron acceptor in the enzyme reaction (9, 16). This plateau is the result of the substrate diffusion limited situation which occurs when the enzyme reaction proceeds rapidly, depleting the concentrations of the substrates locally. Since the glucose oxidase enzyme requires both oxygen and glucose as substrates for the reaction, oxygen is the substrate which limits the reaction at higher glucose concentrations due to the low solubility of oxygen in aqueous solution. The oxygen delivery via diffusion to the reaction zone becomes the rate limiting step and the resulting response of the sensor is insensitive to changes in glucose concentration. Ways to

Table III. Response Times as Function of Glucose Oxidase Concentration in Monomer Solution

External Change (in Glucose or pH)	Glucose Oxidase Concentration (mg/mL)	$t_{1/2}$ (min.)
0.129 mm -thick membrane		
pH 7.4 - 7.1	10	3.1
	25	2.5
	50	3.2
10 - 25 mg%	2.5	5.5, 5.9
	4.0	5.2, 5.9
	10	4.2
	25	3.9
	50	3.6
	100	3.0
500 - 0 mg%	0.125	11.0
	0.5	10.1
1000 - 0 mg%	0.125	10.1
	0.25	12.2
	0.5	9.7
0.307 mm-thick membrane		
pH 7.4 - 7.1	2.5	8.9
	25	8.5, 9.8, 10.1
0 - 30 mg%	2.5	13.3
	25	6.6, 7.5
0 - 100 mg%	2.5	11.8
	25	5.1, 6.4, 7.6

External Dacron mesh support used and flowrate was 2.0 mL/min.

overcome this oxygen limitation are currently being investigated for glucose sensors as well as our insulin delivery systems which use glucose oxidase (*16, 17,* Klumb, L. A., *et al., J. Controlled Rel.,* in press). The membranes with the lowest enzyme loadings in Figure 2, 0.05 and 0.125 mg/mL, exhibit a progressive response as the glucose concentration is increased, although the actual pH drops are small. These membranes are therefore not limited by the rate of substrate delivery to the enzyme, but rather by the rate of reaction which is a function of the enzyme loading and the local concentrations of glucose and oxygen.

The previous discussion also explains why thickness is such an important determinant of response time for both the solution pH and glucose concentration changes. Both the phenol red color change and the enzyme reaction, for this high enzyme loading of 25 mg/mL, are extremely fast compared to the diffusion of the species through the membrane and therefore are substrate diffusion limited. Increasing the internal diffusion path length for the reacting species by increasing the membrane thickness results in an increase in the time to achieve steady state.

The flowrate dependence of response time is more difficult to explain since, as suggested in the Results, this can not be attributed solely to the mixing behavior of the flow cell. Other origins for the flowrate effects include the depletion of substrate concentrations in the flow cell at low flowrates and the effect of convective mixing within the membrane itself. The membrane used in these studies had a high enzyme loading (25 mg/mL) and was therefore substrate diffusion limited. Convective flow within the membrane would speed the delivery of the substrates to the gel interior beyond what could be expected from diffusion through an unstirred gel interior and thus decrease the response times at higher flowrates. Such convective flow could occur due to pressure drops from the front (or flow side) to the back of the membrane, or micro-channels and imperfections in the membrane structure.

The data in Table II for the 0.129 mm-thick membranes subjected to a 10-25 mg% glucose challenge are shown for enzyme loadings (2.5-100 mg/mL) for which the response is substrate diffusion limited as was discussed above. Therefore the sensor response for these cases should exhibit no dependence on enzyme loading. The results for these membranes did demonstrate only slight decreases in response time with increases in enzyme loading. These small changes in response time may be the result of either differing diffusional characteristics between the membranes or the actual amount of active enzyme. For the higher glucose concentrations shown in Table II for the 0.129 mm-thick membranes (500 and 1000 mg%), the enzyme loadings are in the range (0.125-0.5 mg/mL) where the rate of the reaction is the limiting step.

These membranes should exhibit more dramatic differences in response times with increasing enzyme loadings. However, the three enzyme loadings shown do not demonstrate great response time differences, possibly because the enzyme loadings are not sufficiently different from each other.

The thicker membranes are also substrate diffusion limited for both of the enzyme loadings shown in Table II, and should not exhibit the response time dependence on enzyme loading which was observed. It is possible that there are convective effects within the membrane with the higher enzyme loadings which would decrease the response time. This would be the case if the membranes with the higher GO loadings are actually more porous due to the immobilization of greater amounts of enzyme. It is also possible that the actual concentration of active enzyme within these membranes is such that the lower enzyme loading (2.5 mg/mL) is actually a reaction rate limited and not a substrate diffusion limited system. An increase in enzyme loading in a reaction rate limited situation would show a corresponding decrease in response time.

In summary, the steady state and transient performance of the poly(acrylamide) hydrogel with immobilized glucose oxidase and phenol red dye (pAAm/GO/PR) demonstrates phenomena common to all polymer-based sensors and drug delivery systems. The role of the polymer in these systems is to act as a barrier to control the transport of substrates/products and this in turn controls the ultimate signal and the response time. For systems which rely upon the reaction of a substrate for example via an immobilized enzyme, the polymer controls the relative importance of the rate of substrate/analyte delivery and the rate of the reaction. In membrane systems, the thicker the polymer membrane the longer the response time due to substrate diffusion limitations as demonstrated with our pAAm/GO/PR system. However a membrane must not be so thin as to allow convective removal of the substrates before undergoing reaction, or removal of the products before detection. The steady state as well as the transient response of the pAAm/GO/PR system was used to demonstrate these considerations with the more complicated case in which two substrates are required for the reaction.

The design of a polymer-based system requires understanding both the steady state and transient behavior in response to the substrate or analyte of interest. For sensor applications, this information is obtained during the operation of the sensor. However, for other applications of environmentally sensitive polymers, such as drug delivery systems, the polymer response to the substrate/analyte is not usually studied directly. Our work with the pAAm/GO/PR system illustrates the usefulness of an in situ probe to measure what governs the membrane's performance in response to the substrate/analyte and how to analyze it. We continue to use this valuable information in the further modeling and analysis of improved insulin delivery systems without the oxygen diffusional limitations.(Klumb, L. A., *et al., J. Controlled Rel.*, in press).

Acknowledgments

The financial support of the National Institute of Arthritis, Diabetes, Digestive and Kidney Diseases through grant DK30770 is gratefully acknowledged.

Literature Cited

1. Turner, A. P. F. In *Biosensors: Fundamentals and Applications*; Turner, A. P. F.; Karube, I.; Wilson, G. S., Eds.; Oxford Science Publications; Oxford University Press: New York, N.Y., 1987; pp v-vii.
2. Clark, L. C. In *Biosensors: Fundamentals and Applications*; Turner, A. P. F.; Karube; I., Wilson, G. S., Eds.; Oxford Science Publications; Oxford University Press: New York, N.Y., 1987; pp 3-12.
3. *Biosensors: Fundamentals and Applications*; Turner, A. P. F.; Karube, I.; Wilson, G. S., Eds.; Oxford Science Publications; Oxford University Press: New York, N.Y., 1987.
4. *Reversible Polymeric Gels and Related Systems*; Russo, P. S., Ed.; ACS Symposium Series No. 350; American Chemical Society: Washington, D. C., 1987.
5. *Pulsed and Self-Regulated Drug Delivery*; Kost, J., Ed.; CRC Press: Boca Raton, FL, 1990.
6. Albin, G.; Horbett, T. A.; Ratner, B. D. *J. Controlled Rel.* **1985**, *2*, pp 153-164.
7. Kost, J.; Horbett, T. A.; Ratner, B. D.; Singh, M. *J. Biomed. Mater. Res.* **1985**, *19*, pp 1117-1133.
8. Miller, S. R. thesis: *The Development and Characterization of a Glucose Sensing Polymer*, University of Washington: Seattle, WA, **1986**.
9. Albin, G. W.; Horbett, T. A.; Miller, S. R.; Ricker, N. L. *J. Controlled Rel.* **1987**, *6,* pp 267-291.
10. Layman-Spillar, L. M. thesis: *Development of a pH-Sensitive Membrane for use as a Flow Controller*; University of Washington: Seattle, WA, **1990**.
11. Albin, G. W.; Horbett, T. A.; Ratner, B. D. In *Pulsed and Self-Regulated Drug Delivery*; Kost, J., Ed.; CRC Press: Boca Raton, FL, 1990; pp 159-185.

12. Siegel, R. In *Pulsed and Self-Regulated Drug Delivery*; Kost, J., Ed.; CRC Press: Boca Raton, FL, 1990; pp 129-157.
13. Iwata, H; Matsuda, T. *J. Membr. Sci.* **1988**, *38*, pp 185-199.
14. Ito, Y.; Casolaro, M.; Kono, K.; Imanishi, Y. *J. Controlled Rel.* **1989**, *10*, pp 195-203.
15. Peterson, J. I.; Goldstein, S. R.; Fitzgerald, R. V. *Anal. Chem.* **1980**, *52*, pp 864-869.
16. Gough, D.A.; Lucisano, J. Y.; Tse, P. H. S. *Anal. Chem.* **1985**, *57*, pp 2351-2357.
17. Davis, G. In *Biosensors: Fundamentals and Applications*; Turner, A. P. F.; Karube, I.; Wilson, G. S., Eds.; Oxford Science Publications; Oxford University Press: New York, N.Y., 1987; pp 247-256.

RECEIVED October 22, 1991

Chapter 23

Water and the Ion-Selective Electrode Membrane

D. Jed Harrison, Xizhong Li, and Slobodan Petrovic

Department of Chemistry, University of Alberta, Edmonton, Alberta
T6G 2G2, Canada

Water uptake in plasticized polyvinylchloride based ion selective membranes is found to be a two stage process. In the first stage water is dissolved in the polymer matrix and moves rapidly, with a diffusion coefficient of around 10^{-6} cm^2/s. During the second stage a phase transformation occurs that is probably water droplet formation. Transport at this stage shows an apparent diffusion coefficient of 2×10^{-8} cm^2/s at short times, but this value changes with time and membrane addititives in a complex fashion. The results show clear evidence of stress in the membranes due to water uptake, and that a water rich surface region develops whose thickness depends on the additives. Hydrophilic additives are found to increase the equilibrium water content, but decrease the rate at which uptake occurs.

Ion selective membranes are the active, chemically selective component of many potentiometric ion sensors (1). They have been most successfully used with solution contacts on both sides of the membrane, and have been found to perform less satisfactorily when a solid state contact is made to one face. One approach that has been used to improve the lifetime of solid state devices coated with membranes has been to improve the adhesion of the film on the solid substrate (2-5). However, our results with this approach for plasticized polyvinylchloride (PVC) based membranes suggested it is important to understand the basic phenomena occurring inside these membranes in terms of solvent uptake, ion transport and membrane stress (4,6). We have previously reported on the design of an optical instrument that allows the concentration profiles inside PVC based ion sensitive membranes to be determined (7). In that study it was shown that water uptake occurs in two steps. A more detailed study of water transport has been undertaken since water is believed to play an important role in such membranes, but its exact function is poorly understood, and the quantitative data available on water in PVC membranes is not in good agreement (8-10). One key problem is to develop an understanding of the role of water uptake in polymer swelling and internal stress, since these factors appear to be related to the rapid failure of membranes on solid substrates.

By adding water sensitive dyes to the PVC based membranes we are able to determine the distribution of water as a function of position inside the membrane, measured from the water interface, as a function of time (7). We present here a detailed analysis of the transport of water which shows that water uptake occurs in two stages at very different rates, and that this is related to a phase transformation inside the membrane. Surprisingly, when the dye added is a hydrophilic salt the rate of water diffusion during the initial wetting process is an inverse function of the dyes concentration. Internal stress as a result of water uptake is demonstrated by the fact that there are two stages of water uptake, the diffusion constant for the second, slower process is a function of time, and the rate of evaporation of water is faster than the rate of uptake. The presence of a water rich surface region is also demonstrated.

Experimental

A diagram of the instrument used to probe the internal concentration profile of a PVC based membrane is shown in Figure 1. A pulsed N_2 pumped dye laser and two HeNe lasers (633 and 544 nm) are used as sources. The magnification power of the optics is 7 to 20 times. Further details of the instrument design are presented elsewhere (7). The cell dimensions in the figure are exaggerated for clarity. The optical path length is actually quite short, 0.15 mm, while the membrane thickness in the direction of transport (vertical direction in the figure) is in the range of 0.8 to 1.5 mm. The membranes studied were composed of 33% PVC, 66% dioctyladipate (DOA), and several other additives such as valinomycin, $KB(C_6H_5)_4$ (KTPB), anhydrous $CoCl_2$, and 2,6-diphenyl-4-(2,4,6-triphenyl-N-pyridinio)phenolate (Et30) in varying proportions up to 1%.

Results and Discussion

Figure 2 shows the uptake of water into a membrane containing 0.05% $CoCl_2$ after a short period of time. The water enters from the right edge and evaporates into the atmosphere at the left edge. The decrease of absorbance observed in the bulk of the membrane arises from bleaching of the dye on reacting with water. At the surface the absorbance increase corresponds to the growth of light scattering centers in the membrane as the concentration of water increases, leading to an increase in absorbance. This effect indicates a phase transformation has occurred inside the polymer. It is most likely that this is a phase separation due to the formation of water droplets, as has been observed in other hydrophobic polymers in which salts or fixed charge sites are located (11). The light scattering centers penetrate much more slowly into the membrane than the water associated with bleaching of the dye.

   The phenomenon shown in Figure 2 may be understood in terms of there being at least two chemical states of water within the membrane. The first rapid ingress of water is due to the uptake of water that becomes dissolved in the membrane matrix. The second, slower step is associated with separation of the water phase into droplets. The processes can be modeled using Fick's laws of diffusion, and by separating the two stages in the analysis. A closed form solution is obtained if the diffusion coefficients are assumed to be described by a step function at the boundary between regions. Solutions are presented in a number of sources (12,13). If the data is analyzed at the early stages when semi-infinite boundary conditions are applicable then diffusion in the bulk region can be expressed by equation 1 (12,13).

$$A_1 = A_{1max}[1 - erf(x/\sqrt{4D_1t})] \qquad (1)$$

where A_1 is the decrease in absorbance due to bleaching of the dye. A_{1max} is the apparent maximum change in absorbance (13) due to the equilibrium concentration of water, erf is the error function, x is the distance from the edge of the membrane, D_1 is the diffusion coefficient for the dissolved water, and t is the time.

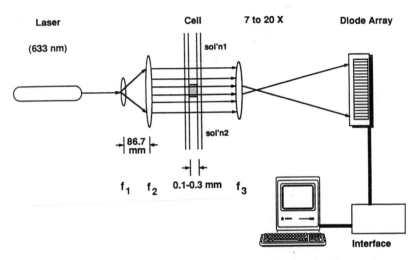

Figure 1. Optical arrangement to probe water distribution inside membranes. The magnification is 7 to 20 times.

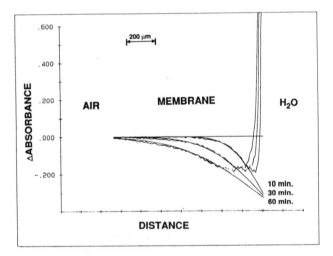

Figure 2. Absorbance profile inside CoCl$_2$ containing membrane during first 60 min. Water enters from the right and evaporates at the left edge. Smooth curves are fits to Eq 1 for dye bleaching.

The second stage of uptake associated with the onset of light scattering can be described by equation 2.

$$A_2 + A_s = A_{2max} - B\ erf(x/\sqrt{4D_2t}) \tag{2}$$

where A_2 is the absorbance increase due to light scattering alone, A_{2max} is the maximum value of A_2, and D_2 is the diffusion coefficient associated with the apparent motion of the light scattering centers. It must be recognized that the scattering centers themselves are not moving, but H_2O molecules move from scattering site to site at this apparent rate. Referring to Figure 3, $-A_s$ is the maximum decrease in absorbance that arises due to bleaching. It is smaller than A_{1max} since the total concentration of soluble water in the membrane exceeds the dye concentration, so that the dye is essentially saturated. The critical water concentration at which light scattering begins to occur, C_k, is different than this value, and the point at which this occurs in the membrane can be approximated as a moving boundary that occurs at position x_k, which extends more deeply into the membrane with time. The parameter B is a complex function of D_1, D_2, x_k and t and has a value close to 1 when D_1 and D_2 are very different. A detailed derivation and treatment of this function is given in ref. 13.

The model curve in Figure 3 clearly resembles the data in Figure 2. To treat the experimental data the bleaching is first analyzed in the bulk region according to eq. 1. The absorbance due to light scattering alone is then obtained by correcting for the absorbance due to the dissolved water and for A_s. From this x_k is obtained, and the diffusion coefficient for light scattering, D_2, is estimated using eq. 2. Eq. 1 assumes no concentration perturbations occur near the surface, but this is not the case since the water concentration continues to change once light scattering centers begin to form. To correct for this the bleaching at x_k is determined, $A_1(x_k)$, and this value is taken as unchanging in time. Then as the light scattering front moves deeper into the membrane A_{1max} is allowed to increase so that $A_1(x_k)$ is constant at the new value of x_k. An equivalent correction is obtained by shifting the value of $x = 0$ inwards and holding A_{1max} constant as x_k shifts. Values of D_1, D_2, and x_k are then recalculated until the analysis is self-consistent, which usually requires one or two iterations. This method is clearly an approximation since the original boundary conditions did not allow for A_{2max} to change over time.

Effect of Dye Concentration on Diffusion. Several curve fits resulting from these calculations are shown in Figure 2. For the $CoCl_2$ dye the value D_1 decreases as the dye concentration increases, as is shown in Figure 4. The value in the absence of dye can be estimated by extrapolation to be 1×10^{-6} cm^2/s, decreasing to 0.2×10^{-6} cm^2/s at a concentration of 1.2 wt%. This result is surprising as adding a hydrophilic dye will increase the water uptake of a membrane, and it is unexpected that the rate of that process would decrease. It may result from the fact that the dye binds water to it and these hydrophilic sites must be saturated before water can pass deeper into the membrane. Whatever the cause, the results show that improving the rate of membrane wetting and water equilibration may not be achieved by what seems the most obvious approach.

When the less hydrophilic dye Et30 is added to a membrane an absorbance profile similar to that in Figures 2 and 3 is obtained, since this dye also bleaches in the presence of water. For the initial uptake of dissolved water a diffusion coefficient of 1 to 2×10^{-6} cm^2/s is obtained, indicating the validity of the extrapolation for the $CoCl_2$ dye.

The formation of light scattering centers is observed in membranes containing only PVC and DOA, and with the additives valinomycin, KTPB, $CoCl_2$, Et30 or

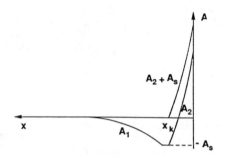

Figure 3. Model of Absorbance profile inside membrane.

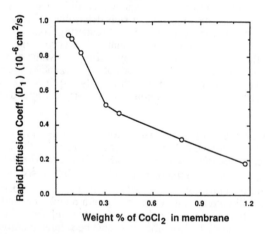

Figure 4. Diffusion coefficient vs concentration of $CoCl_2$ (wt%) in membrane for the first rapid diffusion step (bulk region).

some combination of these. Although the effects are not identical the differences are a matter of degree, as suggested in Figure 5, showing a dependence of D_2 on the additive and its concentration. The behaviour of membranes with $CoCl_2$ added has been examined in the most detail, and is presented here.

The transport behavior associated with the light scattering centers depends on $CoCl_2$ concentration and time. The diffusion process is psuedo-Fickian. The maximum concentration expressed as A_{2max} increases with time, while the value of D_2 decreases with time at low dye concentrations, and increases with time at higher concentrations. In Figure 5 the water is penetrating from the left edge and the magnification is increased so that only a portion of the membrane is visible. It can be seen that the water has penetrated a much greater depth into the membrane after 24 h for the membrane with a higher $CoCl_2$ concentration. Figure 6 shows that as the concentration of dye increases the apparent rate of diffusion of the light scattering centers decreases, when measured 10 minutes after exposure to water. The same is true after 20 min. or 1 h. However, if the extent of penetration of this water rich region is determined after about 24 h the trend is reversed, as seen in Figure 6. At low dye concentration the light scattering centers remain fixed in a region near the surface, and D_2 decreases to less than 10^{-9} cm^2/s. At concentrations of 0.4% or more the water rich region continues to penetrate the membrane and D_2 increases with time. It should be noted that eq. 2 is derived using assumptions that are clearly poor approximations for this complex system. In fact the surface concentration represented by A_{2max} increases with time, and D_2 is likely to be a function of concentration, time or position in the membrane. These factors mean the magnitudes of D_2 calculated here may not be accurate, but they should reflect the real trends in D_2 with $CoCl_2$ concentration.

The behavior of water in the membrane once the phase transformation is induced is complex, but can be understood in the following general terms. When the concentration of water reaches a critical value droplet formation will occur. At this point the polymer begins to be stressed due to volume changes and the diffusion coefficient for water transport decreases. {The dependence of diffusion rates on stress in polymers is a well recognized phenomenon *(13-15)*.} This would predict that at short times both D_1 and D_2 should follow the same trend with dye concentration, and this is the case as shown in Figures 4 and 6. At low dye concentration the increasing stress with water uptake causes a further decrease in D_2 until the penetration of water is essentially halted. At high concentrations of the salt $CoCl_2$ the increased hydrophilicity of the membrane overcomes the effect of stress and so it continues to absorb large amounts of water in the bulk region. The net result is that adding a hydrophilic salt to the membrane does increase the uptake of water as would be expected, but at the same time the initial rate of uptake is decreased.

Effect of Indicator Dye Selected. The surface region that can be seen in Figure 5 has a width that depends on the dye concentration and the additive itself. For 0.05% $CoCl_2$ this region is about 90 µm thick, while for the zwitterion Et30 or the hydrophobic NO_2^- carrier bromo(pyridine){5, 10, 15, 20 tetraphenylporphyrinato}-cobaltate it is about 50 µm thick. For a standard K^+ sensitive membrane containing valinomycin and KTPB this water rich region is about 20 µm thick. The trend observed within these compounds is a decreasing thickness with decreasing hydrophilicity of the additive.

Figure 7 shows the light scattering in a valiomycin membrane with KTPB after 40 h of exposure to water. Weak light scattering is observed in the central region, while a large change is seen at the edges. Developing this distribution requires about 24 h, and it is stabilized within 40 h, indicating the transport process is as slow as with the water sensitive dye additives. The most interesting feature is shown by the

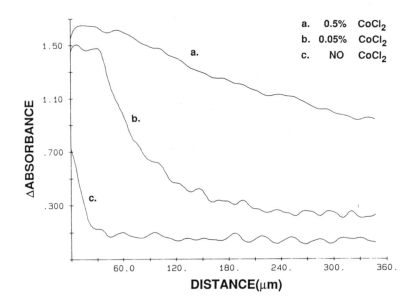

Figure 5. Absorbance profile in membrane after 24 h exposure to water with a) 0.5% CoCl₂ , b) 0.05% CoCl₂ , c) no additives.

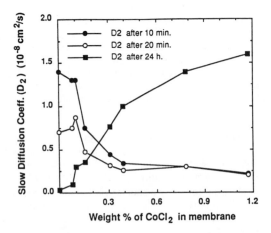

Figure 6. Diffusion coefficient vs concentration of CoCl₂ (wt%) in membrane for the second step associated with light scattering. The values depend on time, and are shown for the same membranes after 10 min. and 24 h.

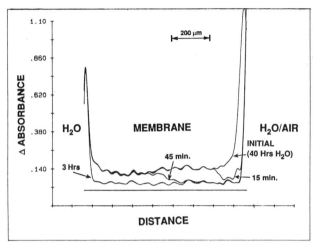

Figure 7. Light scattering profile in valinomycin membrane after 40 h equilibration with H2O on both sides. A decrease in absorbance with time is then initiated when water is removed from the right side.

changes that occur at one side of the membrane after the water is removed at that interface. The water rich region at the surface disappears very rapidly. Further, the decrease in light scattering in the bulk of the membrane appears as a moving front, rather than as a Fickian diffusion process distributed across the entire bulk region. Only an hour or so is required to reverse the effects of 24 hours of water uptake. This effect has not yet been analyzed in detail. However, such differences in ingress and loss of solvent from a polymer matrix are well known and arise from changes in the diffusion coefficient (*13-15*). In this case the data indicate the diffusion coefficient increases as the water content decreases. This is consistent with the measurements presented above, and also with the hypothesis that there is stress in the membrane following significant uptake of water.

Summary

Based on our analysis of the water uptake phenomena in PVC based membranes we can make several statements about the behavior of water.

(1) There must exist at least two chemical states of water inside the membrane. This is inferred from the fact that two stages of water uptake occur, and that there is a phase transformation leading to the formation of light scattering centers. Although not proven, we take these centers to be water droplets. There is a vast difference in the rates of penetration of these two types of water into the membrane. The first, dissolved form enters rapidly with a diffusion coefficient of about 10^{-6} cm^2/s, while the second form associated with the phase transformation shows a complex dependence on time and additive concentration, but is in the range of 10^{-8} cm^2/s initially, and this decreases with time leaving a water rich surface region.

(2) Stress develops within the membrane with the uptake of water. This is demonstrated by the fact that the diffusion coefficient for the second stage of water uptake changes in time, and by the differences in the rate of loss and entry of water into the membrane (Figure 7). The stress field inside may explain the development of a water rich surface region over time, instead of what would be the expected uniform distribution.

(3) The nature and concentration of membrane additives strongly affect the behavior of water. Even for the first stage of water uptake that involves dissolved

water, increasing the concentration of the hydrophilic salt $CoCl_2$ has an effect. The rate of diffusion is seen to decrease as more dye is added. The formation and distribution of a water rich surface region in which light scattering centers form depends on the additive and its concentration. It appears that the more hydrophilic the additive the greater the depth this surface layer penetrates into the membrane bulk. Also, at least for $CoCl_2$, the initial rate of water uptake decreases as the dye concentration increases. This effect is inverted after long equilibration times, and essentially means that the water rich region penetrates throughout the bulk of the membrane. Consequently, adding hydrophilic dye has the expected effect of increasing the water content, but its effect on the rate of uptake may be undesirable for fast wet-up times.

Acknowledgments. We thank the Natural Sciences and Engineering Research Council of Canada, and Ciba-Corning Diagnostics for financial support of this research.

Literature Cited

1. Janata, J. In *Solid State Chemical Sensors*; Janata, J., Huber, R. J., Ed.; Academic Press: London, 1985; Chapter 2.
2. Blackburn, G.; Janata, J. *J. Electrochem. Soc.* **1982**, *129*, 2580-2584.
3. Satchwell, T.; Harrison, D. J. *J. Electroanal. Chem.* **1986**, *202*, 75-81.
4. Harrison, D. J.; Cunningham, L. L.; Li, X.; Teclemariam, A.; Permann, D. *J. Electrochem. Soc.* **1988**, *135*, 2473-2478.
5. Moody, G. J.; Thomas, J. D. R.; Slater, J. M. *Analyst* **1988**, *113*, 1703-1707.
6. Harrison, D. J.; Teclemariam, A.; Cunningham, L. L. *Anal. Chem.* **1989**, *61*, 246-251.
7. Li, X.; Petrovic, S.; Harrison, D. J. *Sens. Actuat.* **1990**, *B1*, 275-280.
8. Thoma, A. P.; Viviani-Nauer, A.; Arvantis, S.; Morf, W. E.; Simon, W. *Anal. Chem.* **1977**, *49*, 1567-1572.
9. Morf, W. E.; Simon, W. *Helv. Chim. Acta* **1986**, *69*, 1120-1131.
10. Marian, S.; Jagur-Grozinski, J.; Kedem, O.; Vodsi, D. *Biophys. J.* **1970**, *10*, 901-910.
11. Southern, E.; Thomas, A. G. *ACS Symp. Ser.* **1980**, *127*, 375-386.
12. Buck, R. P.; Berube, T. R. *J. Electroanal. Chem.* **1988**, *256*, 239-253.
13. Crank, J. *The Mathematics of Diffusion*; Oxford Press: Oxford, 1956.
14. Petropoulos, J. H.; Rousis, P. P. *J. Membrane Sci.* **1978**, *3*, 343-356.
15. Rudolph, F.; Peschel, G. *Ber. Bunsenges. Phys. Chem.* **1990**, *94*, 456-461.

RECEIVED October 22, 1991

Chapter 24

Swelling of a Polymer Membrane for Use in a Glucose Biosensor

Marian F. McCurley[1] and W. Rudolf Seitz[2]

[1]National Institute of Standards and Technology, Gaithersburg, MD 20879
[2]Department of Chemistry, University of New Hampshire,
Durham, NH 03824

A membrane for a fiber optic biosensor based on polymer swelling and optical displacement has been investigated. A copolymer of polyacrylamide and 2-(dimethylamino) ethylmethacrylate (DMAEM), lightly crosslinked with N,N-methylenebisacrylamide (MBA) served as the membrane for immobilizing glucose oxidase. The swelling was found to be a function of the glucose concentration in the physiological range. The final degree of swelling varies inversely with the extent of crosslinking and directly with the amine content.

Analyte induced swelling of polymers can serve as a transduction mechanism in biosensors. The changes in polymer size may be measured either electrically or optically and related to analyte concentration. For example, we recently described a fiber optic sensor for salt concentration based on polymer swelling (1). Changes in polymer size were detected as changes in the intensity of light reflected into an optical fiber. Other investigators have described a sensor for organic solvents in which small changes in polymer size are detected interferometrically (2).

Fiber optic sensors based on polymer swelling offer several potential advantages. They can be designed so that the optical measurement is separated from the polymer by a diaphragm so that the measurement can not be affected by the optical properties of the sample. Unlike fiber optic sensors based on indicator absorbance or luminescence, photodegradation is not a potential source of sensor instability. Measurements can be made in the near infrared region of the spectrum and take advantage of inexpensive components available for fiber optic communications.

Membranes that swell in the presence of glucose have been prepared by immobilizing glucose oxidase in an amine-containing

0097–6156/92/0487–0301$06.00/0
© 1992 American Chemical Society

acrylate polymer hydrogel (3,4). The enzyme catalyzes the oxidation of glucose to gluconic acid, which protonates the amine. Electrostatic repulsion between charged amine groups causes the gel to expand. The degree of swelling increases with glucose concentration.

Studies on glucose-induced polymer swelling have focussed on developing membranes that could serve in systems for controlled delivery of insulin to diabetics (3,4). It has been shown that hydrophobic methacrylate copolymers undergo a sharp swelling transition as the pH is decreased from 7 to 6 (3-7). However, the kinetics of the transition are too slow for the proposed application to glucose delivery.

We are interested in further investigating the potential of biosensors based on polymer swelling. We report here experiments with glucose oxidase impregnated membranes prepared from polyacrylamide copolymerized with (dimethylamino) ethylmethyacrylate. Because this polymer is more hydrophilic than the hydrophobic methacrylate gels investigated previously (3-7), it may provide faster response. The pH dependence of the membrane with no enzyme is reported. The response of the membrane with immobilized glucose oxidase to glucose is then described. In addition the effects of percent crosslinking and percent of amine moieties in the membrane on swelling are also investigated.

Experimental

Apparatus. The diameters of gel disks were measured with a Cambridge Instruments Stereoscope with a micrometer reticle. An Ohaus GA200D balance was used to weigh swollen gel disks. The volume was calculated from the weight assuming a density of 1.00. An Orion Research pH meter (Model EA920) and electrode were used to monitor pH. A BioRad electrophoresis chamber was used to form the gel slab.

Reagents. Anhydrous d-glucose and glucose oxidase Type VII Aspergillus Niger were purchased from Sigma. Acrylamide 97%, potassium persulfate 99%, 2-(dimethylamino)ethylmethacrylate 98% (DMAEM), N,N-methylenebisacrylamide 99% and NaOH were from Aldrich. NaCl was from Mallinckrodt. HCl was purchased from Fisher.

Procedures.

Polymer Preparation. The polymer gel was prepared by combining specified volumes of a filtered 40% acrylamide solution, a 2% N,N-methylenebisacrylamide solution and 98% DMAEM. Total volumes ranged from 3 to 5 mL. The mixture was deoxygenated by bubbling N_2 through it for 15 minutes. Ten to twenty microliters of a saturated potassium persulfate solution were added to initiate polymerization. At this point 0.0010 g of glucose oxidase, dissolved in 0.010 mL saline, was added to those gel solutions that were used to measure glucose concentration. The gel solution was then transferred to the mold which consisted of two glass plates separated by 0.50 mm spacers. The solidified gel slab was removed from the glass plates and allowed to swell in distilled water. A cork borer was used to cut gel disks from the slab.

Polymer Swelling. The swelling of the gel as a function of solution pH, percent crosslinking and percent DMAEM was measured as follows. Gel disks were placed in citrate/phosphate buffer solutions ranging in pH from 2.6 to 7.0. These buffers were prepared according to published methodology (8). In addition, they each contained 0.010 M NaCl. At specific time intervals the gel was taken from the solution, excess water was removed with a tissue, and the change in diameter was measured using a stereoscope. This was repeated until the gel diameter reached a constant value.

The gel disks used to measure glucose contained 40% (w/w) DMAEM and 3.3% (w/w) N,N-methylenebisacrylamide. The gel was placed in aqueous glucose standard solutions, and the change in diameter was measured as above.

Results and Discussion

Water in contact with freshly prepared polymer has a pH close to 7. This is lower than expected for aqueous solutions of aliphatic amines which typically have pK_b values close to 9. Since the polymer preparation procedure does not involve acids which could protonate the amine groups, we conclude that the amine moieties must not be in full contact with the water.

Figure 1 plots the percent increase in disk diameter vs. time at various pH values for gels with 5% crosslinking agent and 40% amine. The data show that swelling is slow; the disks take an hour or more to reach a constant volume. Also, decreasing the pH from 4 to 2.6 causes a large increase in the constant volume observed after two hours of contact with buffer. Since a dissolved aliphatic amine would be fully protonated over the range from 2.6 to 7, the data in Figure 1 also suggest that the amine is not in full contact with the solvent. confirm that the gel is sensitive to changes in pH.

Figure 2 shows volume of the gel vs. pH for a polymer gel with 2% crosslinking agent and 40% amine. In Figure 2a the gel was initially exposed to base, then titrated with 0.10 M HCl. In Figure 2b the gel was initially exposed to acid, then titrated with 0.10 M NaOH. The pH effect has a high degree of hysteresis. If polymer pH is lowered by adding acid, then swelling occurs between pH 4 and pH 2.6. However, if the direction of the pH change is reversed by adding base, then shrinking occurs between pH 6 and 10, a range much closer to that expected based on amine pK_b values.

Figure 3 plots percent increase in disk diameter vs. time after exposure to a pH 2.6 phosphate/citrate buffer for polymers containing 1% crosslinking agent and 10, 20 and 40% amine. It illustrates that the final degree of swelling is highly dependent on the percent of amine, in the form of DMAEM, in the membrane.   As the amine content

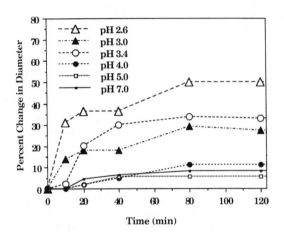

Figure 1.   Effect of pH on the swelling of a 5% crosslinked
polyacrylamide gel with 40 % DMAEM. The percent increase in disk
diameter vs time at various pH values.

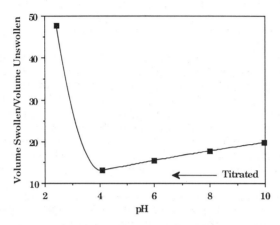

Figure 2a. Swelling ratio of a 2% crosslinked polymer gel with 40% DMAEM as a function of pH. The gel was initially exposed to base, then titrated with 0.10 M HCl.

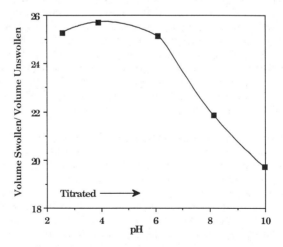

Figure 2b. Swelling ratio of a 2% crosslinked polymer gel with 40% DMAEM as a function of pH. The gel was initially exposed to acid, then titrated with 0.10 M NaOH.

increased the disk diameter increased. The polymer containing 40% amine undergoes substantial swelling, while formulations with 10 and 20% amine swell only slightly. According to theory, in a swollen polymer gel the electrostatic component of swelling should vary with the number of charges to the 1.2 power(9). The data in Figure 3 do not show this relationship. Instead, there appears to be something unique about the polymer with 40% amine that causes it to swell as the pH is reduced from pH 7 to 2.6.

Figure 4 plots the swelling ratio vs. time after exposure to a pH 2.6 phosphate/citrate buffer for polymers containing 1, 2 and 5% crosslinking agent and 40% amine. As would be expected the final swelling ratio is greatest for the polymer with the lowest amount of crosslinker and smallest for the polymer with the highest crosslinker percentage. However, the differences in the final swelling ratio are relatively small. Furthermore, the rate of swelling is significantly faster for the more highly crosslinked polymer. Normally, one would expect mass transfer processes to be faster in the more loosely crosslinked material.

Figure 3 and 4 show that when exposed to a buffered solution, polymer volume reaches a constant value after approximately 20 minutes with little or no further change observed.

Figure 5 shows percent increase in diameter of a 3.3% crosslinked polyacrylamide gel with 32% DMAEM as a function of glucose concentration vs. time. After exposure to glucose, gel placed in pH 7.4 phosphate buffer took about 60 minutes to shrink to its original size. The data in Figure 5 confirm that the amine-modified polyacrylamide membranes swell in response to glucose and thus are potentially useful for biosensors. The enzyme catalyzes the oxidation of glucose to gluconic acid, which protonates the amine. Electrostatic repulsion between charged amine groups causes the gel to expand. The degree of swelling increases with glucose concentration. Our data is similar to that reported for membranes prepared using 2-hydroxyethyl methacrylate (HEMA) rather than acrylamide (3). Response is reversible but the rate of deswelling is quite slow.

Proposed Model. We proposed the following model for gel behavior. The hydrophobic aminoesters associate with each other, forming a separate phase within the polymer. However, the absence of any visible cloudiness in the membrane implies that the aminoester domains must be significantly smaller than the wavelengths of visible light. According to this model, the reason that the solution in contact with freshly prepared polymer doesn't become basic is because the amine groups are tied up in a separate phase. Furthermore, it is known that the crosslinking reagent, N,N'-methylenebisacrylamide, also forms a separate phase in polyacrylamides (10). It may be miscible with the amine phase.

The pH has to be adjusted below 4 to protonate the amine to the extent that the amine phase is disrupted and the amines enter the aqueous phase. This explains the swelling observed as the pH is reduced below 4.

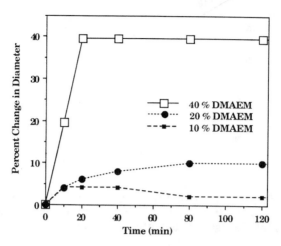

Figure 3. Effect of amine concentration on swelling. Percent increase in diameter as a function of time after exposure to a pH 2.6 phosphate/citrate buffer for 1% crosslinked polyacrylamide gels with 10, 20 and 40% DMAEM.

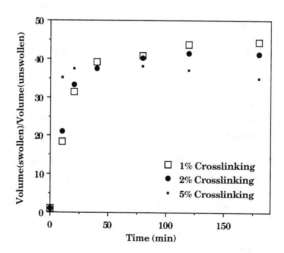

Figure 4. Effect of percent crosslinking on swelling. Swelling ratio vs. time after exposure to a pH 2.6 phosphate buffer for 1, 2 and 5% crosslinked polyacrylamide gels with 40% DMAEM.

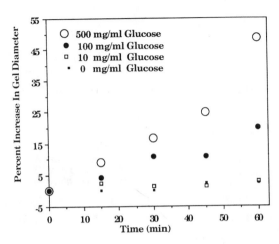

Figure 5. Effect of glucose on the swelling of a 3.3% crosslinked polyacrylamide gel with 32% DMAEM. Percent increase in diameter as a function of glucose concentration vs. time.

Once formed, the protonated amine groups remain in the aqueous phase until the pH is increased to the point that the proton is removed. At this point the microdomains of amine reform. This accounts for the hysteresis in the titration curves that is shown in Figure 5.

This model is consistent with the observation that only the polymer with 40% amine undergoes significant swelling at low pH. In the polymers with lower amine percentages, the amine level may be too low to allow formation of a separate phase. Only at the highest level is there enough amine present for the formation of separate amine domains.

The phase structure of the polymer may also be affected by the degree of crosslinking. The observation that the highly crosslinked polymer swells more rapidly may indicate that the amine domains are smaller in this material.

Implications for Biosensor Design. This study indicates that aliphatic amine containing polymers are not appropriate for biosensing. The proposed model suggests that formation of a separate amine phase is necessary to see swelling as the pH decreases from 7 to lower values. However, the formation of a separate amine phase is also linked to slow rates of response and hysteresis in the variation of volume vs. pH when measured after equilibration for 40 minutes.

The original motivation for using amines in glucose sensitive polymers was because of the intended application to drug delivery. This requires that polymer permeability <u>increase</u> with glucose concentration. This requires a functional group that becomes charged as the pH decreases due to gluconic acid formation. In biosensors, however, this is not required. Therefore, it may prove advantageous to use other pH sensitive functional groups to prepare glucose sensitive polymeric membranes. One possibility is to hydrolyze the amide group into a carboxylic acid moiety. The swelling characteristics of this system as a function of pH have been well established by Tanaka et al.(*11*) At high pH the carboxylate group is uncharged and the gel is swollen. The decrease in volume observed as the solvent pH is lowered is due to protonation of the carboxylate group which leaves the polymer uncharged. Similar behavior was observed in this laboratory.

The amine polymers used to prepare glucose sensitive polymers are heterogeneous. The rate of response is slow because of the slow rate of amine transfer from a hydrophobic phase to the aqueous phase. Faster rates of response should be possible with systems that do not require phase transformations.

Certain commercial equipment, instruments or materials are identified in this report to specify adequately the experimental procedure. Such identification does not imply recommendation or endorsement by the National Institute of Standards and Technology, nor does it imply that the equipment, instruments or materials are necessarily the best available for the purpose.

Literature Cited

1.   McCurley, M.F.; Seitz, W.R. *Anal. Chim. Acta*, **1991**, *249*, 373-380
2.   Butler, M.A.; Ricco, A.J.; Buss, R. *J. Electrochem. Soc.* **1990**,*137*, 1325-1326
3.   Albin, G; Horbett, T.A.; Ratner, B.D. *J. Controlled Release* **1985**, *2*, 153-164
4.   Siegel, R.A.; Firestone, B.A. *J. Controlled Release* **1990**, *11*, 181-192
5.   Siegel, R.A.; Falamazian, M.; Firestone, B.S.; Moxley, B.C. *J. Controlled Release* **1988**, *8*, 179-182
6.   Siegel, R.A.; Firestone, B.A.; Johannes I.; Cornejo, *J. Polymer Preprints* **1990**, *31*, 231-232
7.   Siegel, R.A.; Firestone, B.A. *Macromolecules* **1988**, *21*, 3254-3259
8.   *Methods In Enzymology*; Colowick, S.P.; Kaplan, N.O., Eds.; **1955**, Vol. 1
9.   Flory, P.J. *Principles of Polymer Chemistry*; Cornell University Press, Ithaca, NY, 1953
10. Parreno, J.; Pierola, I.F. *Polymer* **1990**, *31*, 1768-71
11. Tanaka, T.; Fillmore, D.; Sun, Shao-Tang; Nishio, I.; Swislow, G. and Shah, A. *Phys. Rev. Lett* **1980**, *45*, 1636-1639

RECEIVED November 26, 1991

Chapter 25

Optical Immunosensors Using Controlled-Release Polymers

Steven M. Barnard and David R. Walt[1]

Max Tishler Laboratory for Organic Chemistry, Chemistry Department, Tufts University, Medford, MA 02155

Fiber-optic sensors based on controlled-release polymers provide sustained release of indicating reagents over long periods. This technique allows irreversible chemistries to be used in the design of sensors for continuous measurements. The sensor reported in this paper is based on a fluorescence energy transfer immunoassay. The sensor was cycled through different concentrations of antigen continuously for 30 hours. Although the sensor was not optimized, the data indicate the viability of the technique. Improvements in the sensor design to provide shorter response times are discussed.

In the last decade, fiber-optic chemical sensors (FOCS), also known as optrodes, have emerged as alternatives to conventional methods of analysis. FOCS development for a particular analyte depends on the availability of reversible indicating schemes to detect the analyte of interest. Typically, the indicating schemes use commercially available colorimetric or fluorometric indicators (e.g. fluorescein to measure pH (1)). However, the utility of these indicators is limited. Furthermore, indicators may not exist for many analytes. Several reviews discuss the scope of this approach (2,3,4).

Alternatively, irreversible indicating schemes can be used that form irreversibly-colored products or involve complex formation with high binding constants. In general, this approach is important because it may provide the only indicating chemistry available to detect an analyte. Irreversible indicating schemes can be used to develop probes, which by definition, are irreversible and cannot make continuous measurements. However, probes can be adapted to make sequential measurements if the indicating reagent is replenished or regenerated or if the sensor is exposed only to subsaturating concentrations of analyte. One approach to replenishing the probe is to pump fresh reagent to the vicinity of the fiber tip (5). This approach suffers from the need for a constant pumping rate, power to drive the pump, and the added complexity of the system. Immunoprobes have also been fabricated which obviate the irreversibility of the antigen-antibody complex by using chaotropic reagents to disrupt binding (6). Even with this approach, the probe has a limited lifetime.

---

[1]Corresponding author

Furthermore, the necessity to expose the probe tip to chaotropic reagents makes the measurement both cumbersome and discontinuous.

The selectivity and specificity of antibodies make them particularly attractive for sensor use. Although the binding between an antibody and antigen is reversible and noncovalent, most immunological reactions essentially are irreversible because of their large association constants (K_a's), ranging typically between 10^5 to 10^9 M^{-1} (7). The K_a's are composed of large forward (k_1) and small reverse (k_{-1}) rate constants, ranging from 10^7 to 10^9 $M^{-1} s^{-1}$ and 10^2 to $10^{-4} s^{-1}$, respectively. These kinetic parameters make antibodies extremely analyte specific and selective. Since antibodies can be generated to almost any size antigen, an unlimited arsenal of receptors is available to develop indicating schemes. Researchers have developed various immunosensors (8,9,10) but all suffer from some limitation: a) the sensor is irreversible and nonregenerable (8); b) the design relies on the antigen's intrinsic fluorescence, thus is not a general approach (9); c) the design is based on antibodies with fast off-rates, making the technique applicable only to analytes for which fast off-rate antibodies can be obtained (10).

To utilize immunoreactions effectively in sensor design, the problem of irreversibility can be circumvented by creating a reservoir that passively releases immunoreagents to the sensing region of an optical fiber. In previous work (11), we demonstrated how controlled-release polymers could be used to release indicators continuously and passively for extended times. We further demonstrated how to offset the effects of changes in release rate while maintaining calibration. In the first design discussed below, the reservoir is constructed by incorporating immunoreagents in a controlled-release polymer that delivers reagents to an internal chamber. In a preliminary second design, microparticles of controlled-release polymer deliver reagent within a porous hydrophilic polymer layer.

Selection of Immunoassay

The immunoassay selected is based on a competitive fluorescence energy transfer mechanism. This type of assay is general and can be adapted to any antibody and antigen system. The competing reactions are given in equation (1) and (2):

$$F\text{-}Ab + Ag \rightleftharpoons F\text{-}Ab{:}Ag \qquad (1)$$

$$F\text{-}Ab + TR\text{-}Ag \rightleftharpoons F\text{-}Ab{:}TR\text{-}Ag \qquad (2)$$

where F-Ab:Ag is the labeled-antibody:unlabeled antigen immunocomplex and F-Ab:TR-Ag is the labeled-antibody:labeled-antigen immunocomplex. When F-Ab:TR-Ag is formed, the distance between the fluorophors falls within the region where nonradiative energy transfer occurs (12). As fluorescein (F) is excited, a portion of its emission energy is transferred nonradiatively to Texas Red (TR), quenching the fluorescein and enhancing the Texas Red fluorescence intensities. If F-Ab:Ag forms, no energy transfer occurs because Texas Red is not present to accept energy from fluorescein. Therefore, the amount of energy transfer is proportional to the concentration of unlabeled antigen.

Selection of Controlled-Release Polymer

There are a variety of different polymer systems available to deliver a predictable and reproducible release of chemical compounds (13,14). Two of the most versatile systems, adaptable to sensor design, release incorporated reagent by chemically-controlled or diffusion-controlled processes. Figure 1 shows examples of these two systems.

Chemically-Controlled Systems. In these systems, the polymer matrix contains chemically-labile bonds. On exposure to water or enzymes the bonds hydrolyze, erode the three dimensional structure of the polymer and release the incorporated reagent into the surrounding medium. Depending on the polymer used, the erosion products may act as interferences, such as by altering the pH of the solution. Examples of these systems are polyglycolic acid (PGA) and a polyglycolic acid - polylactic acid (PGA/PLA) copolymer. PGA hydrolyzes to hydroxyacetic acid, and PGA/PLA hydrolyzes to lactic acid and hydroxyacetic acid. Other chemically-controlled systems are based on polyorthoesters, polycaprolactones, polyaminoacids, and polyanhydrides.

Diffusion-Controlled Systems. Generally, these systems are based on reagent incorporated uniformly throughout a polymer matrix. On exposure to an aqueous environment, the normally nonporous polymer structure forms a tortuous interconnected pore network. It has been proposed that reagent diffuses through the porous network (15). Unlike the chemically-controlled systems, no breakdown products are formed during the release. These systems are usually rubbery polymers such as silicone rubber, ethylene vinyl acetate, or polyurethanes. We selected ethylene vinyl acetate to construct our polymer reservoirs due to its diffusion-controlled mechanism of release, the lack of breakdown products, and the ease of incorporating various reagents. In addition, EVA has been used to release molecules as large as 2×10^6 daltons for over 100 days (16).

Construction of Immunosensor

The sensors are constructed from Plexiglass cubes that have been drilled to accommodate a 600 μm optical fiber, polymer, and to provide a pathway for the analyte (Figure 2). The physical dimensions of the sensor are 15 mm by 12 mm by 10 mm. The volumes of the polymer reservoir and the reaction chamber are 100 μl and 40 μl, respectively. The lyophilized immunoreagents are incorporated into EVA at 1.5% (dryweight) for fluorescein-labeled anti-IgG antibody (F-Ab) and 8.0% Texas Red-labeled IgG (TR-Ag). Pieces of the polymers are then packed in the polymer reservoirs.

Determination of Antigen Concentration

As immunoreagents are released from the polymer and as unlabeled antigen diffuses into the reaction chamber of the sensor, a competition reaction occurs between labeled and unlabeled antigen for the available binding sites on the labeled antibody. The fluorescence intensity depends on the presence of unlabeled antigen as shown in Figure 3. As the sensor is exposed to a high concentration (500 μg/ml, 3.6 μm) of unlabeled antigen (500μg/ml), less energy transfer occurs, resulting in larger fluorescein and smaller Texas Red emission intensities. In contrast, low concentrations of unlabeled antigen (0μg/ml) result in a large amount of energy transfer, since most of the F-Ab is bound to TR-Ag, producing smaller fluorescein and larger Texas Red intensities. Alternatively, the concentration of unlabeled antigen can be solved by using the equilibrium equations for reactions (1) and (2):

$$Keq_1 = [\text{F-Ab:Ag}] \, / \, ([\text{F-Ab}][\text{Ag}]) \qquad\qquad (3)$$

$$Keq_2 = [\text{F-Ab:TR-Ag}] \, / \, ([\text{F-Ab}][\text{TR-Ag}]) \qquad\qquad (4)$$

Solving for the antibody concentration:

$$[\text{F-Ab}] = [\text{F-Ab:Ag}] \, / \, ([\text{Ag}]Keq_1) \qquad\qquad (5)$$

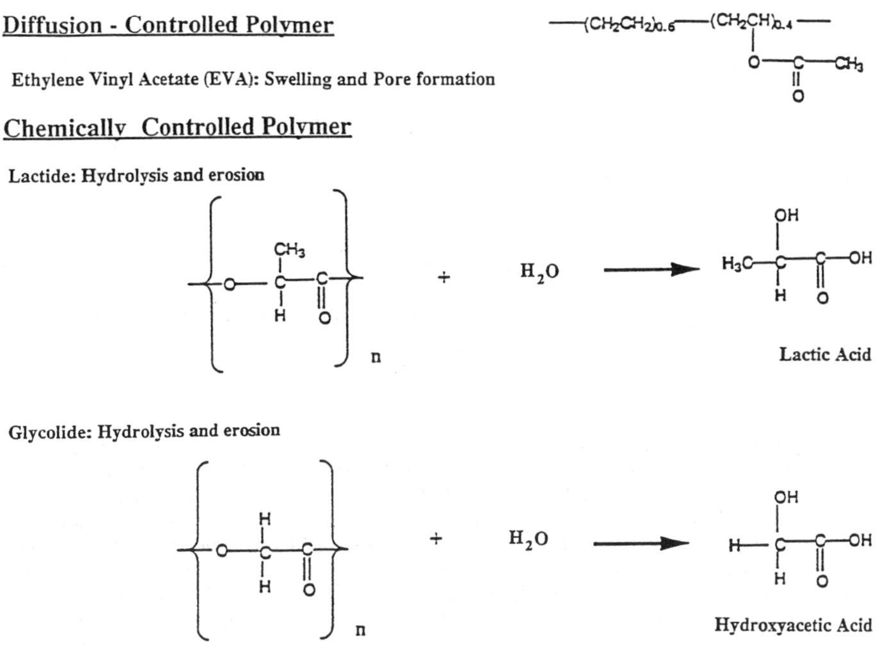

Diffusion - Controlled Polymer

Ethylene Vinyl Acetate (EVA): Swelling and Pore formation

$$-(CH_2CH_2)_{0.6}-(CH_2CH)_{0.4}-$$

Chemically Controlled Polymer

Lactide: Hydrolysis and erosion

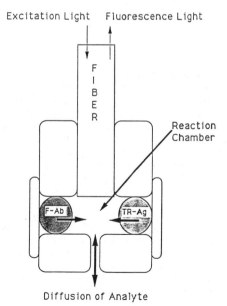

Glycolide: Hydrolysis and erosion

Figure 1. Controlled-Release Polymers

Figure 2. Immunosensor Configuration. Reagent is released from the polymer reservoirs into the reaction chamber where a competition reaction takes place with analyte diffusing in from the bulk solution.

$$[F-Ab] = [F-Ab:TR-Ag] / ([TR-Ag]Keq_2) \qquad (6)$$

Equating (5) and (6) and solving for [Ag]:

$$[Ag] = ([F-Ab:Ag][TR-Ag]Keq_2) / ([F-Ab:TR-Ag]Keq_1) \qquad (7)$$

Equation (7) can be written in terms of fluorescence intensities as equation (8):

$$[Ag] = \frac{([I_{480/520}][I_{570/610}])}{([I_{480/610}]-0.06[I_{570/610}])} \qquad (8)$$

where $I_{x/y}$ is the fluorescence intensity of the excitation wavelength at x nanometers and emission wavelength at y nanometers.

Results and Discussion

Figure 4 shows data collected over 14 days, cycling between 0 and 500 μg/ml of IgG. The values were calculated using equation (8). The data show that the sensor responds reversibly to the two concentrations. Although the sensor is not quantitative, it qualitatively indicates the presence or absence of antigen. Response time is dictated by the rate of analyte diffusion from the bulk solution to the reaction chamber or by the release rate reagent from the polymer. Analyte diffusion can be regulated by controlling the length and diameter of the diffusion channel. Release rate is dependent on the concentration of reagent and the size and shape of the polymer reservoir. It is important that the release rates of F-Ab:Ag or F-Ab:TR-Ag maintain sufficient concentrations of these complexes in the reaction chamber to obtain a measurable fluorescence signal. The basic limitation of the present design are the large diffusional distances for both the analyte and reagent due to the physical separation of the reservoirs and reaction chamber. To overcome this limitation, a second approach places the polymer reservoir intimately in contact with the reaction chamber.

Second Design Strategy

To improve the response time and quantitative behavior of the immunosensor, a second strategy is being investigated that combines specific properties of two polymers: a controlled-release polymer - EVA; and a porous hydrophilic polymer - polyacrylamide. Microparticles of controlled-release polymer containing reagent are entrapped in a crosslinked polyacrylamide polymer layer on the distal face of an optical fiber. The microparticles act as tiny reservoirs, releasing reagent into the polyacrylamide layer. Analyte diffuses into the polyacrylamide layer and reacts with the reagent released from the microparticles. After the reaction is complete, the complex diffuses out into the bulk solution. As the complex diffuses away, fresh reagent is replenished in the internal solution of the polyacrylamide layer.

Using this approach, a pH sensor was constructed using a common pH indicator, 8-hydroxypyrene 1,3,6 trisulfonic acid, HPTS, (Figure 5). Although HPTS is a reversible indicator, it was chosen for the following reasons: it is a common pH indicator, it is highly water soluble, it has a molecular weight that will not hinder diffusion within the polyacrylamide layer, it has a high quantum yield making detection easy, and it can be ratioed for quantitation. In addition, HPTS undergoes a change in emission spectrum upon analyte binding.

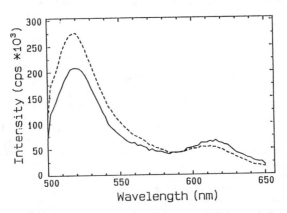

Figure 3. The dependence of the sensor's emission spectra on antigen concentration: spectrum 1 (---) high concentration (500 μg/ml); spectrum 2 (-) low concentration (0 μg/ml). Placing the sensor in a solution of high free antigen concentration causes less energy transfer to occur, decreasing the Texas Red emission maximum (610nm) and increasing the fluorescein emission maximum (520nm peak).

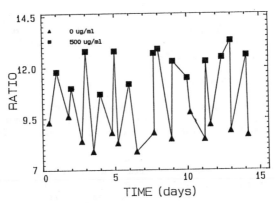

Figure 4. Immunosensor response to 0 μg/ml and 500 μg/ml of IgG over 14 days.

The polyacrylamide layer is a critical feature in the design. It has three functions. First, it acts as a support structure by physically immobilizing the reservoir particles in the sensing region of the optical fiber. Second, it acts as a reaction layer allowing the released reagent to react and mix with released reagents within the porous network of the polyacrylamide polymer. Third, the polyacrylamide layer acts as a diffusional barrier allowing the released reagents to react without being diluted and quickly washed away into the bulk solution. HPTS's incorporation into the polymer, its release from the microparticles and its subsequent reaction, have validated the design strategy.

Fabrication of Sensor. The sensor is constructed by physically entrapping dye-containing-EVA microparticles in a polyacrylamide layer on the distal end of an optical fiber. The microparticles are made by incorporating a pH sensitive dye, 8-hydroxypyrene, 1,3,6 trisulfonic acid (HPTS), into EVA at a dry-weight-loading of 30% (17). A 1.0 cm^3 dry piece of polymer was grated against a 450 μm wire mesh screen, resulting in particles of random size and shape. The particles were then sequentially graded through 250, 150, and 75 μm wire mesh sieves, producing reasonably homogeneous particles as observed through a microscope. The supporting polyacrylamide layer was grown from the distal end of a 600 μm silanized optical fiber by photopolymerization. One gram of microparticles, 100 mg of benzoin ethyl ether, 1.0 ml of propanol and 1.0 ml of stock acrylamide solution (TOXIC) (40.0 grams of acrylamide and 2.0 grams of bisacrylamide (TOXIC) dissolved in 100 mls of pH 6.6 phosphate buffer), were added to a 3.0 ml vial and deoxygenated with nitrogen for 15 minutes. A 600 μm optical fiber, silanized with 3-(trimethoxysilyl) propyl methylacrylate, was positioned in the vial and light (400 μW) from a xenon arc lamp was launched down the fiber. After irradiation for 30 seconds, a polyacrylamide layer can be seen growing from the distal face of the fiber.

Results. The cumulative release of HPTS from the microparticles is shown in Figure 6. During the first hour, an initial fast-release rate is observed, followed by a slower sustained-release rate. The initial rate is attributed to surface diffusion of HPTS from the microparticles. The sustained-release rate results from diffusion of HPTS through an interconnected pore network.

The response time of the sensor is shown in Figure 7. The sensor has approximately a 7.0 minute response time for a change of 1 pH unit. Although the response is still slow it is much faster than the bulky design of Figure 2 and proves the validity of the sensor design. The sensor's titration curve is graphed in Figure 8.

Application to Irreversible Indicators

The response of the HPTS sensor based on the two polymer configuration indicates that an indicator can be released from microparticles and react within the internal solution of the polyacrylamide layer. Figure 9 shows possible irreversible indicators that approximate the molecular weight of HPTS and that undergo a change in spectrum upon binding. Figure 10 shows a proposed adaptation of a fluorescence energy transfer immunoassay to the design. In this configuration, the immunoreagents are released from microparticles, instead of large polymer reservoirs. The reaction chamber is the internal solution of the polyacrylamide layer instead of a large chamber at the end of the optical fiber. These improvements should minimize diffusional barriers that impair the response time and quantitation of the sensor. It is important to recognize that immunosensors are inherently more complicated to design as they pose challenges to solve incorporation of immunoreagents into microparticles, loss of activity during microparticle fabrication, the slow diffusion of large molecule weight reagents, and the selection of appropriate indicating chemistry.

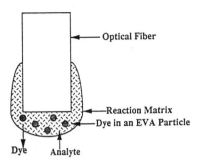

Figure 5. Microparticle Sensor Design. In this design, microparticles release reagent into a polyacrylamide layer, reacting with analyte that has diffused in from the bulk solution.

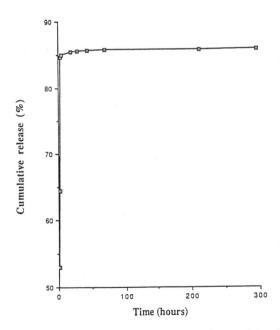

Figure 6. Cumulative Release of HPTS from EVA Microparticles. Sensors were fabricated and soaked in pH 7.8 phosphate buffer. The release of HPTS from microparticles entrapped in polyacrylamide on sensor tips was monitored by measuring the increase in fluorescence intensity over 300 hours.

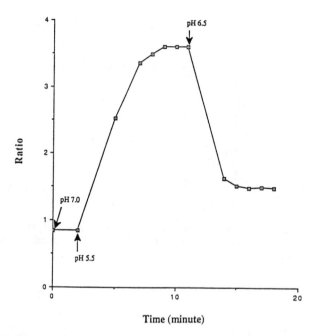

Figure 7. Microparticle Sensor Response Time. The arrows show when the sensor was placed in the respective pH buffer. Measurements were taken at various intervals over 20 minutes.

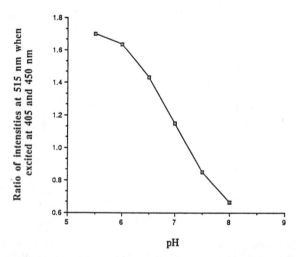

Figure 8. Microparticle Sensor Titration Curve. A sensor was place in a series of six buffers of different pHs. After the sensor signal stabilized, the intensity at the respective excitation and emission wavelengths was collected: 405nm and 515nm; 450nm and 515nm. The ratio at these intensities was plotted against pH.

Fluorescent Irreversible Synthetic Indicators			
	8-hydroxyquinoline-5-sulfonic acid	*p*-tosyl-8-aminoquinoline	Lumogallion
Analyte	Mg^{+2}	Cd^{+2}	Al^{+3}

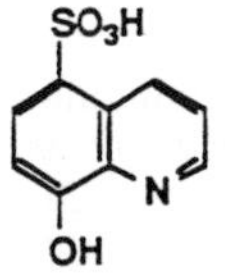

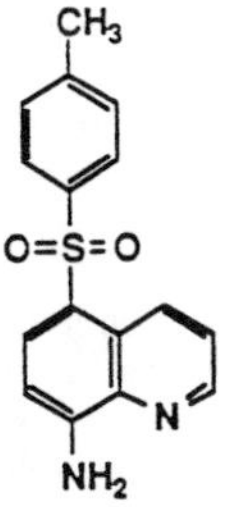

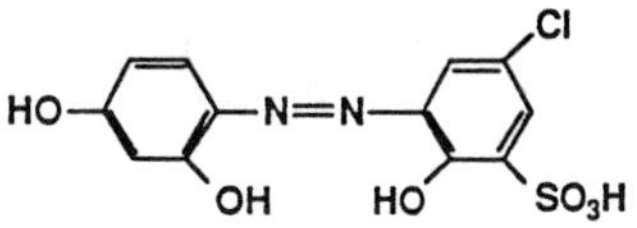

Figure 9.   Fluorescent Irreversible Indicators that are Adaptable to the Microparticle Sensor Design.

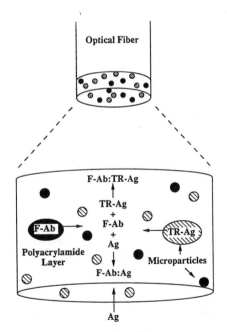

Expanded View of Layer

Figure 10. Proposed Adaptation of a Fluorescence Energy Transfer Immunoassay to the Microparticle Sensor Design. A mixture of two different microparticles, each containing different reagents, are entrapped physically in the polyacrylamide layer. The reagents released from the microparticles set up a competition reaction between the free and labeled antigens for the available binding sites of labeled-antibody. The immunocomplexes formed have different emission spectra, allowing quantitation of free antigen concentration.

Conclusion

Our approach of using controlled-release polymers to construct reservoirs of irreversible indicating chemistries allows sensors to be developed that operate continuously and for long periods. The approach has several advantages:

> 1) Receptors that have large binding constants may be used in a continuous fashion without chemical immobilization or confinement at the fiber's distal tip.

> 2) Indicators that do not have pendent groups for immobilization or that would be altered by immobilization may be used as sensing reagents.

> 3) Multiple indicators may be released that function as internal references, pH indicators, or ionic strength indicators.

In addition, by combining specific properties of polymers in sensor design, new configurations can be constructed that may have advantages over traditional designs. By entrapping controlled-release microparticles in a thin hydrophilic porous polymer, the miniature dimensions of the fiber are preserved, the released reagent and the analyte are in intimate contact, and the diffusion barrier of the analyte is minimized. The most immediate applications of these designs are likely to be in monitoring pollutants at toxic waste sites, groundwater aquifers, and agricultural areas.

LITERATURE CITED

(1)   Munkholm, C.; Walt, D. R.; Milanovich, F. P.; Klainer, S. M. *Talanta* 1988, **35**, 109.
(2)   Wolfbeis, O. S. In *Molecular Luminescence Spectroscopy: Methods and Applications - Part II*; Winefordner J. D. and Kolthoff, I. M., Eds.; Chemical Analysis; John Wiley & Sons: New York, NY, 1988, Volume **77**; pp. 129-283.
(3)   Seitz, R. *CRC Crit. Rev. Anal. Chem.* 1988, **19**, 135.
(4)   Angel, S. M. *Spectroscopy,* 1987, **6**, 38.
(5)   Inman, S. M.; Stromvall, E. J.; Lieberman, S. H. *Anal. Chim Acta* 1989, **217**, 249.
(6)   Bright, F. V.; Bett, T. A.; Litwiler, K. S. *Anal. Chem.* 1990, **62**, 1065.
(7)   Nisenoff, A. *Introduction to Molecular Immunology*; Sinauer Associates, Inc.; Sunderland, MA, 1985, pp. 29-45.
(8)   Vo-Dinh, T; Tromberg, B. J.; Griffin, G. D.; Ambrose, K. R.; Sepaniak, M. J.; Gardenhire, E. M. *Appl. Spectrosc.* 1987, **41**, 735.
(9)   Tromberg, B. J.; Sepaniak, M. J.; Vo-Dinh, T; Griffin, G. *Anal. Chem.* 1987, **56**, 1226.
(10)   Anderson, F. P.; Miller, W. G. *Clin. Chem.* 1988, **24**, 1417.
(11)   Barnard, S. M.; Walt, D. R. *Science* 1991, **251**, 927.
(12)   Schulman, S. G. *Molecular Luminescence Spectroscopy Methods and Applications: Part I*; Elung, P. J. and Winefordner, J. D., Eds.; Wiley: New York, 1985, Vol. **77**, pp. 1-28.
(13)   Langer, R. *Science* 1990, **249**, 1527.
(14)   Chandrasenkaran, S. K.; Wright, R. M.; Yuen, M. J. In *Controlled Release Delivery Systems;* Roseman, T. J. and Mansdorf, S. Z., Eds; Marcel Dekker, Inc.: New York, 1983, pp. 1-25.
(15)   Bawa, R.; Siegel, R. A.; Marasca, B.; Karel, M.; Langer, R. *Journal of Controlled Release* 1985, **1**, 259.
(16)   Langer, R.; Folkman, J. *Nature* 1976, **263**, 797.
(17)   Luo, S.; Walt, D. R. *Anal. Chem.* 1989, **61**, 174.

RECEIVED December 10, 1991

INDEXES

Author Index

Affiliation Index

Subject Index

Production: Paula M. Bérard
Indexing: Deborah H. Steiner
Acquisition: Anne Wilson
Cover design: Amy Hayes

Printed and bound by Maple Press, York, PA